BARRON'S
VISUAL LEARNING
Human
Anatomy

Published by arrangement with UniPress Books Ltd

Publisher: Nigel Browning

Project manager: Kate Duffy

Page design: Sleeperdesign.co.uk

Illustrations: Sarah Skeate

Editorial consultant: Cynthia Pfirrmann

First edition published in North America by Kaplan, Inc.,
d/b/a Barron's Educational Series

Published by Kaplan, Inc., d/b/a Barron's Educational Series

750 Third Avenue
New York, NY 10017

www.barronseduc.com

ISBN: 978-1-5062-8095-0

Printed in China

10 9 8 7 6 5 4 3 2 1

Professor Ken Ashwell, B. Med. Sc., M.B., B.S., Ph.D., teaches anatomy to medical and science students and is particularly interested in the study of brain development. He has contributed to scientific journals and written many books about human anatomy and brain structure. He is professor of anatomy at the University of New South Wales, Sydney, Australia.

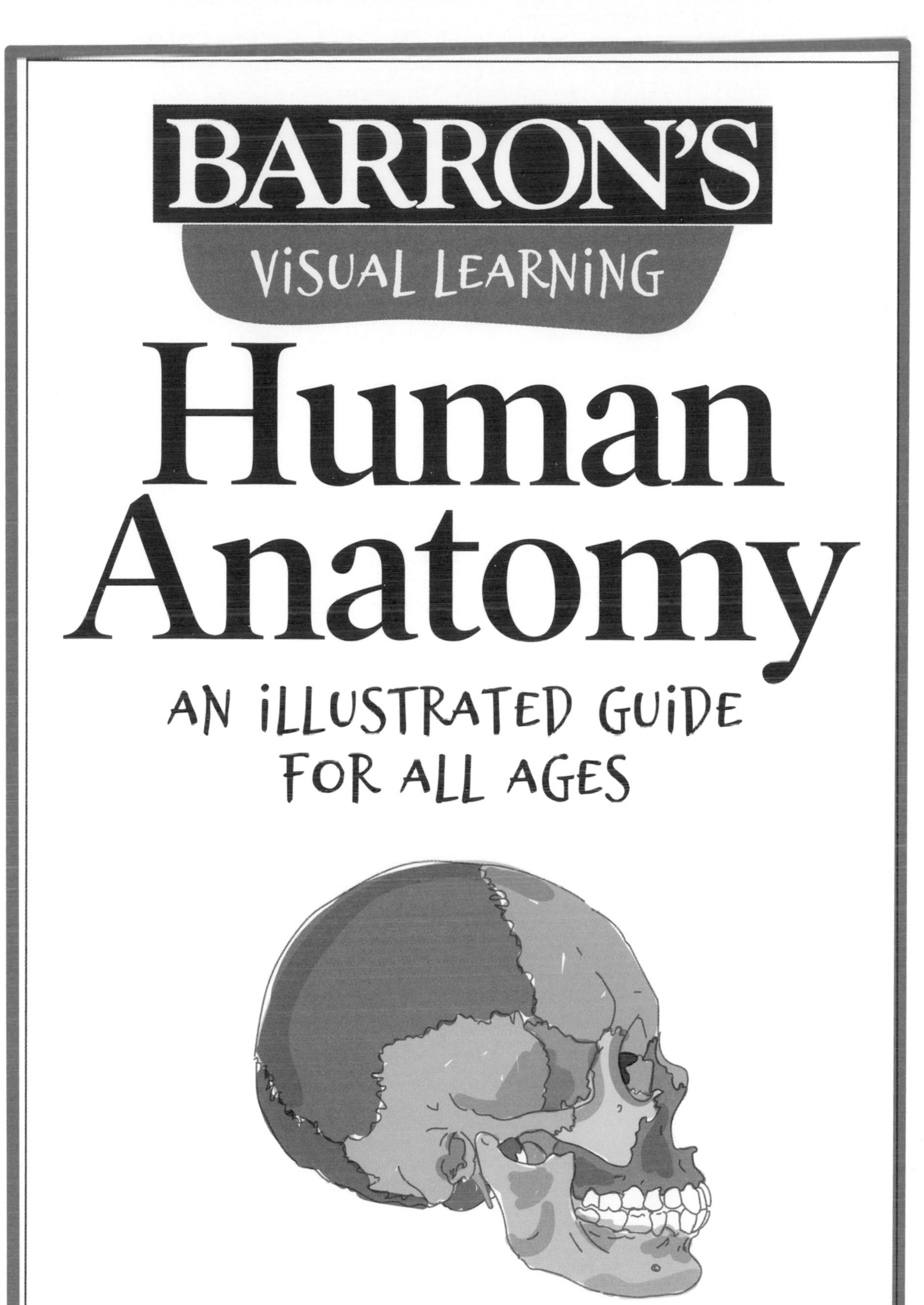
BARRON'S
VISUAL LEARNING
Human
Anatomy
AN ILLUSTRATED GUIDE
FOR ALL AGES

PROFESSOR KEN ASHWELL B.Med.Sc., M.B., B.S., Ph.D.

CONTENTS

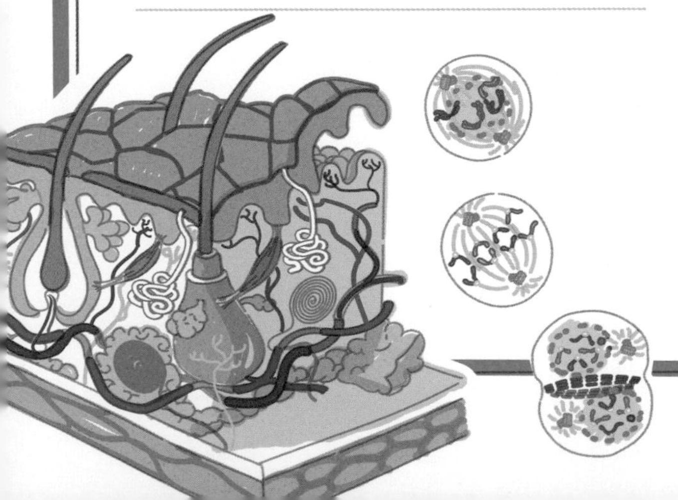

6 Cardiovascular System 106

7 Immune/Lymphatic System 126

8 Respiratory System 136

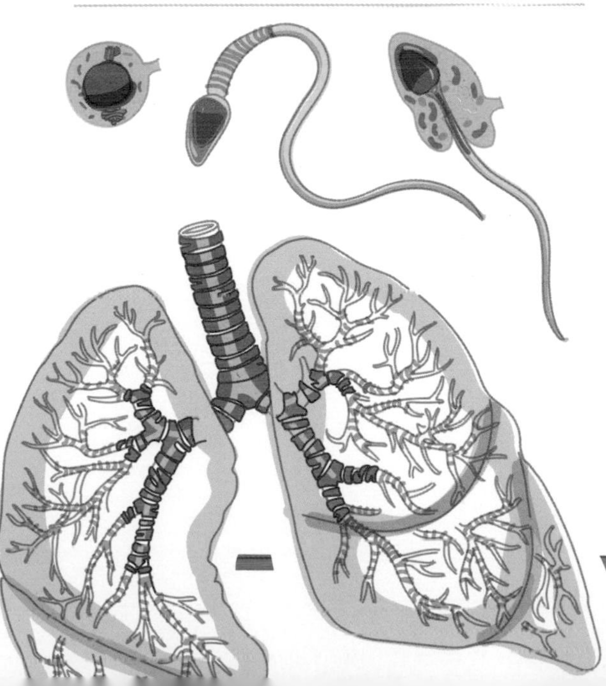

9 Digestive System 146

10 Urinary System 156

11 Reproductive System 164

12 Endocrine System 176

INTRODUCTION

This extensively illustrated book explores the wonderful world of the human body, from the more than 50 trillion tiny cells that make up body tissues to the large, complex organs of the many systems that keep us alive.

Did you know that the word *anatomy* comes from the Greek for "cutting up"? It's rather like that other important word in anatomy—*dissection*—that comes from the Latin for "cutting apart." That, of course, is where anatomy started—by dissecting bodies, observing, and describing the organs and body cavities that were visible to the naked eye. Even when the light microscope came along in the seventeenth century, anatomists continued to use sharp knives to slice the body tissues thinly enough to see the fine detail of cells and tissues.

The invention of the electron microscope in the twentieth century enabled thinner slices to be studied. Now, we have powerful laser microscopes that can focus on optically sliced human tissue. So, anatomy is all about dissecting body parts to see and understand them more clearly.

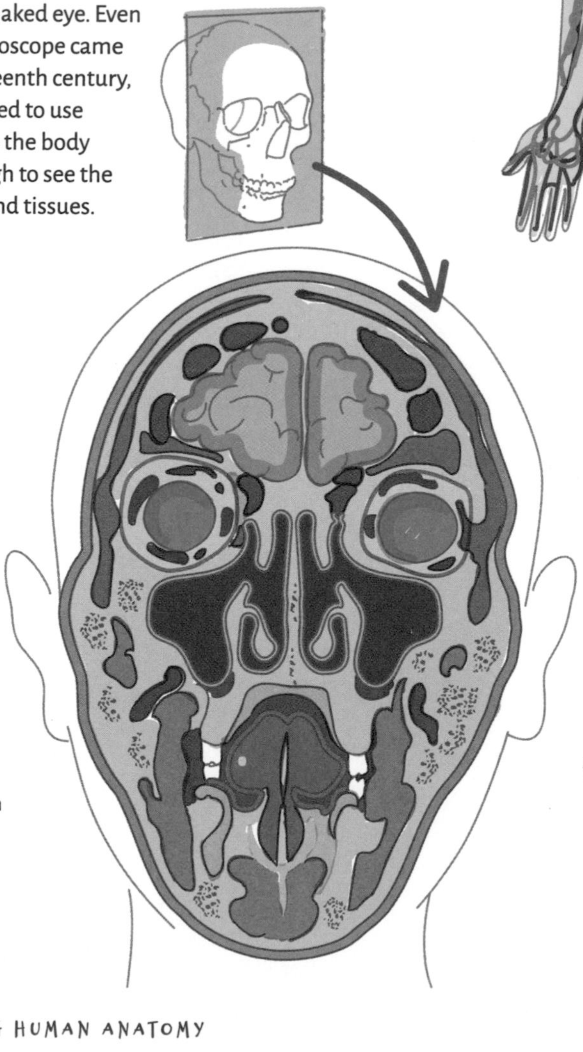

The skull bones contain the paranasal sinuses.

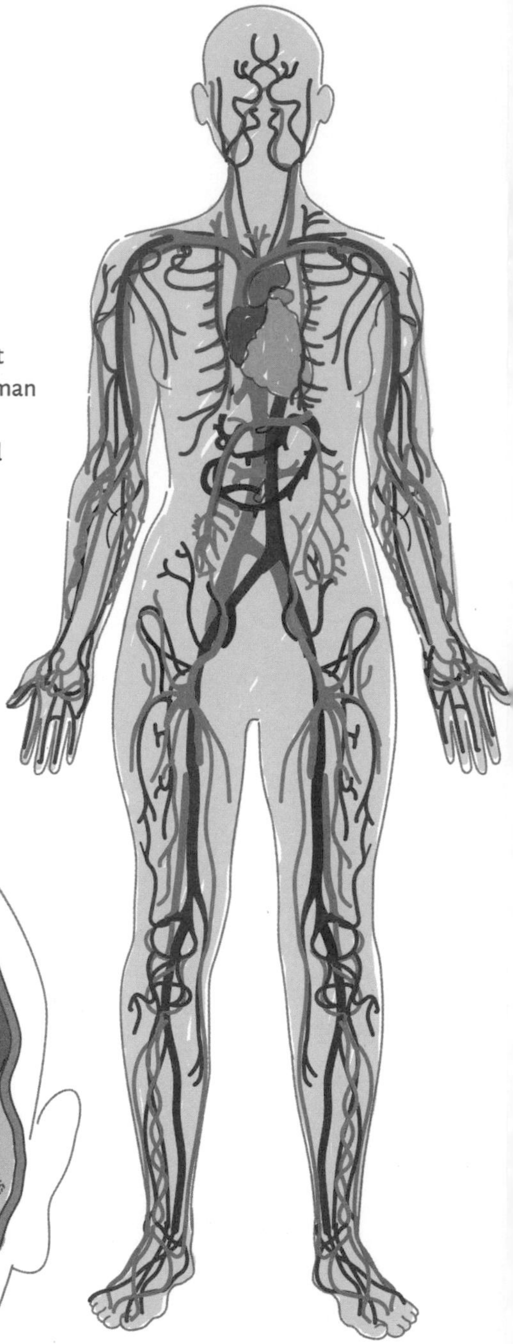

The circulatory system carries gases, nutrients, proteins, and waste around the body.

The first accurate account of anatomy started during the early Renaissance, when pioneer scientific anatomists, such as Andreas Vesalius, started to dissect unclaimed bodies and precisely record what they saw. Vesalius's 1543 work *De Humani corporis fabrica libri septem* ("Seven books on the structure of the human body") is considered one of the greatest works of science. Before then, most knowledge of the human body was derived from dissection of pigs, monkeys, and dogs from the time of Galen, physician to the Roman emperor Marcus Aurelius. The doctrines of Galenic anatomy had been passed unquestioningly from classical times, with many errors that hindered the development of scientific medicine and surgery. By combining dissection with accurate illustration, Vesalius broke out of the constraints of ancient thought.

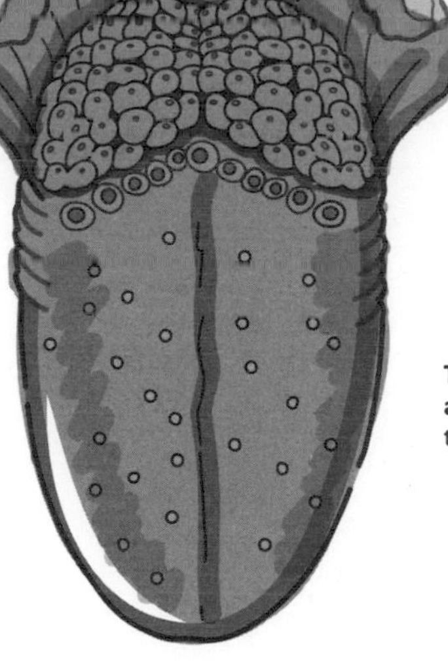

The tongue has about 10,000 taste buds.

Anatomy has always been a visual science, so this book emphasizes the key elements of the human structure seen in vividly colored but simple illustrations. This book is particularly useful for visual learners and reinforces facts and ideas with pictures and diagrams. Look not just at the shape of structures but also their spatial relationship to other structures. When you have mastered the two-dimensional relationships on a single plane, take your understanding to the next level by thinking about depth through layered illustrations. Practice your knowledge by first copying the diagrams and then see if you can reproduce them just from memory.

As Vesalius showed, good science requires the observer to see things for themselves, so try to relate the pictures in this book to the features you can find on your own body. Many bones, muscles, vessels, and nerves come close to the skin's surface, and you can check their position by touch and sight. Your body will be your teacher, this book will be your guide. Welcome to *Visual Learning Human Anatomy*!

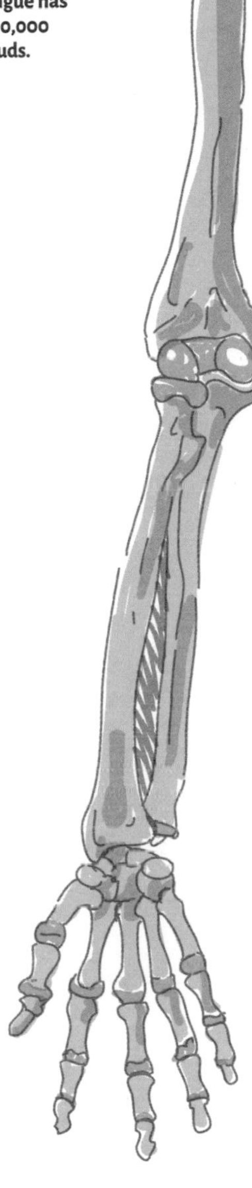

The bones of the upper limb include the pectoral girdle, the humerus in the arm, the ulna and radius in the forearm, the carpals of the wrist, and the phalanges of the fingers.

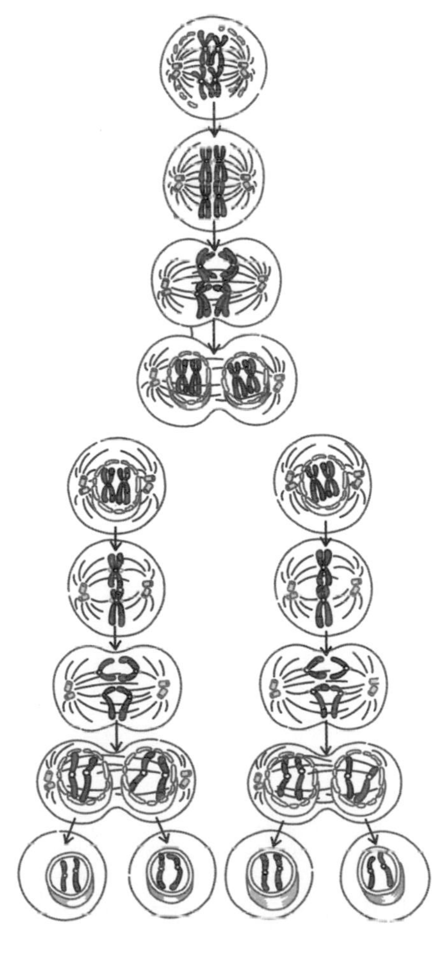

Meiosis cell division happens during the production of sex cells.

CHAPTER 1

OVERVIEW OF SYSTEMS

The body is made up of more than 50 trillion cells that are grouped into tissues and that in turn are assembled into organs that function together as body systems.

A related group of organs that collectively perform a particular function is called a body system. For example, different body systems enable digestion, movement, immunity, and reproduction.

The body systems that will be discussed in human anatomy are called skin (integumentary), skeletal, muscular, nervous, circulatory, respiratory, digestive, urinary, reproductive, immune/lymphatic, and endocrine.

THE BASICS OF SYSTEMS

The components of body systems may be large, such as the brain, heart, or liver, or microscopic, such as the immune system cells.

Homeostasis

The body systems serve the basic functions common to all animals. One of the most vital of these is **homeostasis**—namely to maintain a constant internal body environment. Homeostasis is particularly important, because the body systems must work together to keep the internal state of the body all within narrow limits to keep us alive. Homeostasis requires the ability to detect the internal state (e.g., blood sugars, blood pressure, and ionic balance) and also to respond with effective change by mobilizing nutrient stores, increasing lung ventilation, contracting vascular or gut smooth muscle, or activating glands.

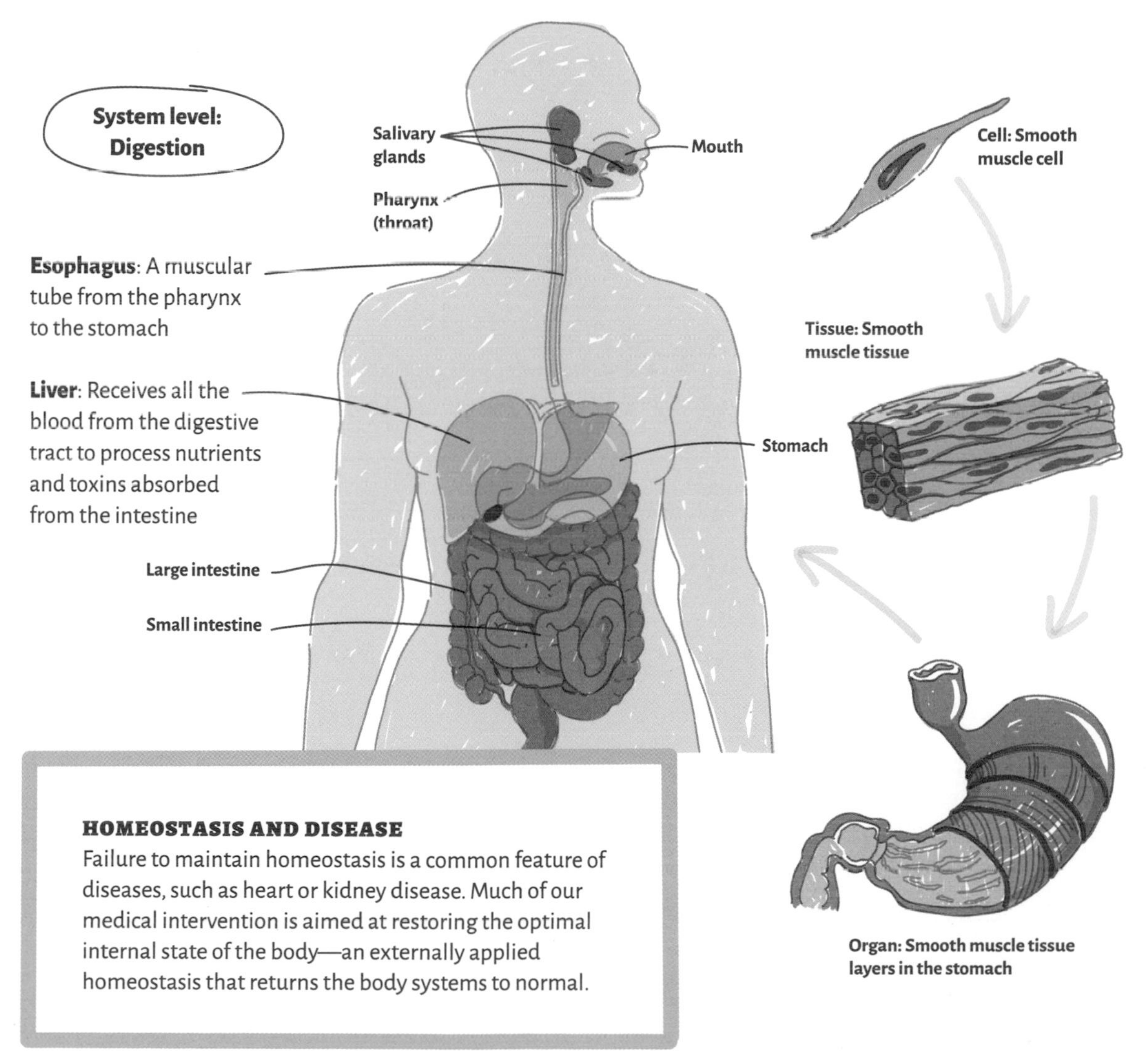

System level: Digestion
Salivary glands
Mouth
Pharynx (throat)
Esophagus: A muscular tube from the pharynx to the stomach
Liver: Receives all the blood from the digestive tract to process nutrients and toxins absorbed from the intestine
Large intestine
Small intestine
Stomach
Cell: Smooth muscle cell
Tissue: Smooth muscle tissue
Organ: Smooth muscle tissue layers in the stomach

HOMEOSTASIS AND DISEASE

Failure to maintain homeostasis is a common feature of diseases, such as heart or kidney disease. Much of our medical intervention is aimed at restoring the optimal internal state of the body—an externally applied homeostasis that returns the body systems to normal.

SKELETAL SYSTEM: BONES AND JOINTS

The skeletal system consists of the bones and joints between them. There are approximately 206 to 213 bones in an adult human. The variation is due to differences in the number of small sesamoid (like a sesame seed) bones in the tendons.

Bone composition

Bone is much like reinforced concrete or fiberglass, in that it is a composite material. This means it is formed from an organic fibrous component (mainly the protein type 1 collagen) and a cellular component (bone cells), both embedded in a crystalline mineral matrix of calcium phosphate called **hydroxyapatite**. The mineral matrix provides hardness, but the organic fibrous component provides elasticity and reduces brittleness. A rib that has had the minerals leached out is flexible and can be tied in a knot.

Bone is much stronger in compression than in tension (pulling) or shear (sideways) forces, so most fractures occur when a force is applied at right angles to the long axis of the bone. The **femur** of the thigh is about four times stronger in compression than a comparable width of wood, and long bones such as the femur are hollow to maximize strength while minimizing weight.

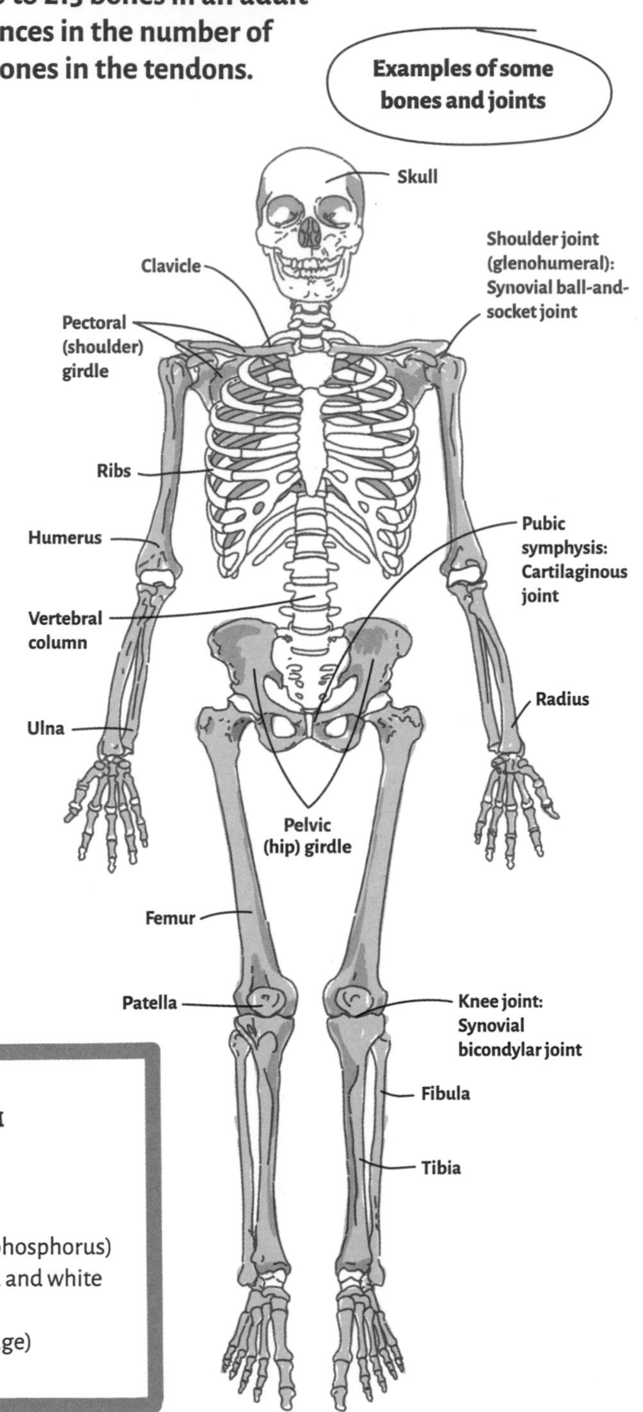

Examples of some bones and joints
Skull
Clavicle
Shoulder joint (glenohumeral): Synovial ball-and-socket joint
Pectoral (shoulder) girdle
Ribs
Humerus
Pubic symphysis: Cartilaginous joint
Vertebral column
Radius
Ulna
Pelvic (hip) girdle
Femur
Patella
Knee joint: Synovial bicondylar joint
Fibula
Tibia

FUNCTION OF THE SKELETAL SYSTEM

★ Provides shape to the body
★ Attachments for muscles
★ Protects internal organs
★ Store for essential minerals (calcium and phosphorus)
★ Supplies red bone marrow (for making red and white blood cells and platelets)
★ Supplies yellow bone marrow (for fat storage)

Bones come in many different shapes: long, short, flat, sesamoid, and irregular.

Many long bones develop in a cartilage model by the formation of ossification centers (endochondral ossification), but some flat bones, such as the mandible and skull plates, form by mineralization inside membrane sheets (intramembranous ossification).

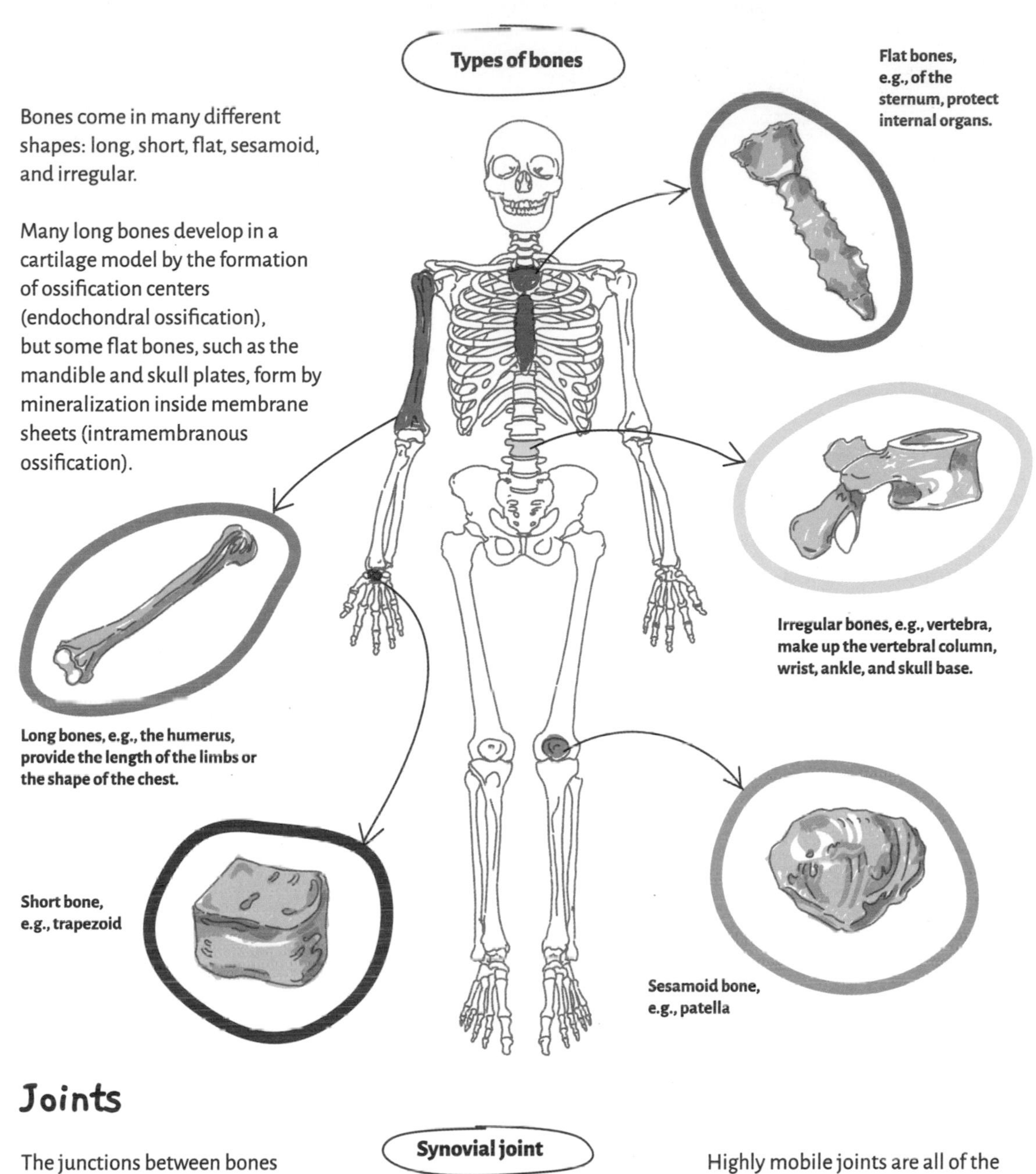

Types of bones
Flat bones, e.g., of the sternum, protect internal organs.
Irregular bones, e.g., vertebra, make up the vertebral column, wrist, ankle, and skull base.
Long bones, e.g., the humerus, provide the length of the limbs or the shape of the chest.
Short bone, e.g., trapezoid
Sesamoid bone, e.g., patella

Joints

The junctions between bones are called **joints**. Joints can be very stable and immobile (e.g., the sutures of the skull) or relatively less stable and more mobile (e.g., the ball-and-socket joints of the shoulder and hip). Joint stability is achieved by close fitting of the adjacent bone surfaces, such as the ball of the femur fitting snugly into the socket of the hip bone, the presence of strong ligaments around the joint, and the existence of strong muscles crossing the joint.

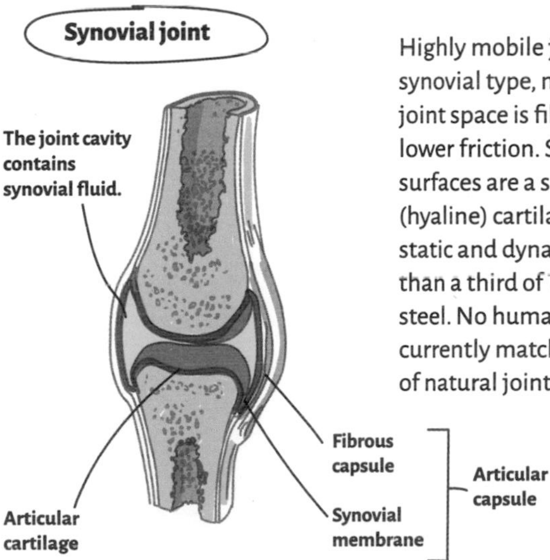

Synovial joint
The joint cavity contains synovial fluid.
Articular cartilage
Fibrous capsule
Synovial membrane
Articular capsule

Highly mobile joints are all of the synovial type, meaning that the joint space is filled with fluid to lower friction. Synovial joint surfaces are a smooth, glassy (hyaline) cartilage that has both static and dynamic friction less than a third of Teflon on stainless steel. No human-made surface currently matches the low friction of natural joint cartilage.

MUSCULAR SYSTEM

Muscles produce movement by shortening and lengthening. Skeletal (voluntary) muscles attach to the skeleton and produce movement at joints by drawing one end of the muscle toward the other.

Muscle structure

Skeletal muscles consist of **muscle fibers** aligned in parallel to make muscle **fascicles.** Muscle fibers must be activated by nerves, which contact muscles at regions called **neuromuscular junctions**.

A muscle's strength depends on how many fascicles are lined up in parallel, i.e., the cross-sectional area. The proteins that produce contraction are called **myofibrils**.

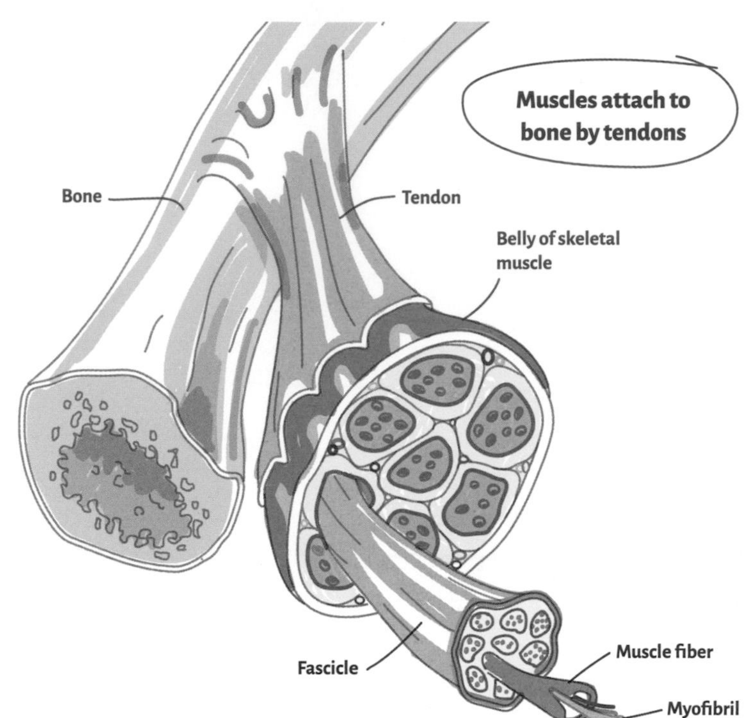

Muscles attach to bone by tendons
Bone
Tendon
Belly of skeletal muscle
Fascicle
Muscle fiber
Myofibril

The three main types of muscles are striated, cardiac, and smooth.

Striated (striped) muscles are voluntary muscle that attach to the skeleton, these make up most of the muscles in the body (about 70% to 80% of lean weight). The striping seen under the microscope is due to the regular alignment of the muscle proteins (actin and myosin) that produce contraction.

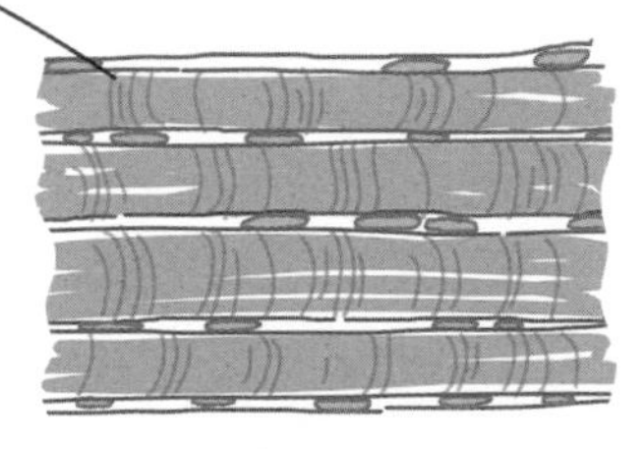

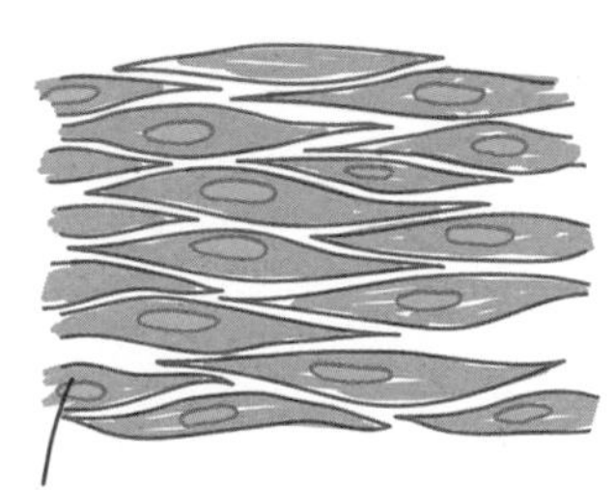

Smooth muscle has no stripes because the proteins that produce contraction don't have regular alignment, but smooth muscle is very important in strengthening the walls and controlling the internal diameters of hollow organs, such as the blood vessels, airways, and alimentary canal.

Cardiac muscle is striated like skeletal muscle but is involuntary and found only in the heart wall.

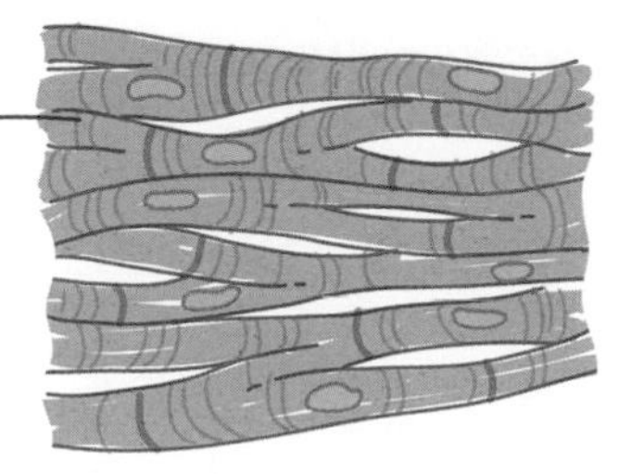

MUSCLE SHAPES AND FUNCTIONS

Muscles have different shapes for different functions.

- ★**Powerful** muscles, e.g., **masseter** and gluteus maximus have a large cross-sectional area for their size
- ★**Convergent**, e.g., pectoralis major: A fan-shaped muscle where the fibers converge on a tendon
- ★**Fusiform**, e.g., biceps brachii: A spindle-like muscle found in the arm
- ★**Sheetlike**, e.g., abdominal wall muscle: Protects the internal organs and moves the trunk
- ★**Parallel**, e.g., sartorius: A long, strap-like thigh muscle that has many muscle fibers
- ★**Circular**, e.g., orbicularis oris: A circular muscle that surrounds the mouth
- ★**Multipennate**, e.g., deltoid: Many featherlike bundles form the rounded curve of the shoulder (pennate means featherlike)
- ★**Bipennate**, e.g., quadriceps femoris: Muscle fibers attach on a central tendon from two directions
- ★**Unipennate**, e.g., tibialis anterior: Muscle fibers attach to a tendon from only one direction

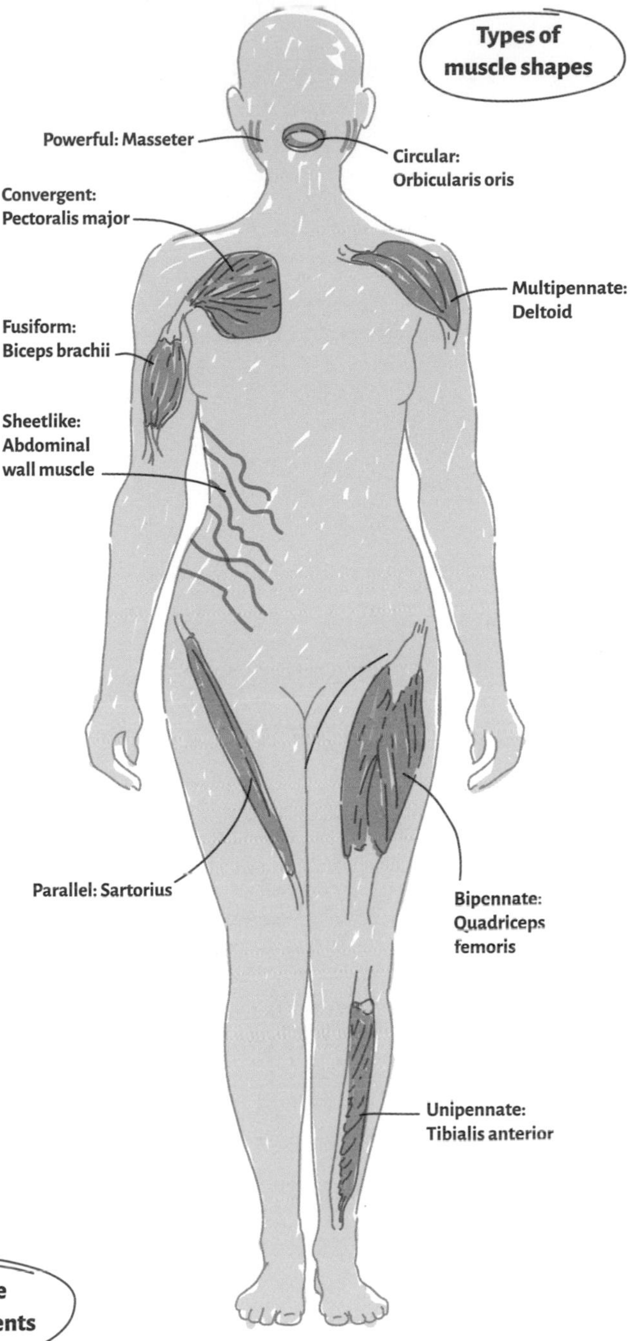

Types of muscle shapes
Powerful: Masseter
Circular: Orbicularis oris
Convergent: Pectoralis major
Multipennate: Deltoid
Fusiform: Biceps brachii
Sheetlike: Abdominal wall muscle
Parallel: Sartorius
Bipennate: Quadriceps femoris
Unipennate: Tibialis anterior

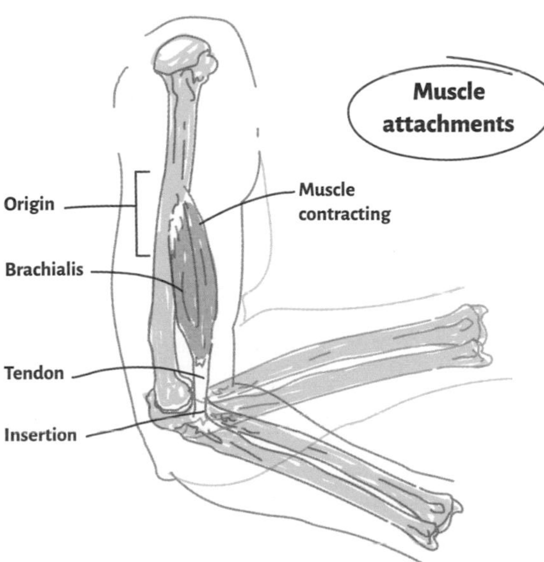

Muscle attachments
Origin
Muscle contracting
Brachialis
Tendon
Insertion

How muscles work

Muscles can only produce movement by shortening, although muscles may lengthen while under power to provide a smooth change in limb position. Most muscles have an **insertion** onto a bone via a **tendon**. The action of a muscle can be deduced by noting which joints it crosses and the side of the joint traversed by the muscle. For example, the **brachialis** that lies on the front of the elbow is an elbow flexor.

NERVOUS SYSTEM AND SENSES

The nervous system serves the functions of sensory perception, information processing, decision-making, and movement control. The most important cell type in the nervous system is the neuron, which actually processes and transmits information, but neurons need support.

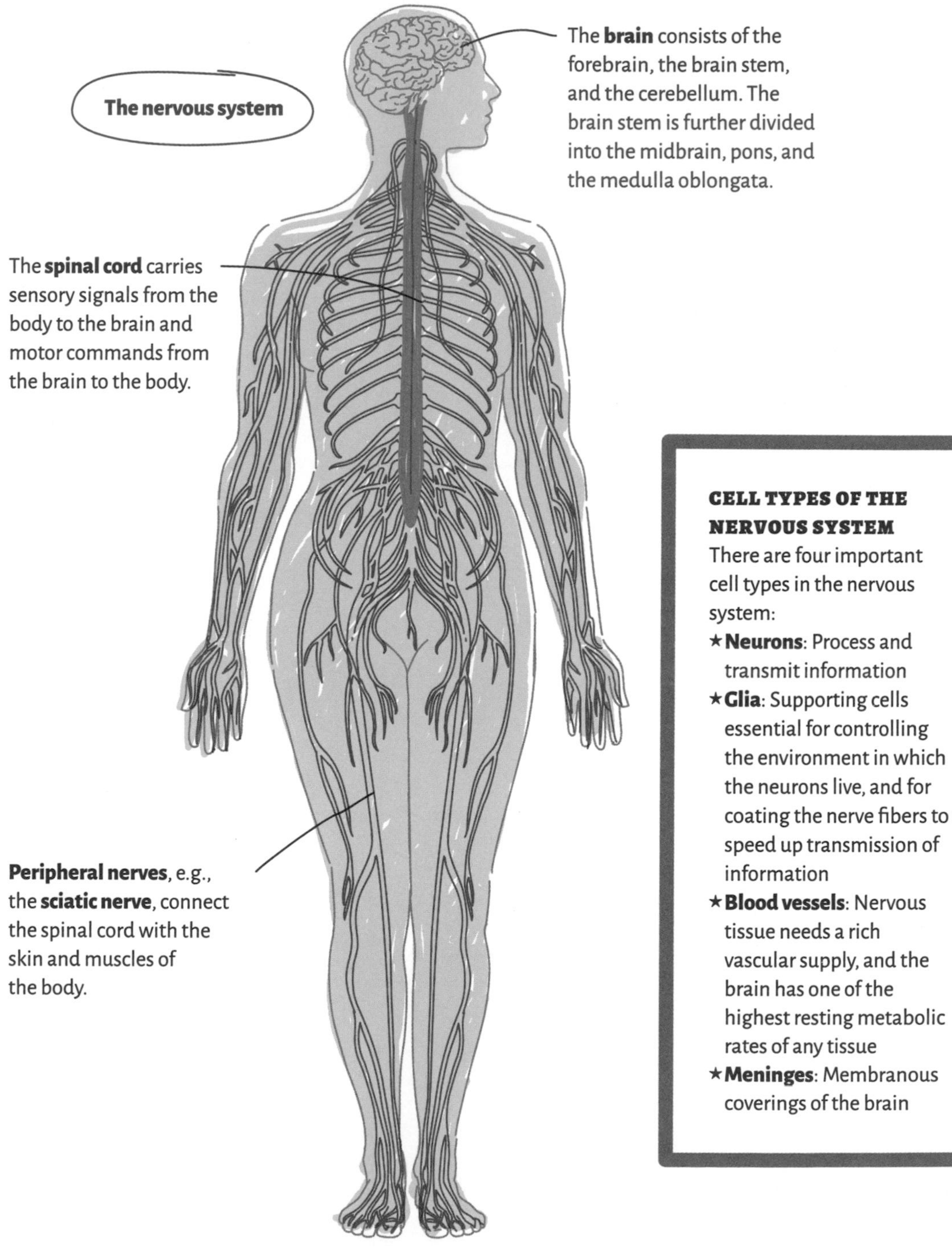

The nervous system
The **brain** consists of the forebrain, the brain stem, and the cerebellum. The brain stem is further divided into the midbrain, pons, and the medulla oblongata.
The **spinal cord** carries sensory signals from the body to the brain and motor commands from the brain to the body.
Peripheral nerves, e.g., the **sciatic nerve**, connect the spinal cord with the skin and muscles of the body.

CELL TYPES OF THE NERVOUS SYSTEM

There are four important cell types in the nervous system:

* **Neurons**: Process and transmit information
* **Glia**: Supporting cells essential for controlling the environment in which the neurons live, and for coating the nerve fibers to speed up transmission of information
* **Blood vessels**: Nervous tissue needs a rich vascular supply, and the brain has one of the highest resting metabolic rates of any tissue
* **Meninges**: Membranous coverings of the brain

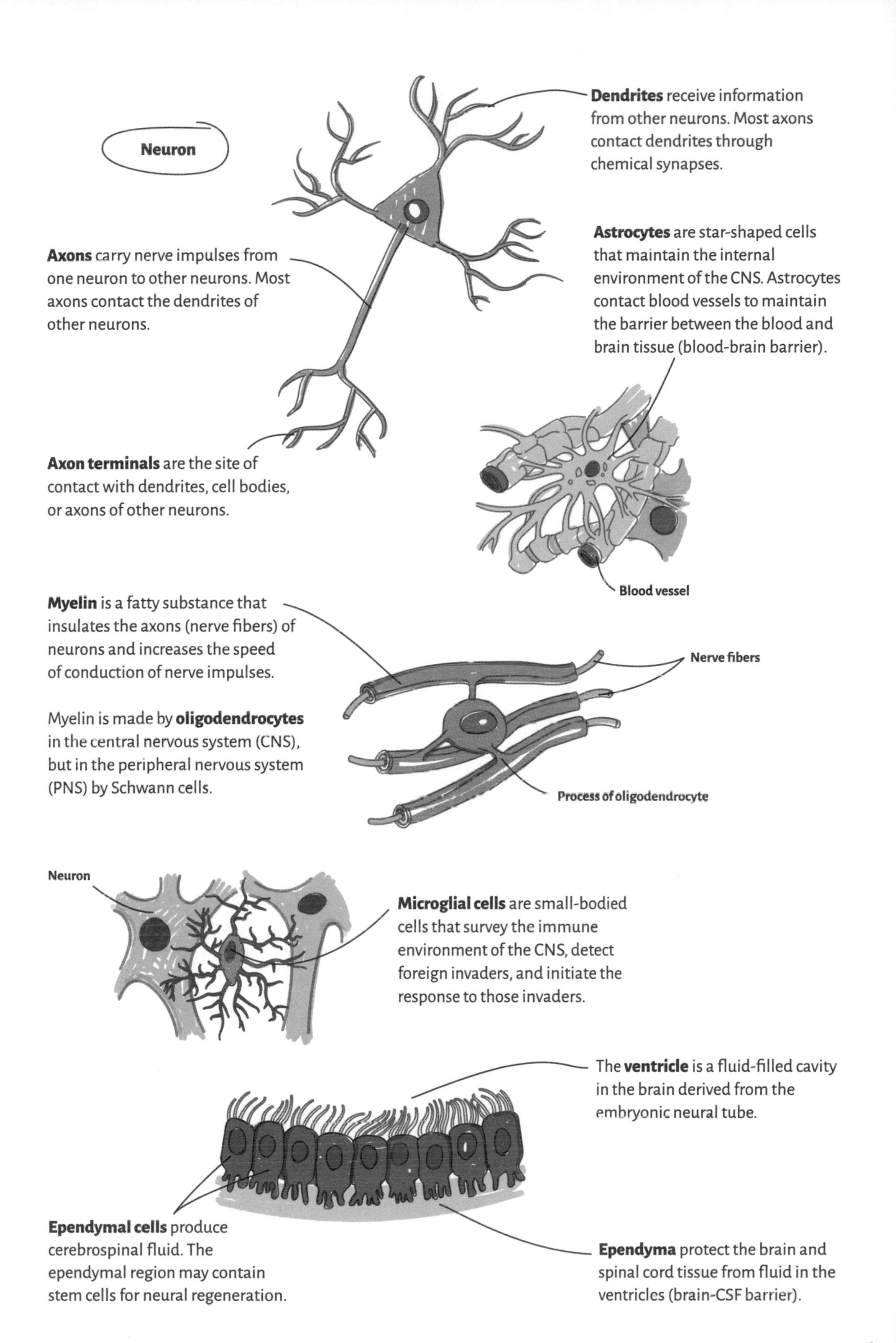
Neuron
Dendrites receive information from other neurons. Most axons contact dendrites through chemical synapses.
Axons carry nerve impulses from one neuron to other neurons. Most axons contact the dendrites of other neurons.
Astrocytes are star-shaped cells that maintain the internal environment of the CNS. Astrocytes contact blood vessels to maintain the barrier between the blood and brain tissue (blood-brain barrier).
Axon terminals are the site of contact with dendrites, cell bodies, or axons of other neurons.
Blood vessel
Myelin is a fatty substance that insulates the axons (nerve fibers) of neurons and increases the speed of conduction of nerve impulses.
Nerve fibers
Myelin is made by **oligodendrocytes** in the central nervous system (CNS), but in the peripheral nervous system (PNS) by Schwann cells.
Process of oligodendrocyte
Neuron
Microglial cells are small-bodied cells that survey the immune environment of the CNS, detect foreign invaders, and initiate the response to those invaders.
The **ventricle** is a fluid-filled cavity in the brain derived from the embryonic neural tube.
Ependymal cells produce cerebrospinal fluid. The ependymal region may contain stem cells for neural regeneration.
Ependyma protect the brain and spinal cord tissue from fluid in the ventricles (brain-CSF barrier).

The nervous systems

The nervous system is divided into the **central nervous system** (CNS), which controls the brain and the spinal cord, and the **peripheral nervous system** (PNS), which consists of the nerves and neurons outside the CNS.

There are about 80 billion neurons in the average human brain and a similar number of glial cells. The human spinal cord has only 70 million nerve cells, but the wall of the alimentary canal (the enteric nervous system) has even more nerve cells than the spinal cord!

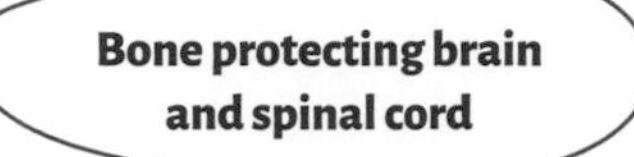
Bone protecting brain and spinal cord

The **skull** is the brain case and consists of bony plates (frontal, parietal, occipital, and temporal) that are locked together by sutures to form a protective dome.

Veins draining the brain can be torn by violent movement of the head (e.g., in motor vehicle accidents).

The **atlas** is the first vertebra of the spinal column. It is a delicate ring that can be fractured by a blow to the head.

The **axis** is the second vertebra, it allows the atlas to rotate around the long axis of the neck.

The **cervical spinal cord** is vulnerable to damage because of the slightly-built cervical vertebral column. It contains motor neurons (phrenic nucleus) that control the diaphragm muscle, which is essential for lung ventilation.

Central nervous system (CNS)

The CNS is very sensitive to physical injury. It is protected by hard bony plates locked together by sutures. The skull base is also very rigid and composed of dense bone, which is usually only fractured by heavy blows or motor vehicle accidents. The spinal cord is protected by the encircling bones (vertebrae) of the vertebral column (backbone).

The vertebrae are smallest and weakest in the neck (cervical) region. This is why whiplash injuries of the neck from motor vehicle accidents are so hazardous, and why stabilizing the neck is a critical step in first aid. Damage to the spinal cord in the cervical region can lead to loss of sensation and paralysis of all four limbs (quadriplegia). Even inside its bony case, the brain is easily damaged by rapid acceleration or deceleration in motor vehicle accidents or blows to the head.

Peripheral and enteric nervous system

The PNS consists of nerves and nerve cells outside the CNS. Ganglia are collections of nerve cell bodies in the PNS. Ganglia may be sensory or autonomic (for control of automatic functions).

Traditionally, the autonomic nervous system has been divided into sympathetic components (used for emergencies) or parasympathetic (used for sustained restorative functions).

The 100 million enteric neurons in the alimentary canal control the movement of the gut smooth muscle and the secretion of fluids from gut glands.

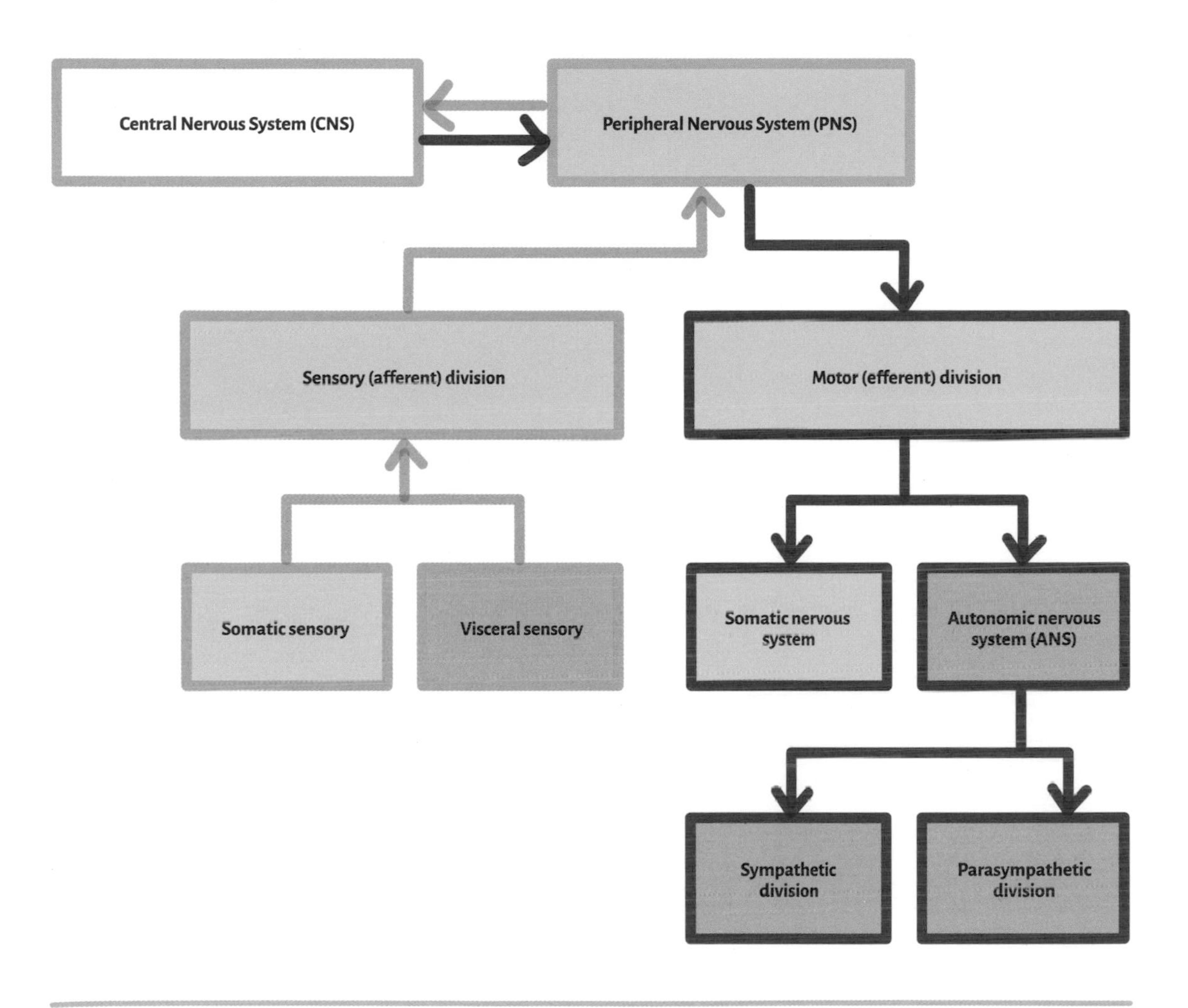

Central Nervous System (CNS)
Peripheral Nervous System (PNS)
Sensory (afferent) division
Motor (efferent) division
Somatic sensory
Visceral sensory
Somatic nervous system
Autonomic nervous system (ANS)
Sympathetic division
Parasympathetic division

Senses

The major sense organs are closely linked with the nervous system. We often think of just five senses (sight, hearing, smell, taste, and touch), but there are many more. Additional senses include:

★ Head position in space
★ Head acceleration or rotation
★ Limb joint position
★ Internal organ filling or emptying (e.g., stomach, bowel, and bladder)
★ Tension on membranes in the abdomen

Even the sense of touch is more complex than we commonly think because it includes:

★ Simple touch (e.g., touch with cotton ball)
★ Pain and itch
★ Pressure
★ Temperature
★ Vibration (actually sense of surface texture)
★ Two-point discrimination (the ability to tell if one has been touched by one point or two points close together)

CIRCULATORY SYSTEM AND BLOOD

The circulatory system is comprised of the heart and blood vessels. Its role is to transport gases, nutrients, waste, and proteins around the body.

Blood vessels

Blood vessels leaving the heart contain blood under high pressure (25 to 150 mm mercury) and are called **arteries**, whereas vessels returning blood to the heart are low-pressure vessels and are called **veins**.

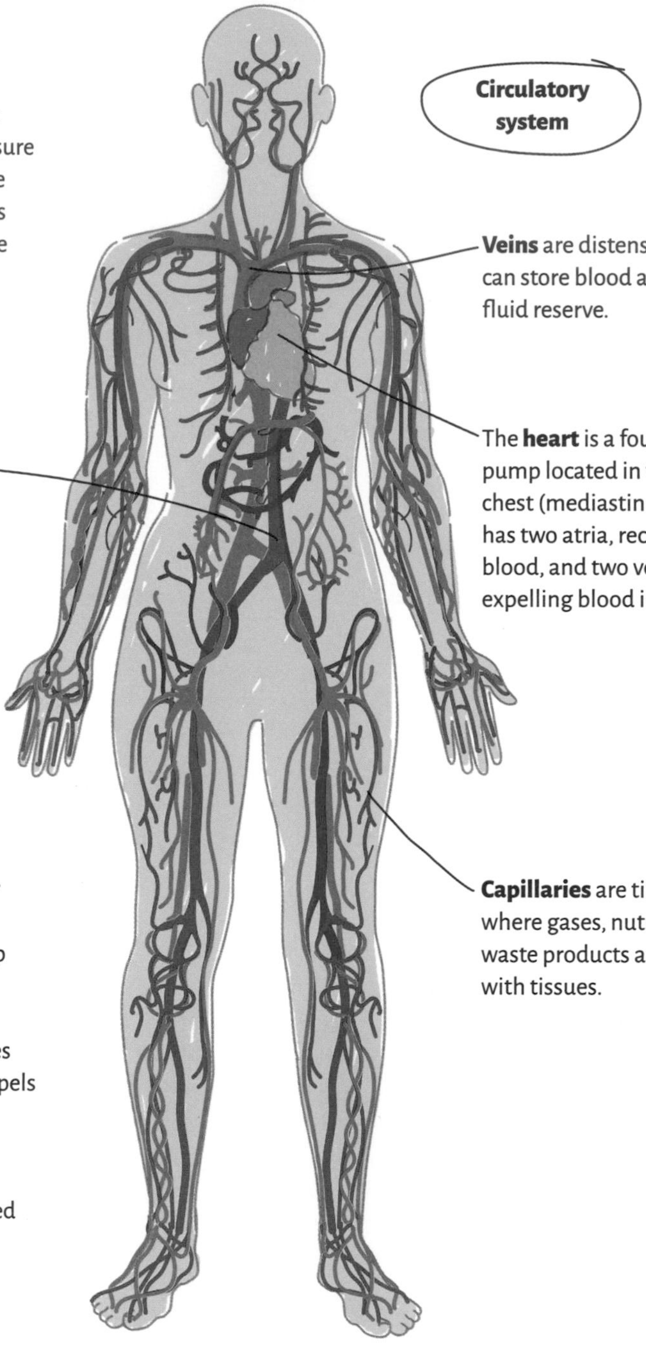

Circulatory system

Veins are distensible vessels that can store blood and provide a fluid reserve.

The **heart** is a four-chambered pump located in the central chest (mediastinum). The heart has two atria, receiving venous blood, and two ventricles, expelling blood into arteries.

Arteries have abundant smooth muscle in their walls to provide elasticity and regulate blood pressure.

Capillaries are tiny vessels where gases, nutrients, and waste products are exchanged with tissues.

Heartbeat

The heart starts beating at four weeks of human embryonic development and does not stop until the moment of death, so the average human heart can beat as many as 2.5 billion times in a lifetime. Each heartbeat expels approximately 70 ml (2.4 fluid ounces) of blood, so more than 150 million liters (40 million gallons) of blood can be pumped in a lifetime.

FUNCTIONS OF THE CIRCULATORY SYSTEM

The circulatory system and the blood it carries transport many important substances around the body.

★ Essential nutrients (sugars, amino acids, fats, nucleic acids)
★ Vitamins
★ Minerals (calcium, iron, copper, magnesium)
★ Chemical messengers (hormones)
★ Clotting factors
★ Immune system cells (white blood cells)
★ Immune system proteins (antibodies, complement)
★ Buffering compounds such as the plasma proteins and bicarbonate ions that maintain optimal pH (acid-base balance) of the body tissues
★ Distribution of blood to the skin in hot weather plays an important role in temperature control (thermoregulation).

Blood cell types

Red blood cells (erythrocytes) are approximately 36% to 50% of blood volume, and carry oxygen and carbon dioxide. Red blood cells are biconcave disks (like a cough lozenge) that have lost their nuclei but are packed with hemoglobin to carry oxygen.

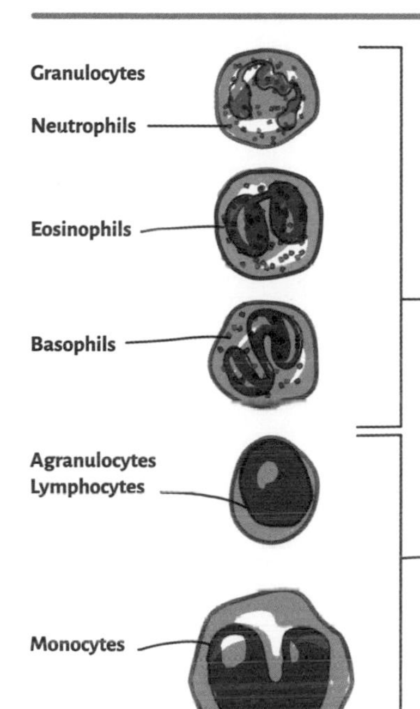
Red blood cells
Granulocytes
Neutrophils
Eosinophils
Basophils
Agranulocytes
Lymphocytes
Monocytes
Granulocytes contain granules in their cytoplasm. They have multilobed nuclei.
Agranulocytes do not have granules in their cytoplasm and have nuclei that fill most of the cell.

White blood cells (leukocytes) are immune system cells that include cells with granules (granulocytes), or without (monocytes and lymphocytes).

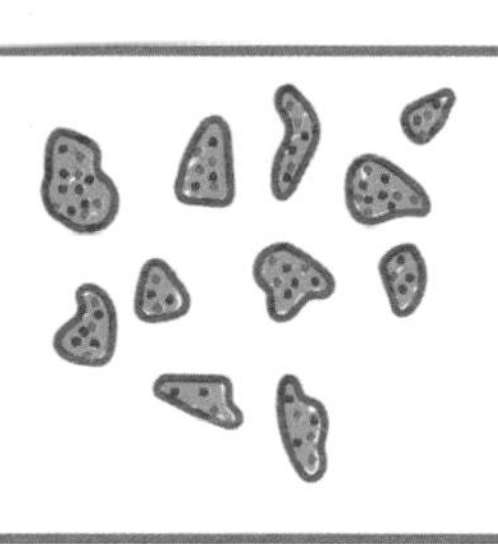

Platelets are cell fragments that are vital in blood clotting (hemostasis).

Proteins in blood

Blood also contains proteins suspended in its fluid. These include plasma proteins that maintain the osmotic pressure of blood (e.g., albumin) and immunoglobulins (antibodies) that protect against foreign proteins, viruses, bacteria, and fungi. Other plasma proteins carry fat molecules (low and high density lipoproteins) and minerals such as iron and copper, help maintain blood pH, or contribute to blood clotting (prothrombin and fibrinogen).

Two circulations

There are two types of circulation: pulmonary and systemic. **Pulmonary circulation** carries blood from the right side of the heart to the lungs for oxygenation and removal of carbon dioxide, and returns that blood to the heart.

Systemic circulation carries blood from the left side of the heart to the rest of the body to provide oxygen to the tissues and gather carbon dioxide for removal from the body. The systemic veins then return this oxygen-depleted, but carbon dioxide-loaded, blood to the heart.

RESPIRATORY SYSTEM

The main job of the respiratory system is to bring oxygen into the body and expel carbon dioxide. It also plays an important role in controlling the pH of the blood and aids in thermoregulation.

On average, humans breathe 12 times each minute, so we take about 400 million breaths in our life. Each breath moves about 500 ml (17 fluid ounces) of air into and out of the lungs, so we breathe about 200 million liters (53 million gallons) of air in our lives.

The lungs and acid-base balance

The respiratory system also contributes to control of the pH of the blood, because carbon dioxide dissolved in blood makes that blood acidic. Increasing loss of CO_2 from the blood makes the blood more alkaline; reducing loss of CO_2 from the blood makes the blood more acid.

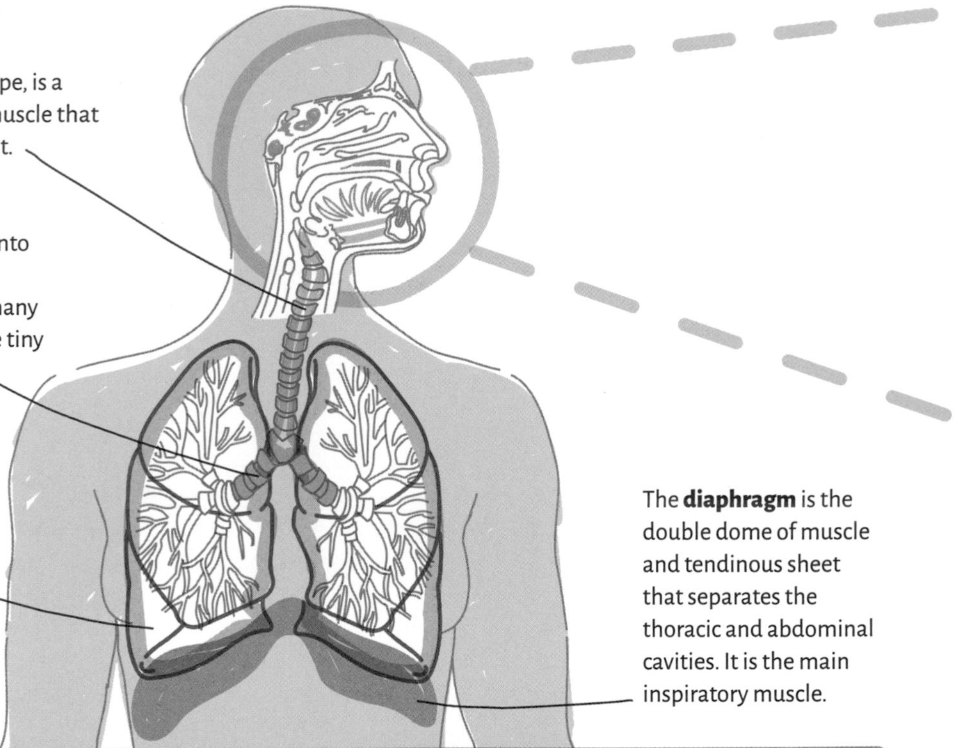

The **trachea**, or windpipe, is a tube of cartilage and muscle that carries air into the chest.

The trachea branches into main **bronchi**, which continue to divide as many as 21 times to reach the tiny air sacs of the lungs.

The **lungs** are the site of gas exchange between blood and inhaled air.

The **diaphragm** is the double dome of muscle and tendinous sheet that separates the thoracic and abdominal cavities. It is the main inspiratory muscle.

LUNGS AND GAS EXCHANGE

Once air reaches the small sacs at the end of the airways, called **alveoli**, gas exchange with the blood of the pulmonary circulation can take place.

The membranes between the alveolar gas and the bloodstream of the capillary bed are only 1 to 2 µm (micrometers) thick (1/25,000 to 1/12,000 of an inch), therefore, gas molecules can freely diffuse from high to low pressure. Oxygen molecules diffuse from the alveoli to the blood, and carbon dioxide diffuses from the blood to the alveoli.

LUNG PROTECTION

The lungs are constantly exposed to the external environment with the risk of infection and damage from toxic inhalants. The alveoli contain alveolar macrophages, a cell type that engulfs debris and microorganisms. Debris is cleared from the lungs by a carpet of tiny hairlike cilia that sweep the debris suspended in mucus toward the larynx, where the debris is swallowed or spit out.

LUNG VENTILATION

Lung ventilation is produced by respiratory muscles pulling on the ribs (intercostal muscles) or by expanding the height of the chest cavity (the muscular diaphragm). The semi-voluntary contraction of these muscles is regulated by the brain stem in response to the concentration of oxygen and carbon dioxide in the blood. We can interrupt the automatic cycle of breathing to speak, cough, and sneeze.

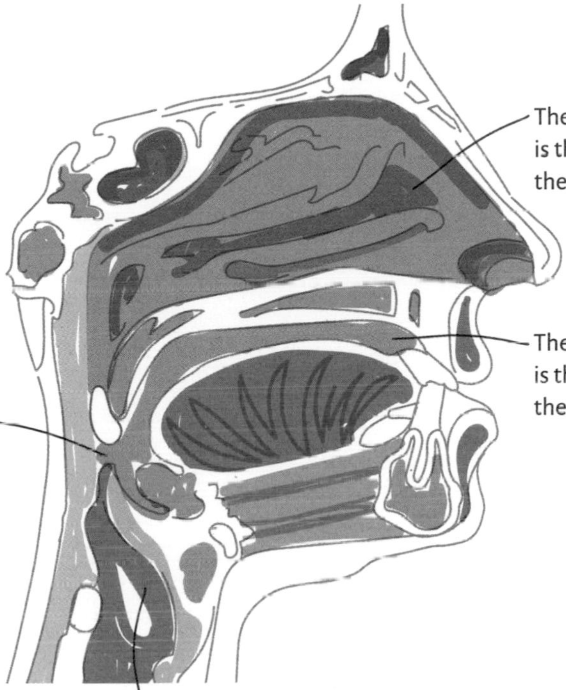

The **nasal cavity** is the inside of the nose.

The **oral cavity** is the inside of the mouth.

The **pharynx** is part of the throat. Inhaled air passes through it to the larynx.

The **larynx** is our voice box. It is vibrated by exhaled air to produce the voice.

Nasal cavity

The **nasal cavity** is the initial part of the respiratory system, and it provides for the sense of smell.

You can breathe through the mouth (**oral cavity**) and through the nose, but the nose is much better adapted to breathing than the mouth. This is because there are fine bony elevations in the nasal cavities that increase the surface area of mucous membranes; these are used to warm and moisten the air, filter out dust and microorganisms, and detect odors.

The human nose is very simple in structure because our sense of smell is so poor. By contrast, animals with a good sense of smell (e.g., dogs) have highly folded nasal cavity walls.

DiGESTiVE SYSTEM

The functions of the digestive system include ingestion (taking food into the mouth and chewing), digestion (breaking food down into component nutrient molecules), absorption (taking nutrients across the wall of the digestive tract into the blood), and excretion (discharging waste through the anus).

The alimentary canal

Did you know that your body is like a doughnut (topologically speaking)? Your alimentary canal extends from your mouth to your anus, much like the hole in a doughnut, with various out-pockets from the tube for glands and their ducts, e.g., the salivary glands, liver, gallbladder, and pancreas.

The digestive tract (gut) forms in the fourth week of embryonic development, when a flat sheet of tissue rolls to form a tube. This forms a mouth at the front end of the embryo and a primitive anus at the back. The jaw, chewing, and tongue muscles form around the primitive mouth, and the glands of the gut develop as buds from the tube.

IMMUNE FUNCTION IN THE ALIMENTARY CANAL

The alimentary canal (gut) has clusters of immune system cells in its wall called **lymphoid nodules**.

- ★ Lymphoid nodules protect against those ingested bacteria, viruses, and fungi that have not been degraded by the stomach acid.
- ★ The gut immune system also controls the bacteria of the natural gut flora and stops them invading the gut wall.

The **esophagus** is a muscular tube (skeletal in its upper parts, smooth in its lower) that carries fluid and food from the pharynx to the stomach.

The **liver** helps digestion by secreting bile salts that break down fats. The alimentary canal is exposed to toxins, foreign proteins, and microorganisms from the external environment and is a potential route for those agents to enter the body. The liver is the main protection against toxins, such as alcohol, ammonia, and other products of microorganisms, and receives all the blood from the gut wall.

The **gallbladder** stores bile from the liver and releases it when needed for a fatty meal.

The **small intestine** is the main site of absorption where amino acids, sugars, fatty acids, glycerol, nucleic acids, vitamins, and minerals pass across the gut wall into the gut bloodstream. Most nutrients pass to the liver where they are processed into proteins and complex sugars or passed through into the general bloodstream.

In the **large intestine**, water and minerals are absorbed to form feces. The large intestine also contains many bacteria of the natural gut flora, which provide nutrients for the host, by breaking down some of the cellulose that cannot be digested by human gut enzymes. As much as 10% of our nutrients is provided by our gut flora.

How the digestive system works

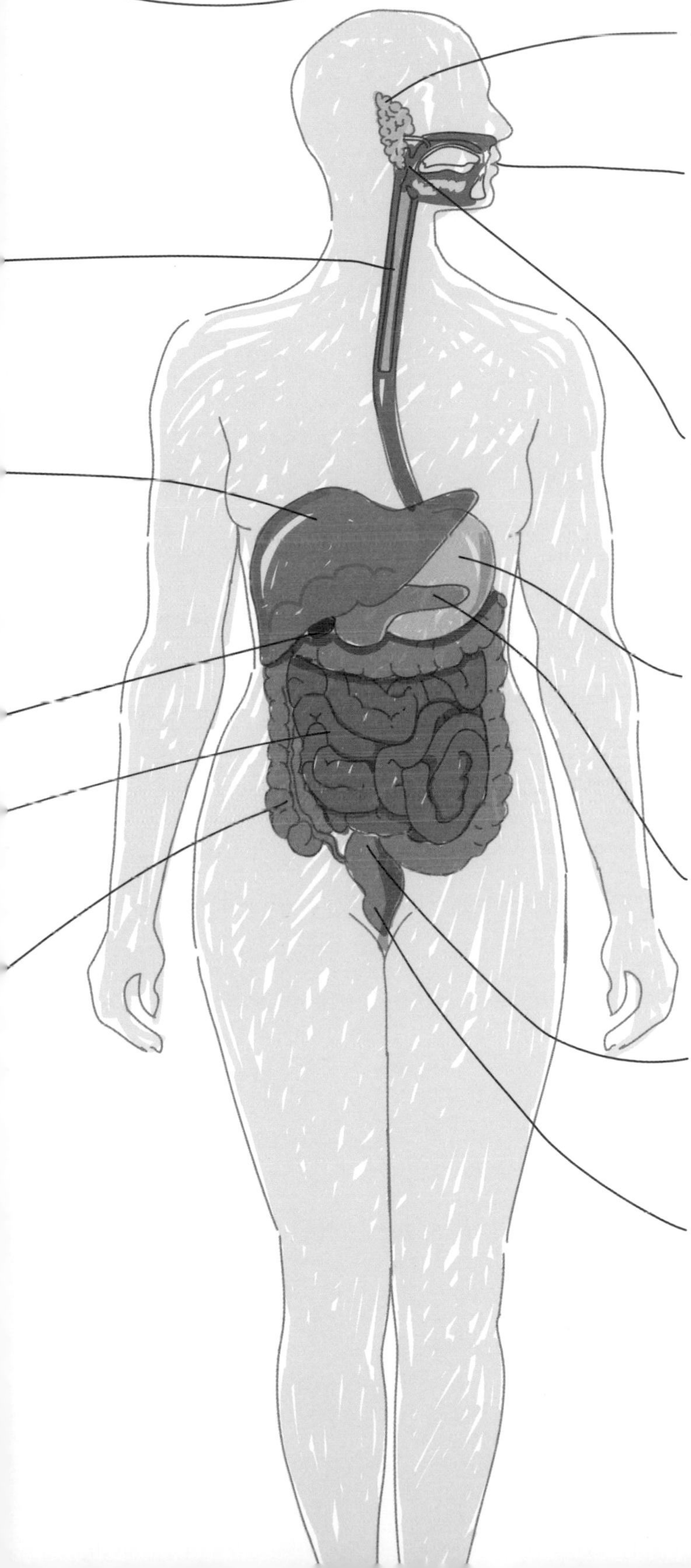

Salivary glands (major and minor) produce saliva to moisten food and begin the digestive process for starches.

In the **mouth**, food is mixed with salivary enzymes to form a soft **bolus**—a lump of chewed food—that is thrown back into the pharynx and moved down to the esophagus, where coordinated waves of muscular contraction (peristalsis) slowly progress the bolus down to the stomach.

The **pharynx** is the common pathway for inhaled air and swallowed fluid and food. The pharynx is a skeletal muscle tube that suspends from the base of the skull. Contraction of the pharyngeal muscles forces the food down to the esophagus.

The **stomach** breaks down food by mechanical, chemical, and biological digestion and passes the products to the small intestine, where digestion continues to produce a soup of nutrients.

The **pancreas** lies behind the stomach and secretes biological catalysts, called **enzymes**, which break down protein, fats, and starches.

Feces is stored in the **rectum** until it is expelled. The presence of feces in the upper part (rectal ampulla) gives the urge to defecate.

The **anus** is the last part of the alimentary canal. It can dilate to allow the passage of feces, and it can distinguish between gas (flatus) and feces.

URINARY SYSTEM

The urinary system consists of the kidneys and ureters, the urinary bladder, and the urethra.

The kidneys need a rich blood supply to function properly—about 20% to 25% of the blood from the heart. Together, the two kidneys produce about 1 ml of urine per minute, or 800 ml to 2 liters (1.7 to 4.2 pints) per day.

The urine flows down the paired ureters to the urinary bladder, which is a smooth muscle bag that can store urine and contract to expel it through the urethra to the external environment.

KIDNEY FUNCTION

- ★Removes nitrogen-containing waste from the body
- ★Regulates the concentration of ions such as sodium, potassium, chloride, and bicarbonate in the blood
- ★Adjustment of blood pressure and pH (acid-base balance)
- ★Controls production of red blood cells

The urinary system develops in close proximity to the male reproductive tract, so urine and semen follow the same pathway to the external world in males.

Kidneys filter the blood to remove nitrogenous waste and control the ionic balance of the blood.

The **ureter** is a smooth muscle tube that carries urine from the kidney to the urinary bladder.

The **urinary bladder** is a muscular bag that stores the urine until it is convenient to release it. It needs to be emptied regularly to avoid a urinary tract infection.

The **urethra** carries urine from the urinary bladder to the external world. It is considerably shorter in females than in males.

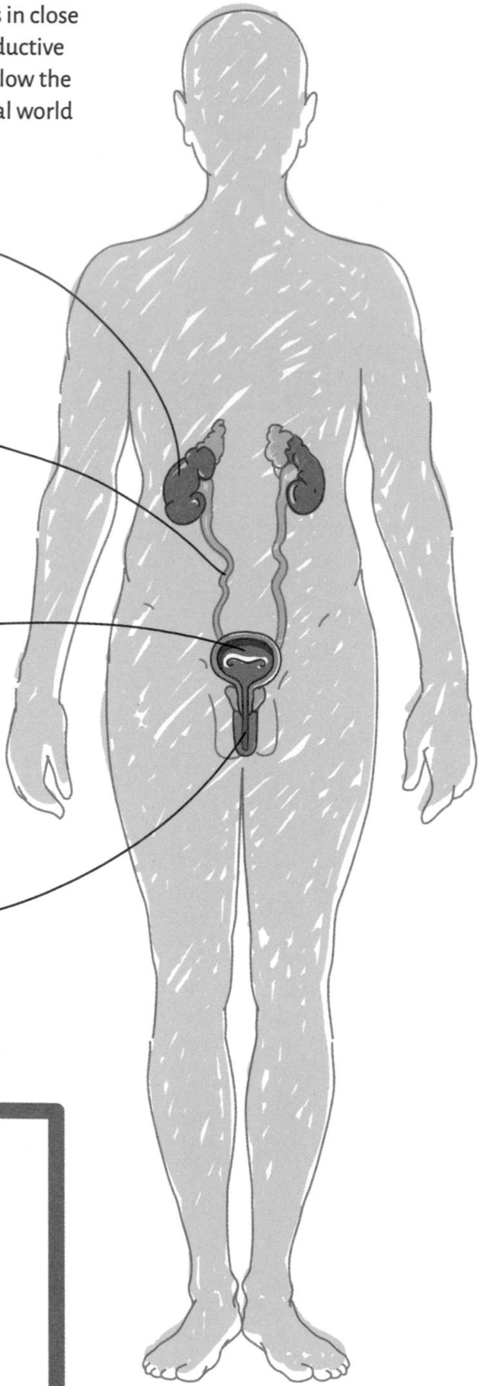

THE NITROGEN PROBLEM

Using excess food proteins as an energy source causes a problem for the body, because the amine part of the amino acids must be sheared off and the resulting toxic ammonium ions disposed of. The liver turns these ammonium ions into urea, which is water-soluble and is discharged from the body in the urine.

REPRODUCTIVE SYSTEM

The reproductive system is concerned with producing and nurturing the next generation. This includes not only producing the sex cells that make new life but also providing a place for the embryo to develop, and a vehicle for delivering nutrition through lactation after birth.

The reproductive systems of both sexes are regulated by hormones produced by the pituitary gland at the base of the brain, allowing the brain to control reproductive cycles and function.

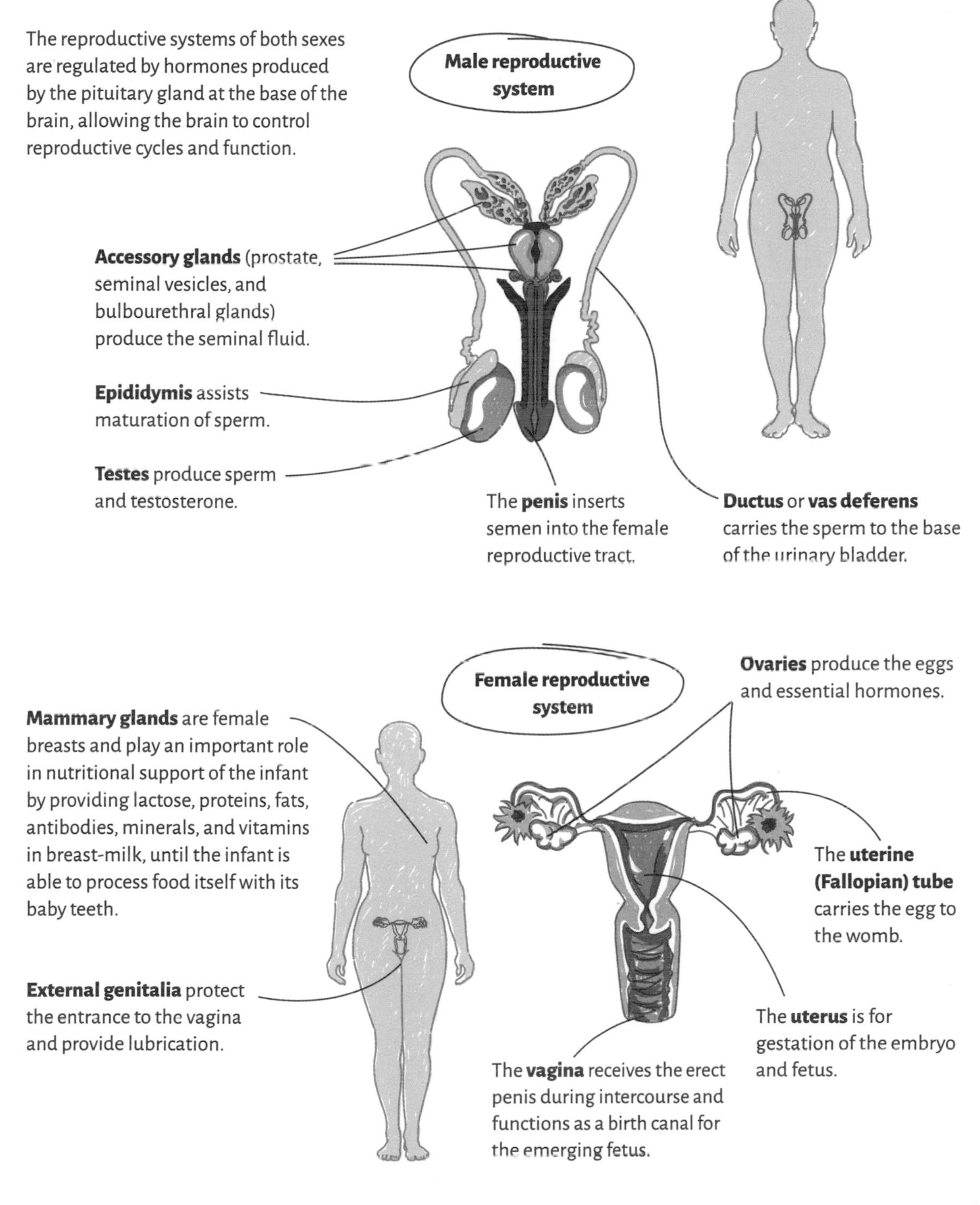

Male reproductive system
Accessory glands (prostate, seminal vesicles, and bulbourethral glands) produce the seminal fluid.
Epididymis assists maturation of sperm.
Testes produce sperm and testosterone.
The **penis** inserts semen into the female reproductive tract.
Ductus or **vas deferens** carries the sperm to the base of the urinary bladder.
Female reproductive system
Ovaries produce the eggs and essential hormones.
Mammary glands are female breasts and play an important role in nutritional support of the infant by providing lactose, proteins, fats, antibodies, minerals, and vitamins in breast-milk, until the infant is able to process food itself with its baby teeth.
The **uterine (Fallopian) tube** carries the egg to the womb.
External genitalia protect the entrance to the vagina and provide lubrication.
The **uterus** is for gestation of the embryo and fetus.
The **vagina** receives the erect penis during intercourse and functions as a birth canal for the emerging fetus.

IMMUNE SYSTEM

We are constantly surrounded by a sea of microorganisms and their toxic products. Without a defense system, our bodies would be quickly invaded and overwhelmed.

The **immune** or **lymphatic system** is a set of small structures distributed throughout the body that collectively drain excess tissue fluid and protect the body from foreign proteins and invaders, such as bacteria, fungi, viruses, rickettsiae, and parasites.

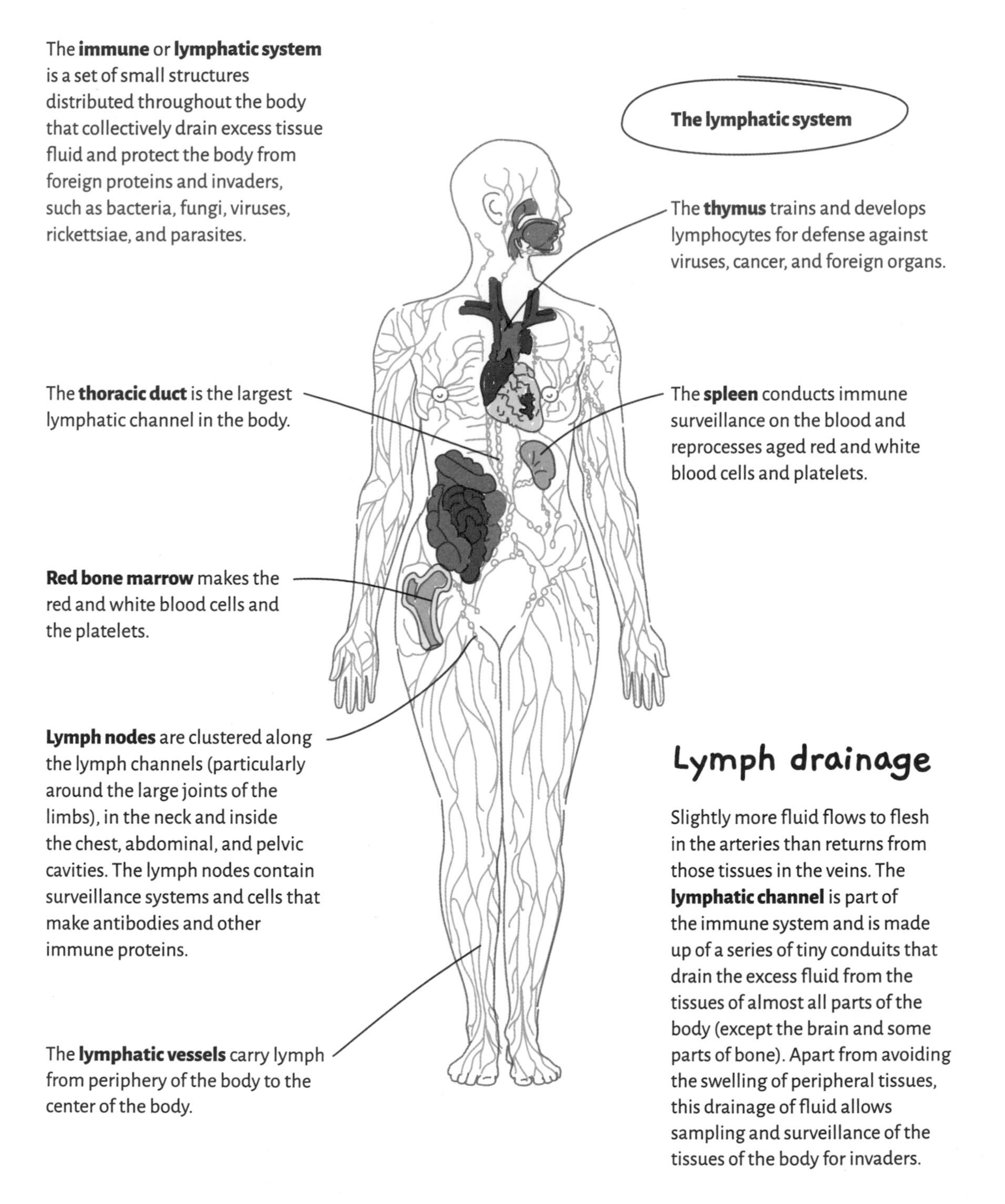
The lymphatic system
The **thymus** trains and develops lymphocytes for defense against viruses, cancer, and foreign organs.
The **thoracic duct** is the largest lymphatic channel in the body.
The **spleen** conducts immune surveillance on the blood and reprocesses aged red and white blood cells and platelets.
Red bone marrow makes the red and white blood cells and the platelets.
Lymph nodes are clustered along the lymph channels (particularly around the large joints of the limbs), in the neck and inside the chest, abdominal, and pelvic cavities. The lymph nodes contain surveillance systems and cells that make antibodies and other immune proteins.
The **lymphatic vessels** carry lymph from periphery of the body to the center of the body.

Lymph drainage

Slightly more fluid flows to flesh in the arteries than returns from those tissues in the veins. The **lymphatic channel** is part of the immune system and is made up of a series of tiny conduits that drain the excess fluid from the tissues of almost all parts of the body (except the brain and some parts of bone). Apart from avoiding the swelling of peripheral tissues, this drainage of fluid allows sampling and surveillance of the tissues of the body for invaders.

ENDOCRINE SYSTEM

The endocrine system is a distributed group of glands that play critically important roles in the control of the body's metabolism and reproductive function.

The term **endocrine** refers to the way that these glands secrete their products (hormones) directly into the bloodstream or body cavities. Both the nervous and endocrine systems regulate internal body function, but the nervous system operates on a much shorter time scale (seconds to minutes) than the endocrine system (hours to weeks to years).

The master endocrine gland

The master gland of the endocrine system is the **pituitary**, which is connected by a stalk to the hypothalamus at the base of the brain. The brain can influence pituitary function by releasing hormones from the hypothalamus (to the anterior pituitary) or by nerve pathways (to the posterior pituitary).

Other endocrine glands include the **adrenal cortex**, which controls the stress response and salt/water balance. The **adrenal medulla** controls the release of epinephrine and norepinephrine when there are emergencies

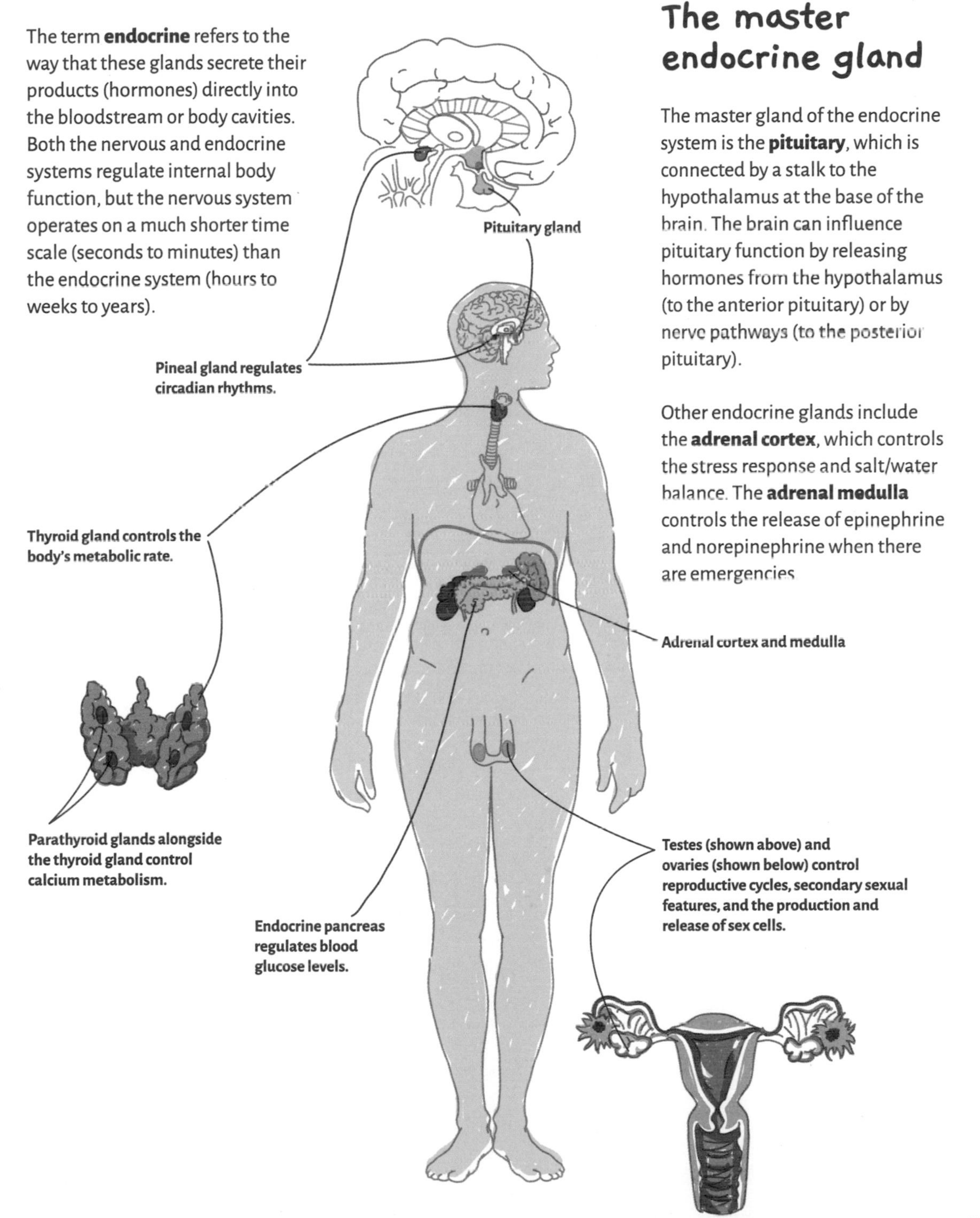

Pituitary gland
Pineal gland regulates circadian rhythms.
Thyroid gland controls the body's metabolic rate.
Adrenal cortex and medulla
Parathyroid glands alongside the thyroid gland control calcium metabolism.
Endocrine pancreas regulates blood glucose levels.
Testes (shown above) and ovaries (shown below) control reproductive cycles, secondary sexual features, and the production and release of sex cells.

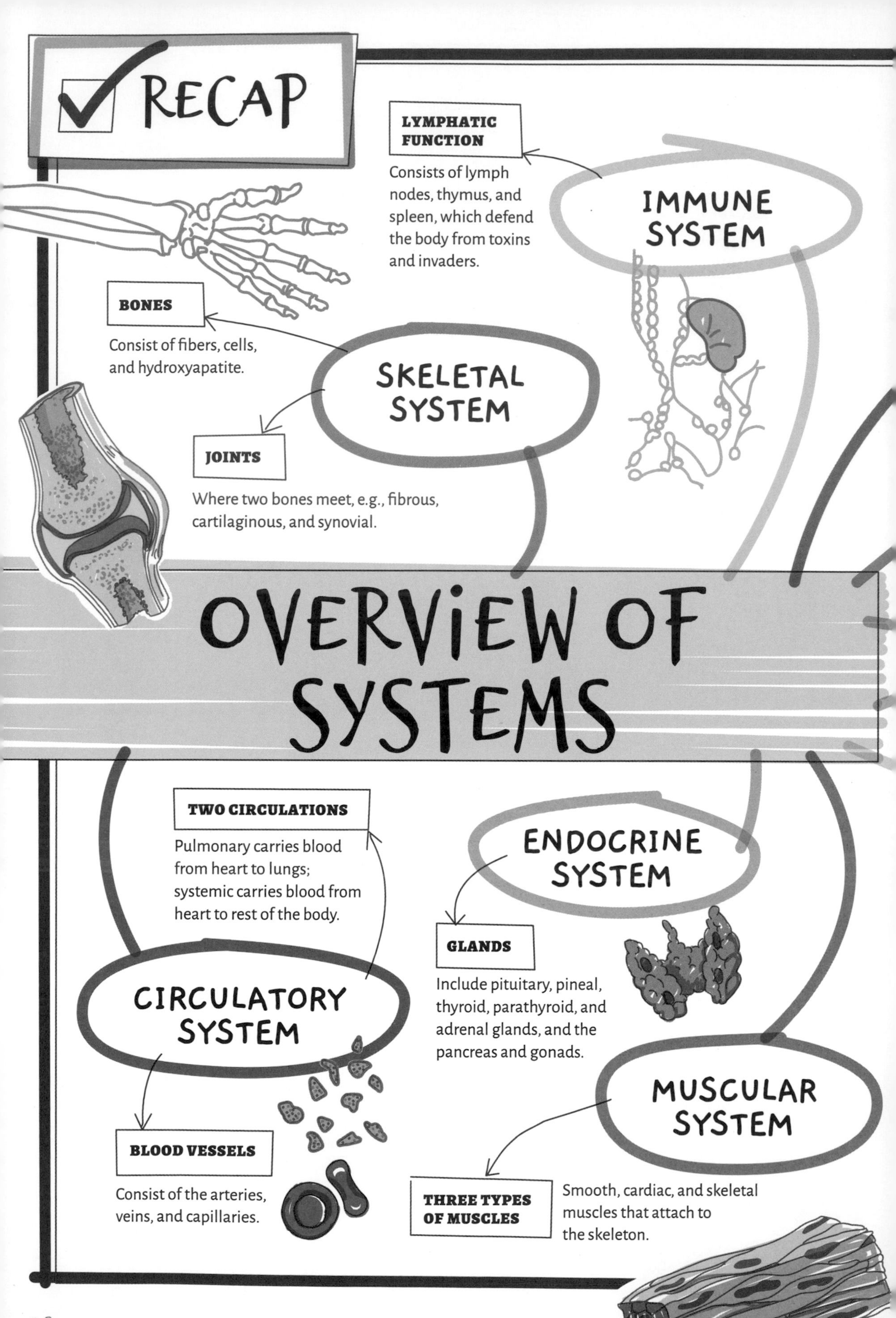
RECAP
LYMPHATIC FUNCTION
Consists of lymph nodes, thymus, and spleen, which defend the body from toxins and invaders.
IMMUNE SYSTEM
BONES
Consist of fibers, cells, and hydroxyapatite.
SKELETAL SYSTEM
JOINTS
Where two bones meet, e.g., fibrous, cartilaginous, and synovial.
OVERVIEW OF SYSTEMS
TWO CIRCULATIONS
Pulmonary carries blood from heart to lungs; systemic carries blood from heart to rest of the body.
ENDOCRINE SYSTEM
GLANDS
Include pituitary, pineal, thyroid, parathyroid, and adrenal glands, and the pancreas and gonads.
CIRCULATORY SYSTEM
MUSCULAR SYSTEM
BLOOD VESSELS
Consist of the arteries, veins, and capillaries.
THREE TYPES OF MUSCLES
Smooth, cardiac, and skeletal muscles that attach to the skeleton.

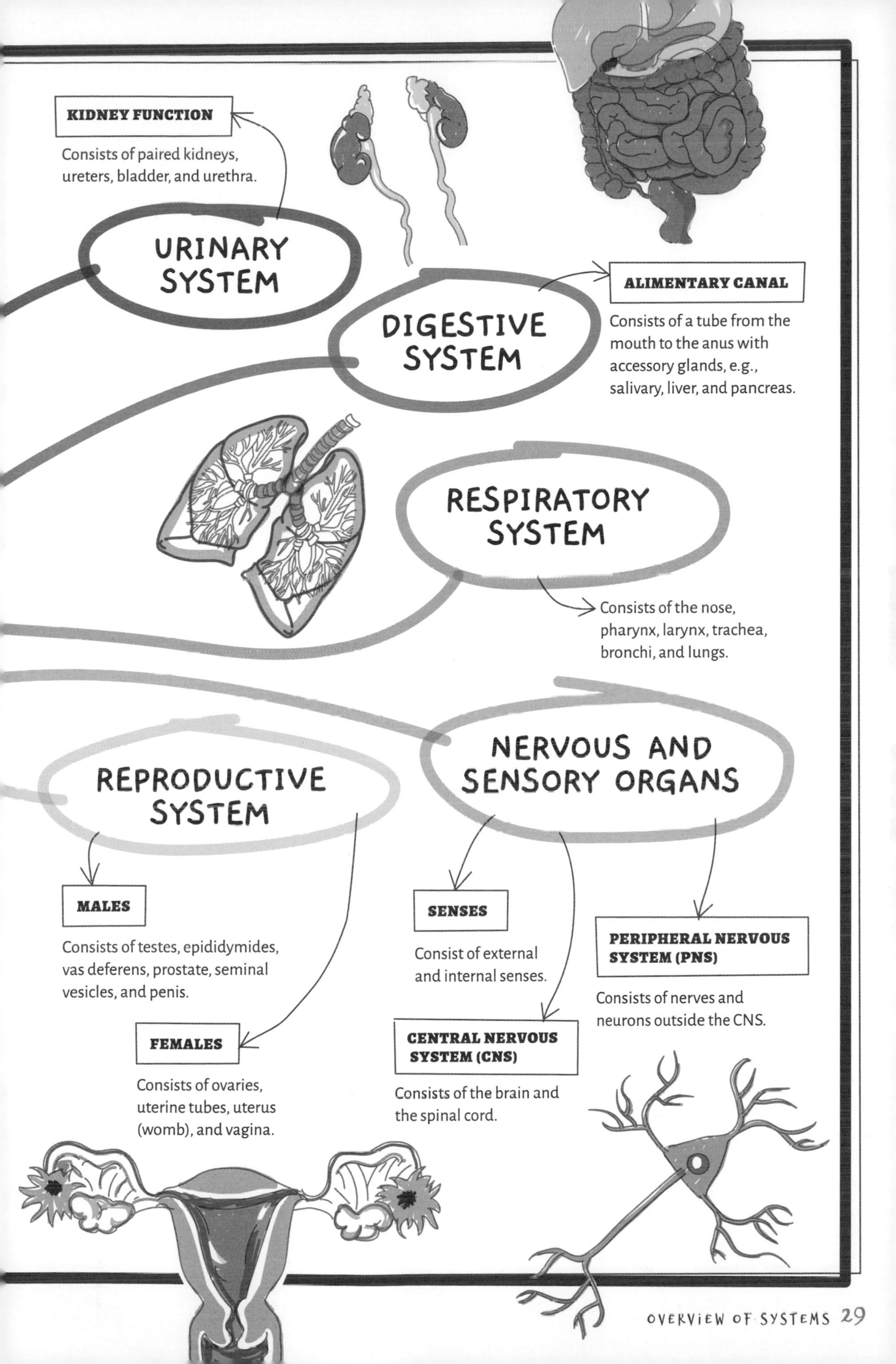
KIDNEY FUNCTION
Consists of paired kidneys, ureters, bladder, and urethra.
URINARY SYSTEM
DIGESTIVE SYSTEM
ALIMENTARY CANAL
Consists of a tube from the mouth to the anus with accessory glands, e.g., salivary, liver, and pancreas.
RESPIRATORY SYSTEM
Consists of the nose, pharynx, larynx, trachea, bronchi, and lungs.
REPRODUCTIVE SYSTEM
NERVOUS AND SENSORY ORGANS
MALES
Consists of testes, epididymides, vas deferens, prostate, seminal vesicles, and penis.
FEMALES
Consists of ovaries, uterine tubes, uterus (womb), and vagina.
SENSES
Consist of external and internal senses.
CENTRAL NERVOUS SYSTEM (CNS)
Consists of the brain and the spinal cord.
PERIPHERAL NERVOUS SYSTEM (PNS)
Consists of nerves and neurons outside the CNS.

CHAPTER 2

THE STRUCTURE OF CELLS AND SKIN

All living things are made up of cells and their products. Cells are the smallest living components of the human body. We all began as a simple unspecialized embryonic cell (the zygote), and cell division and differentiation produced our complex bodies. Differentiation is the process whereby cells acquire special characteristics that adapt them for particular functions (e.g., epithelial cells, neurons, muscle cells, and fat cells). Some cells retain the capacity to undergo cell division (e.g., bone marrow), whereas others lose it (e.g., neurons).

CELLS AND THEIR PRODUCTS

More than 50 trillion cells and their products make up our adult bodies.

Cell types

Cellular products include structural proteins such as collagen, elastin, and reticulin. The first embryonic cell of our body is unspecialized, so a lot of differentiation and cell division must go on to make the adult body. **Differentiation** is the process of changes in gene expression that a cell goes through to become a more specialized cell serving a particular function.

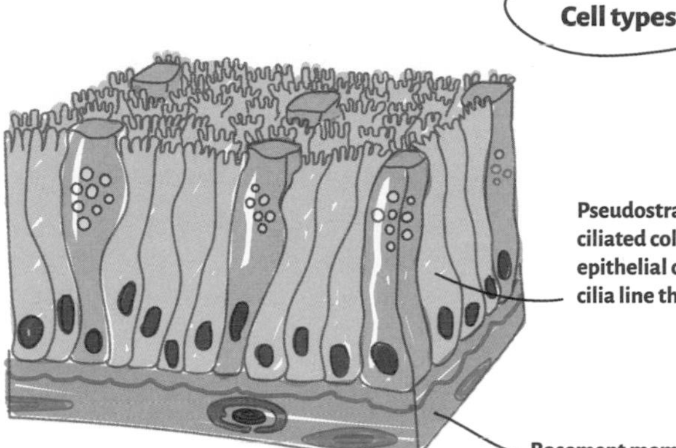

Pseudostratified ciliated columnar epithelial cells with cilia line the airways.

Basement membrane

Cells that line the body surfaces (internal and external) are called **epithelia**. The epithelia of the skin are flattened cells in many layers (stratified squamous).

Those lining the alimentary canal (also called the digestive tract or the gut) are a single layer of columnar cells (simple columnar), while those lining the airways are in a single layer, but with nuclei at different levels and with cilia at their tips to move mucus, called **pseudostratified columnar with cilia**.

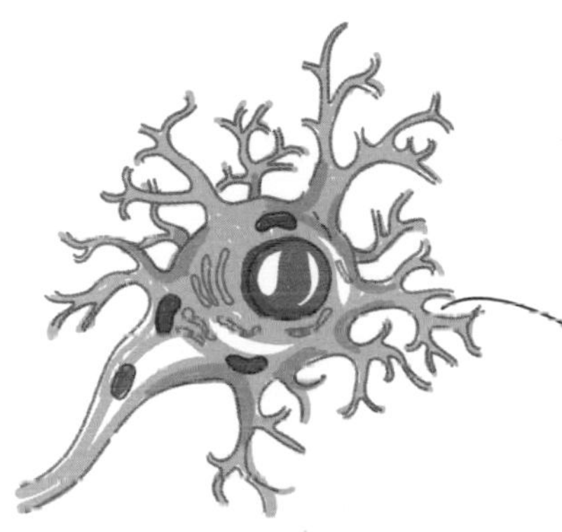

Nerve cells (neurons) process and transmit information.

Muscle cells produce movement and can be voluntary (skeletal muscle) or involuntary (smooth and cardiac muscle).

Connective tissue

Defense cells protect against foreign invaders.

Fibroblasts make tissue fibers.

Fat cells store energy as fat.

Endothelial cells line the blood vessels.

Much of the body is made up of **connective tissue**. Connective tissue is comprised of different components. A characteristic of connective tissue is that cells are suspended in a matrix that can be solid or liquid. Bone, cartilage, tendon, and ligament are examples of solid matrix connective tissue. Mineralization of collagen produces bone, and collagen makes up the bulk of loose and dense connective tissue throughout the body. Blood is a fluid matrix connective tissue, because cells are suspended in plasma.

CELL STRUCTURE AND ORGANELLES

The main components of the cell are the nucleus, where genetic material is stored; a cytoplasm that contains the cellular organelles; and a surrounding plasma membrane that provides a semipermeable barrier between the cytoplasm (intracellular fluid) and the surrounding environment (extracellular fluid).

Nucleus

The structure of a cell

The **nucleus** is the control center for the cell. It is responsible for storing genetic information in DNA (deoxyribonucleic acid) of the chromatin and distributing that information for regulation of the activities of the cell. Cell division (mitosis) requires the chromatin to be wound up into their compact chromosomes and separated into the two daughter cells.

The **nuclear envelope** separates the nucleoplasm from the cytoplasm and provides some management of the movement of substances between the two.

The **nucleolus** is the site where ribosomal RNA (rRNA) and proteins are assembled into ribosomes, and the code of DNA is transcribed into messenger RNA (mRNA). Ribosomes then move to the cytoplasm to produce proteins.

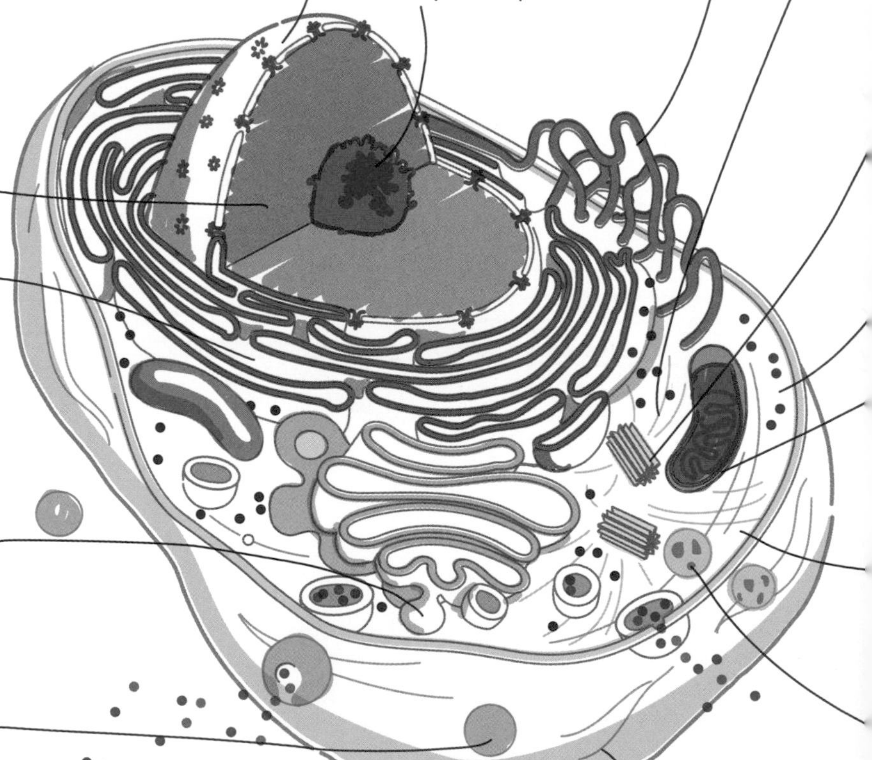

Ribosomes are the protein-manufacturing machinery. They can be free-floating or combined with a membrane system called the **rough endoplasmic reticulum** (**RER**).

The **Golgi apparatus** packages the products of ribosomes and the endoplasmic reticulum for export.

Peroxisomes are responsible for detoxifying ingested substances and hydrogen peroxide produced by oxygen-based metabolism in the cell.

Intermediate filaments resist tension in the cell and also contribute to the cytoskeleton.

The **smooth endoplasmic reticulum** is a ribosome-free set of membranes that makes fats and steroid hormones.

Microtubules are cylindrical structures, composed of the protein tubulin, which provide pathways for movement within the cell as well as strengthening the structure of the cell.

Centrioles are formed by microtubules joining together. Centrioles move the chromosomes during cell division (mitosis) and also provide the bases of cilia and flagella.

The **cytoskeleton** provides structure for the cell and includes microfilaments composed of actin, a key part of the cytoskeleton, which plays a role in muscle contraction and movement within the cell.

A **mitochondrion** (plural mitochondria) produces and stores ATP (adenosine triphosphate) for the cell. It has its own DNA and is believed to have once been a free-living microorganism that has become symbiotic with the mammalian cell.

The **cytoplasm** is made of a fluid, called cytosol, with suspended structures, called organelles, providing the metabolic machinery of the cell.

Lysosomes are membranous sacs in the cytoplasm where unwanted structures are dissolved.

Plasma membrane

The **plasma membrane** of the cell is a double-layered membrane, composed of **phospholipids** (a type of lipid) and **cholesterol**, with **suspended proteins**.

The **lipid layer** is impermeable to water, whereas the embedded proteins serve important membrane functions.

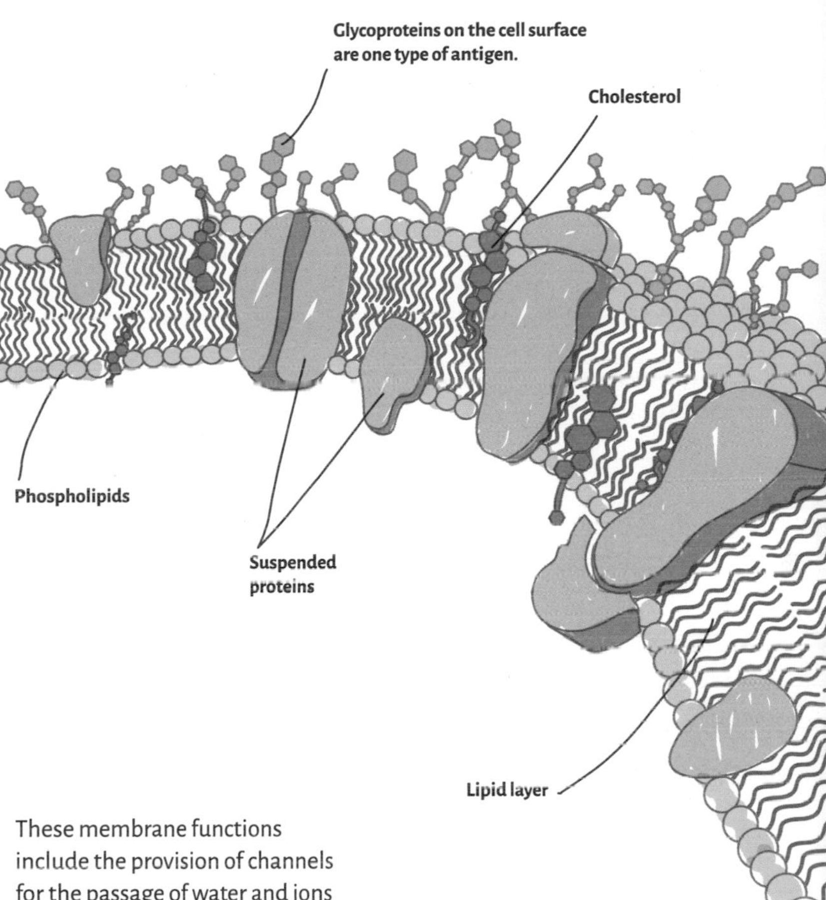

Glycoproteins on the cell surface are one type of antigen.
Cholesterol
Phospholipids
Suspended proteins
Lipid layer

These membrane functions include the provision of channels for the passage of water and ions (e.g., sodium, potassium, and chloride) and nutrients (e.g., glucose and amino acids) through the membrane. They also include the provision of receptors for hormones and neurotransmitters to signal to the cell and the formation of external cell markers (antigens) that can be recognized by the immune system.

Some specialized cells have structural adaptations of their plasma membranes to increase the surface area (e.g., the microvilli of the gut cells); they can also move the surrounding medium (e.g., the cilia of airway cells), or they can swim through the reproductive tract (e.g., the flagellum of sperm).

CELL DIVISION: MITOSIS AND MEIOSIS

We all begin life as a single cell; therefore, many hundreds of trillions of cell divisions, called mitoses, must occur during our lifetime to first build our bodies and then to replace damaged or dying cells.

Where does mitosis happen?

Mitosis is particularly common in tissues that are continuously exposed to the damaging effects of the external environment, physical wear and tear, and when lost cells must be replaced. So, cell division occurs unceasingly in the epidermis of the skin, the lining of the alimentary canal, and in the bone marrow to make red and white blood cells. For some tissues, cell division is largely confined to repair (e.g., broken bones), whereas mitosis to produce other structurally complex cell types (e.g., neurons) is mainly confined to development.

PROCESS OF MITOSIS
Mitosis is primarily a process for division of the nuclear material from one cell between two daughter cells. There are four phases: prophase (includes early and late stages), metaphase, anaphase, and telophase.

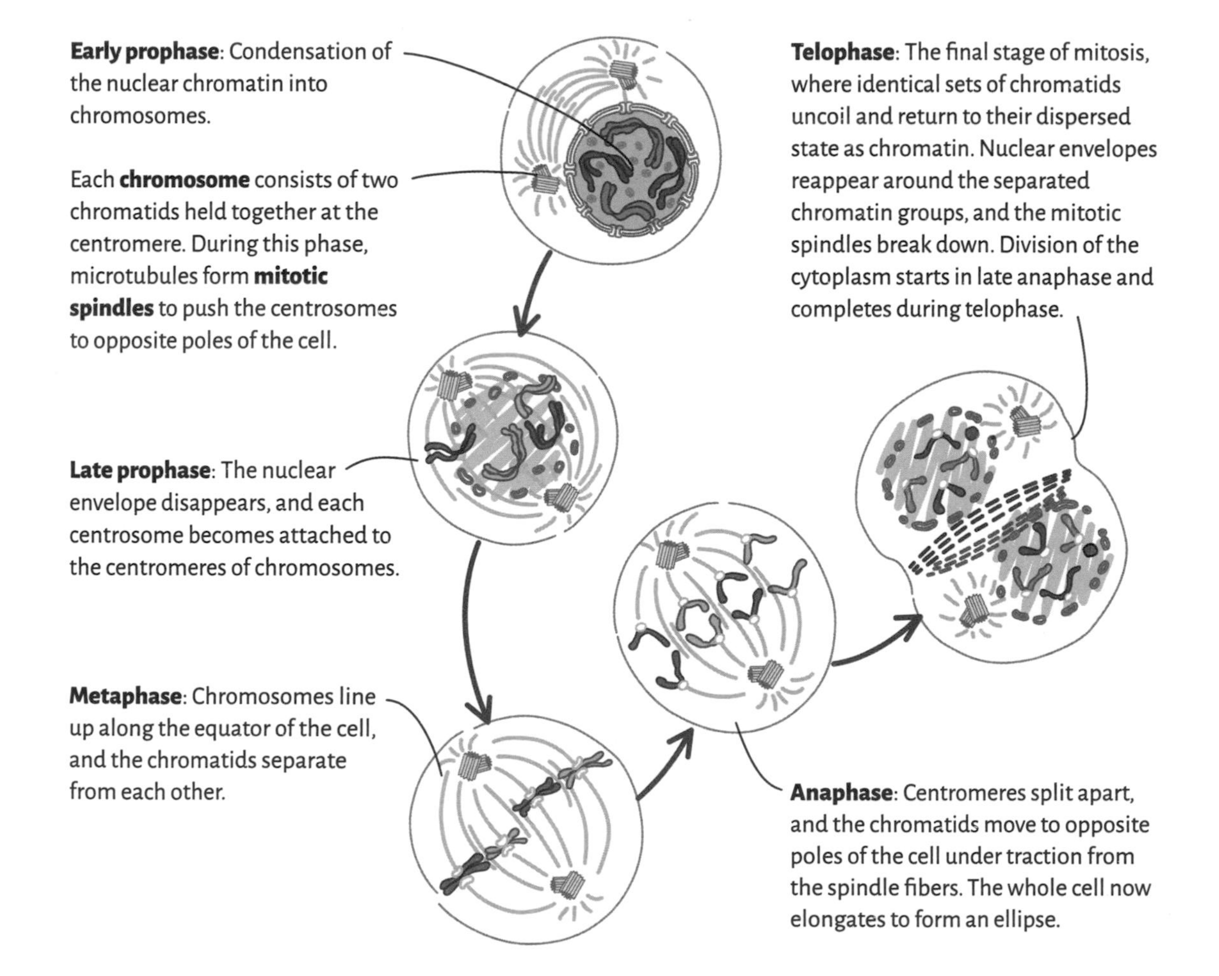

Early prophase: Condensation of the nuclear chromatin into chromosomes.
Each **chromosome** consists of two chromatids held together at the centromere. During this phase, microtubules form **mitotic spindles** to push the centrosomes to opposite poles of the cell.
Late prophase: The nuclear envelope disappears, and each centrosome becomes attached to the centromeres of chromosomes.
Metaphase: Chromosomes line up along the equator of the cell, and the chromatids separate from each other.
Anaphase: Centromeres split apart, and the chromatids move to opposite poles of the cell under traction from the spindle fibers. The whole cell now elongates to form an ellipse.
Telophase: The final stage of mitosis, where identical sets of chromatids uncoil and return to their dispersed state as chromatin. Nuclear envelopes reappear around the separated chromatin groups, and the mitotic spindles break down. Division of the cytoplasm starts in late anaphase and completes during telophase.

RISKS OF MITOSIS
Very active cell division carries risks, because genetic errors can occur in cell lines when mitosis is frequent. This is why many cancers arise in tissues that are frequently exposed to dangerous agents, such as toxins, physical wear, and radiation, and which must undergo frequent cell division. These include the epidermis of the skin, and the lining of the respiratory and digestive tracts. It also includes some tissues under strong hormonal influence, such as the breasts, ovaries, testes, and the prostate gland

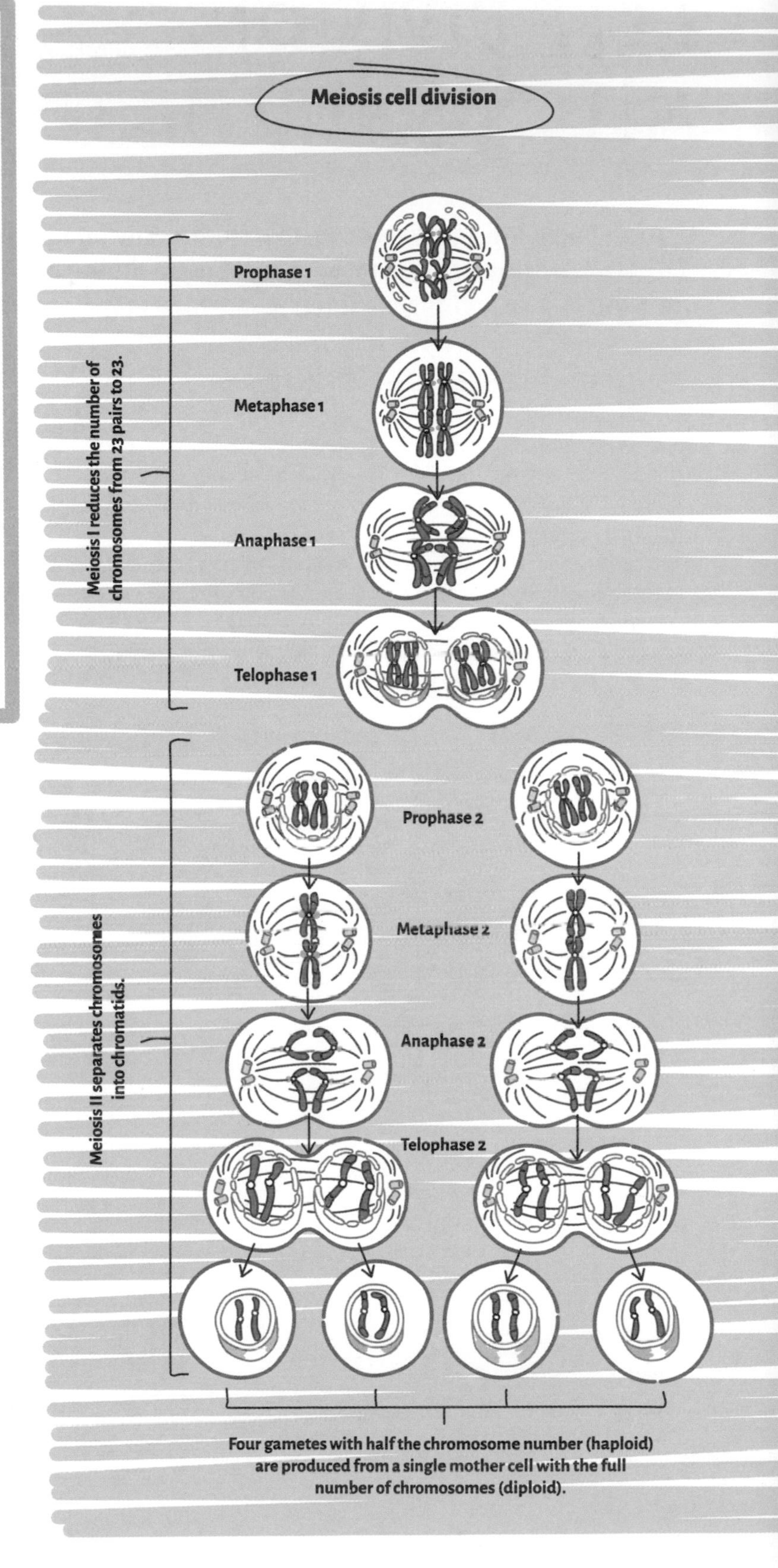

Meiosis cell division
Meiosis I reduces the number of chromosomes from 23 pairs to 23.
Prophase 1
Metaphase 1
Anaphase 1
Telophase 1
Meiosis II separates chromosomes into chromatids.
Prophase 2
Metaphase 2
Anaphase 2
Telophase 2
Four gametes with half the chromosome number (haploid) are produced from a single mother cell with the full number of chromosomes (diploid).

Meiosis

Meiosis is cell division that occurs during production of the sex cells (gametes) and is found in the testes and ovaries. Meiosis differs from mitosis in that the normal number of chromosomes (diploid, 23 pairs) is reduced to half normal (haploid, 23 single chromosomes) when sex cells are produced.

Meiosis involves two successive divisions of the nucleus: meiosis I and meiosis II. Meiosis I is the process that reduces the chromosome number, whereas meiosis II is more like mitosis.

The entire process produces four gametes (spermatid or oocyte) from one cell. Meiosis is essential, so that when the gametes combine during fertilization, the resulting cell (zygote) has the normal number of chromosomes.

ANATOMICAL POSITION AND PLANES

The accurate description of anatomical structures requires a standard position for the body. In this anatomical position, the subject stands with head erect and eyes directed forward, the upper limbs by the side, and the palms facing forward. The feet are together with toes facing forward.

Anatomical planes

Anatomists describe planes that pass through the body in the anatomical position. There are three sets of planes, each aligned in one of the three spatial dimensions: horizontal or transverse, coronal or frontal, and sagittal planes.

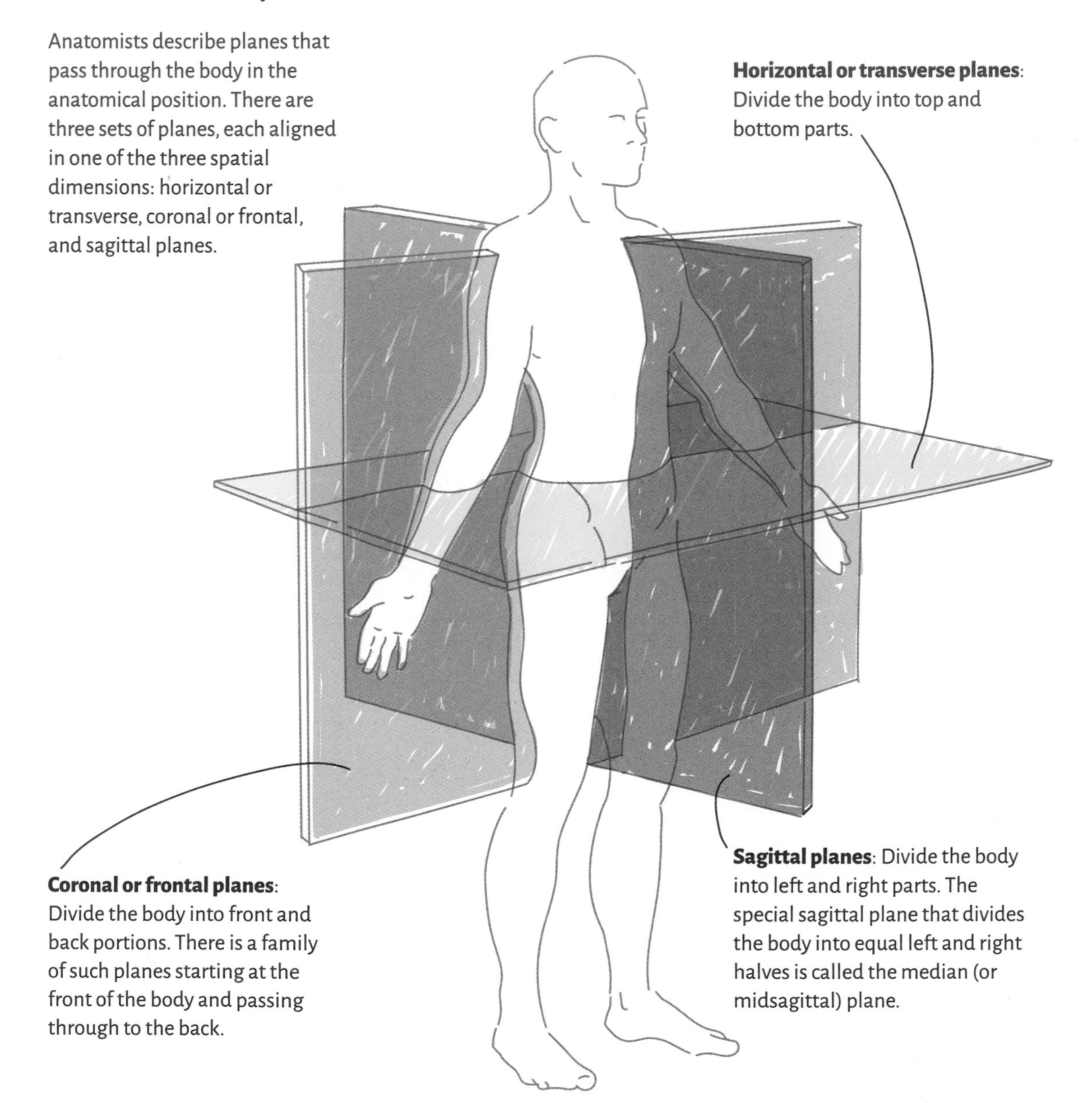

Horizontal or transverse planes: Divide the body into top and bottom parts.
Coronal or frontal planes: Divide the body into front and back portions. There is a family of such planes starting at the front of the body and passing through to the back.
Sagittal planes: Divide the body into left and right parts. The special sagittal plane that divides the body into equal left and right halves is called the median (or midsagittal) plane.

Directions in the body

Anatomists use pairs of terms to describe positions within the body in the anatomical position.

Structures close to the surface of the body are superficial, whereas structures farther from the skin surface are deep. Directions in the hand may be palmar (toward the skin of the palm) or dorsal (toward the back of the hand). Similarly, the foot has plantar (sole) and dorsal (top) surfaces.

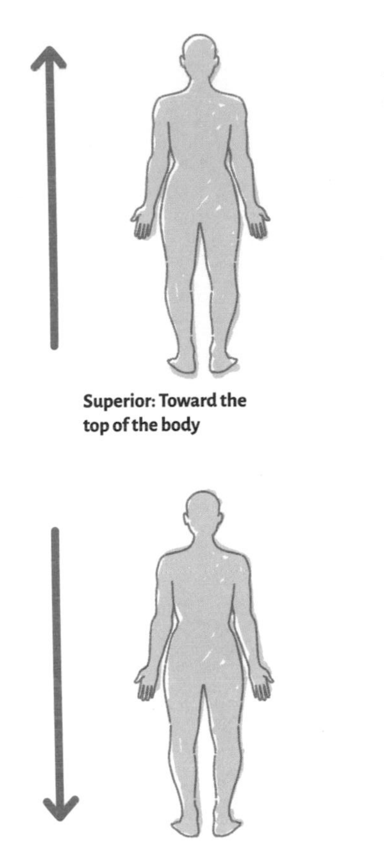

Superior: Toward the top of the body

Inferior: Toward the bottom of the body

Anterior: Toward the front of the body

Posterior: Toward the back of the body

Medial: Toward the midline of the body

Lateral: Toward the side of the body

Proximal: Close to the limb base

Distal: Farther from the limb base

Proximal and distal may also be used in the alimentary canal (gut tube), with proximal toward the mouth and distal toward the anus.

How are anatomical terms used?

Here are some examples of how anatomists use these special terms. For instance, we can say that "the nose is anterior to the eye" or that "the ear is lateral to the eye."

Remember that these terms are defined with the body in the anatomical position, so we would also say that "the thumb is lateral to the index finger."

Detailed anatomical description can become complex, for instance: "the tendon of the flexor carpi ulnaris inserts distally on the palmar surface of the pisiform and hamate bones."

TERMS FROM ZOOLOGY

Anatomists have borrowed some terms from zoology, as in the spinal cord, for instance.

Dorsal: Toward the back (literally toward the dorsum or back)
Ventral: Toward the front (literally toward the ventrum or belly)
Rostral: Toward the brain (literally toward the rostrum or beak)
Caudal: Away from the brain (literally toward the cauda or tail)

These terms are all borrowed from zoological descriptions of a simple vertebrate (e.g., a fish). Like other primates, the axis of the human nervous system takes a bend at about the midbrain, so that rostral is equivalent to anterior in the forebrain, but rostral is equivalent to superior in the spinal cord.

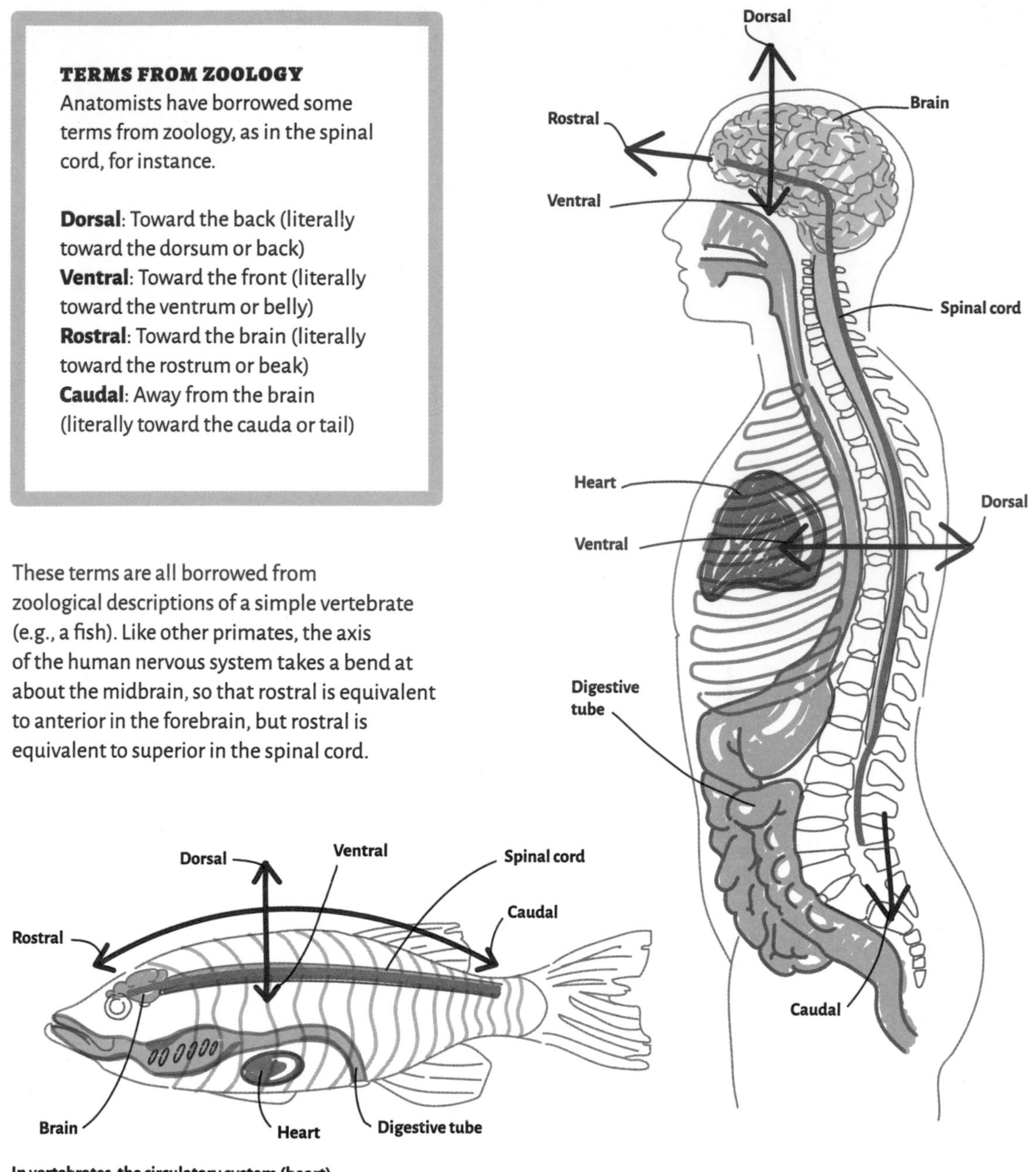
Dorsal
Rostral
Brain
Ventral
Spinal cord
Heart
Ventral
Dorsal
Digestive tube
Caudal
Dorsal
Ventral
Spinal cord
Caudal
Rostral
Brain
Heart
Digestive tube

In vertebrates, the circulatory system (heart) and alimentary canal are ventral to the central nervous system.

Anatomical names not in common usage

Anatomical names for parts of the body may differ slightly from everyday usage. The upper limb extends from the shoulder joint to the fingers of the hand (manus), but the term arm (brachium) is confined to that part of the upper limb between the shoulder and elbow. The forearm (antebrachium) extends from the elbow to the wrist (carpus). The fingers and thumb are collectively called digits.

The lower limb extends from the hip joint to the toes, but the region between the hip and the knee is called the thigh. The leg is that part of the lower limb between the knee and the ankle. The foot is called the pes, the proximal part of the foot is the tarsus, and the toes are digits.

SKIN, NAILS, AND HAIR

The skin is the largest organ of the body, weighing about 3.5 kg (7.8 pounds) and covering 2 square meters (about 22 square feet), but it is only a few millimeters thick.

Skin structure

Human skin is glandular and has sweat and sebaceous glands to control temperature and protect the skin surface with an oily layer.

Hair and nails are keratin-rich appendages made by the skin. The skin consists of a superficial epidermis and a deeper dermis.

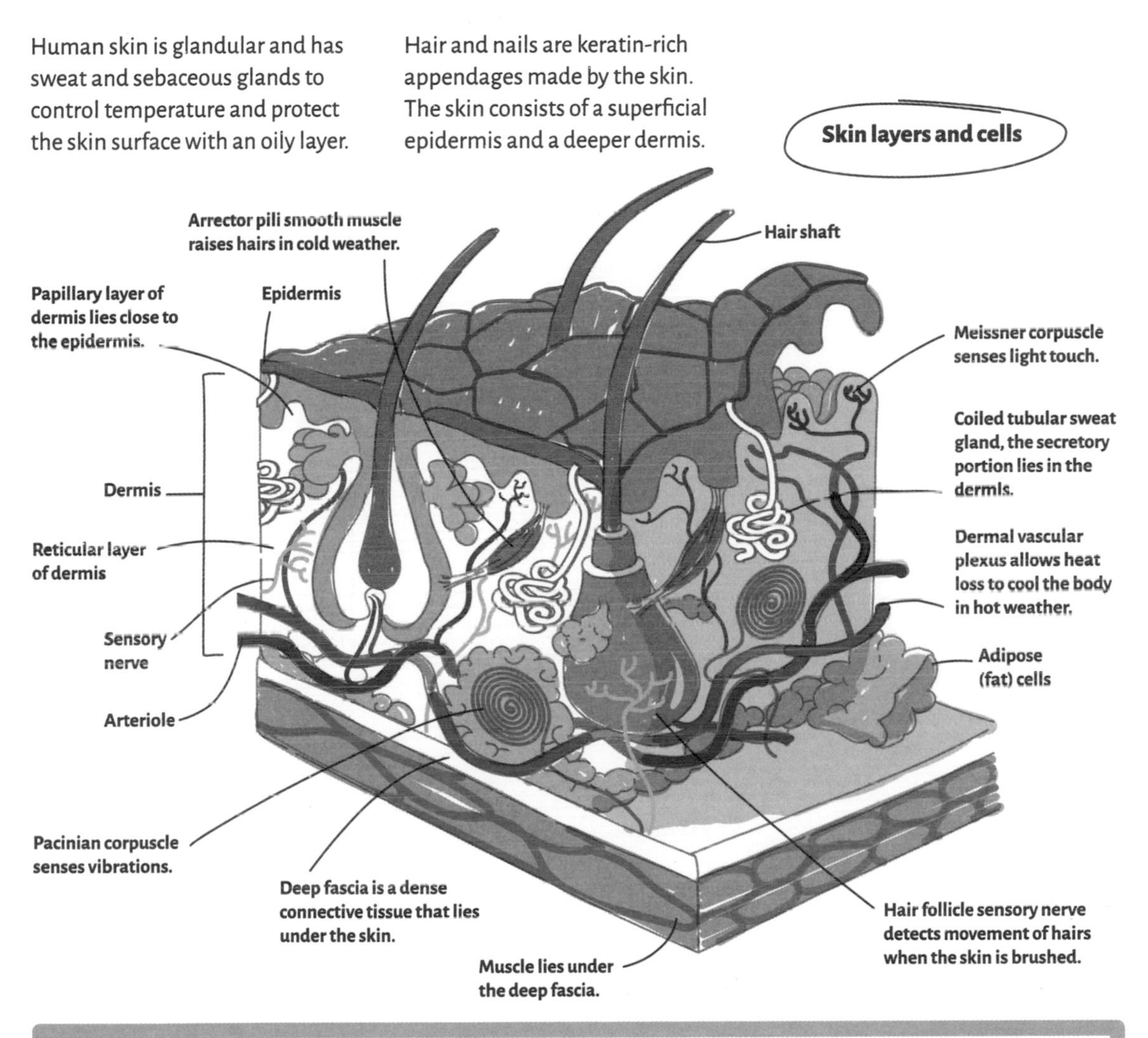

Skin layers and cells
Arrector pili smooth muscle raises hairs in cold weather.
Hair shaft
Papillary layer of dermis lies close to the epidermis.
Epidermis
Meissner corpuscle senses light touch.
Coiled tubular sweat gland, the secretory portion lies in the dermis.
Dermis
Reticular layer of dermis
Dermal vascular plexus allows heat loss to cool the body in hot weather.
Sensory nerve
Adipose (fat) cells
Arteriole
Pacinian corpuscle senses vibrations.
Deep fascia is a dense connective tissue that lies under the skin.
Hair follicle sensory nerve detects movement of hairs when the skin is brushed.
Muscle lies under the deep fascia.

SKIN FUNCTIONS

- ★ Protects the underlying tissues from the hazards of the external environment, e.g., water loss, heat, radiation, microorganisms, and physical wear
- ★ Produces vitamin D, essential for bone formation and regulation of cell division, and differentiation
- ★ Plays a critically important role in temperature control
- ★ Key sense organ for touch, pain, and temperature sensation
- ★ Facial skin's movement serves an important social and communication function.

Epidermis structure

The epidermis is made up of a type of surface tissue, called epithelium, and has a layered structure. The surface layers of the epidermis must be continuously replenished by cell division in the base of the epidermis, called the **stratum basale** or germinativum.

Daughter cells (**keratinocytes**) from this deep zone undergo changes as they move toward the skin surface. They acquire a tough protein called **keratin** in their cytoplasm, and their plasma membranes become stronger. Their nuclei disappear, and the cells of the epidermis are eventually transformed into tough, dead flakes that protect the underlying skin from abrasion and penetration. The dead keratinocytes are eventually shed as squames.

In skin regions, such as the palms of the hands and the soles of the feet, which are constantly exposed to physical wear, the outer dead layer of the epidermis (**stratum corneum**) is particularly thick and tough.

Tactile structures, such as Merkel's disks for light touch and free nerve endings for pain, are distributed among the keratinocytes, but most skin sensation is detected deeper in the dermis.

Other cells in the epidermis include **melanocytes** (about 10% to 25% of cells in the **stratum basale**), which produce the pigment melanin and transfer it to the keratinocytes. Melanin clusters on the superficial side of the keratinocytes provide an umbrella protecting against ultraviolet radiation. People with dark skins have darker melanin, more melanin granules, and more pigment in each melanocyte, but have the same number of melanocytes as people with light skins.

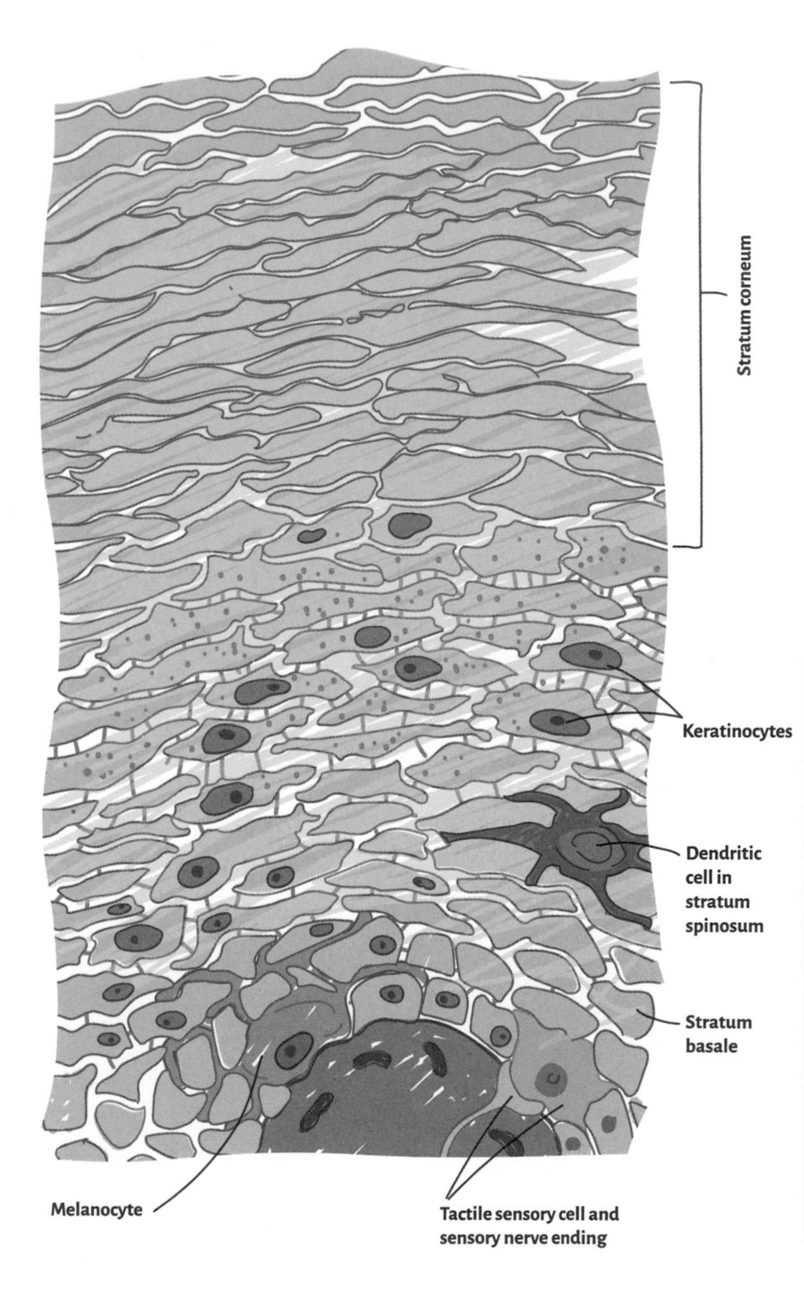
Stratum corneum
Keratinocytes
Dendritic cell in stratum spinosum
Stratum basale
Melanocyte
Tactile sensory cell and sensory nerve ending

IMMUNE FUNCTION
The skin also has an immune function in the form of **dendritic cells** scattered throughout the **stratum spinosum**. These cells take up foreign proteins that have invaded the epidermis, then leave the skin and travel to the nearest lymph node, where they begin an immune response to all foreign cells or viruses that carry the foreign protein.

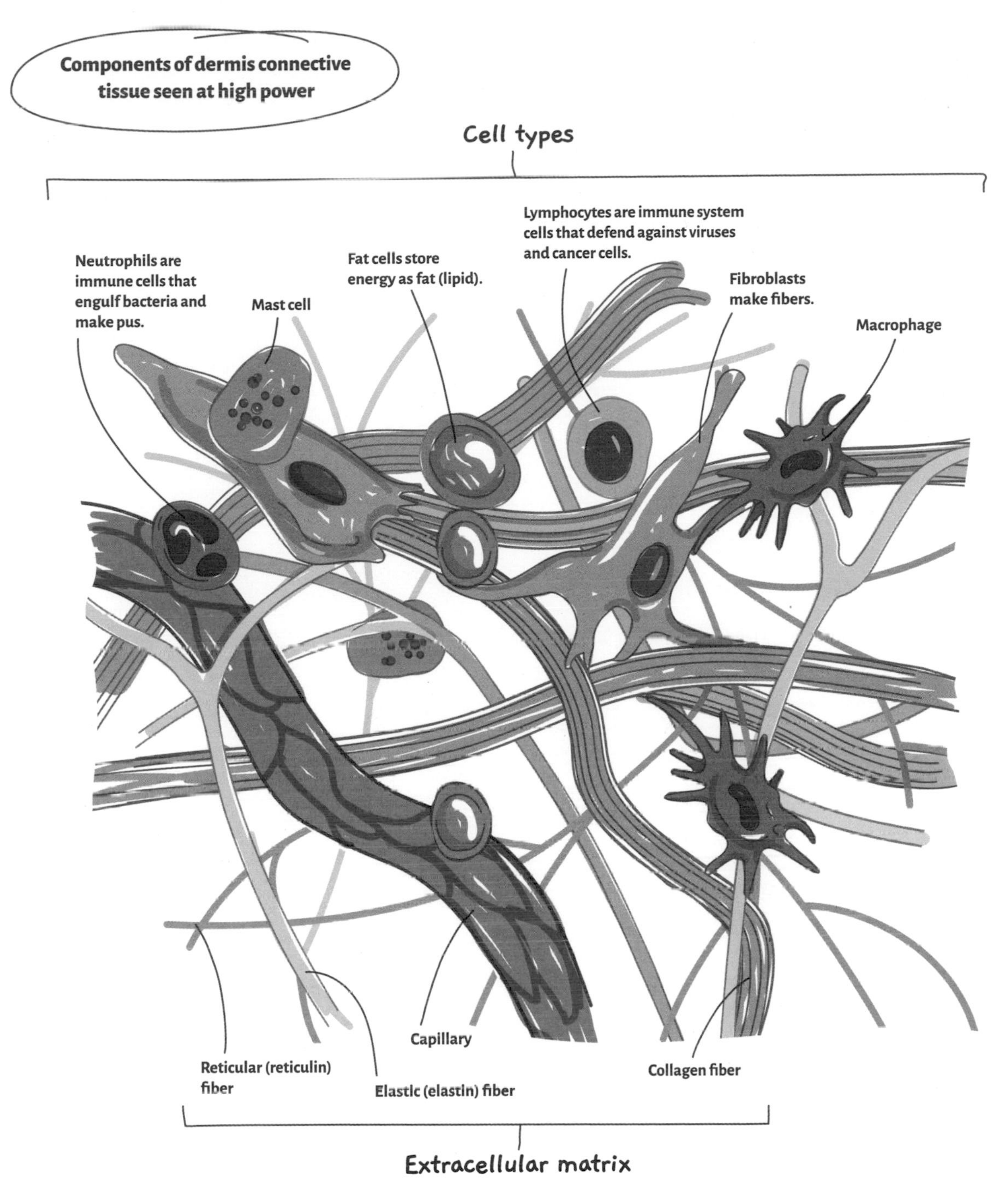
Components of dermis connective tissue seen at high power
Cell types
Neutrophils are immune cells that engulf bacteria and make pus.
Mast cell
Fat cells store energy as fat (lipid).
Lymphocytes are immune system cells that defend against viruses and cancer cells.
Fibroblasts make fibers.
Macrophage
Capillary
Reticular (reticulin) fiber
Elastic (elastin) fiber
Collagen fiber
Extracellular matrix

Dermis

The epidermis sits on a deeper layer called the **dermis**, which is divided into an outer papillary and a deeper reticular layer.

The dermis is connective tissue filled with cells called **fibroblasts** and their fibrous products: **collagen**, **reticulin**, and **elastin**.

This tissue also contains **macrophages**, which are cells used to engulf invaders; **mast cells** that regulate allergic responses; and blood vessels.

The dermis also contains many sense organs. These include free nerve endings that sense pain and temperature, bulb-like Meissner corpuscles that sense light pressure and discriminative touch, and onion-like Pacinian corpuscles that sense vibration.

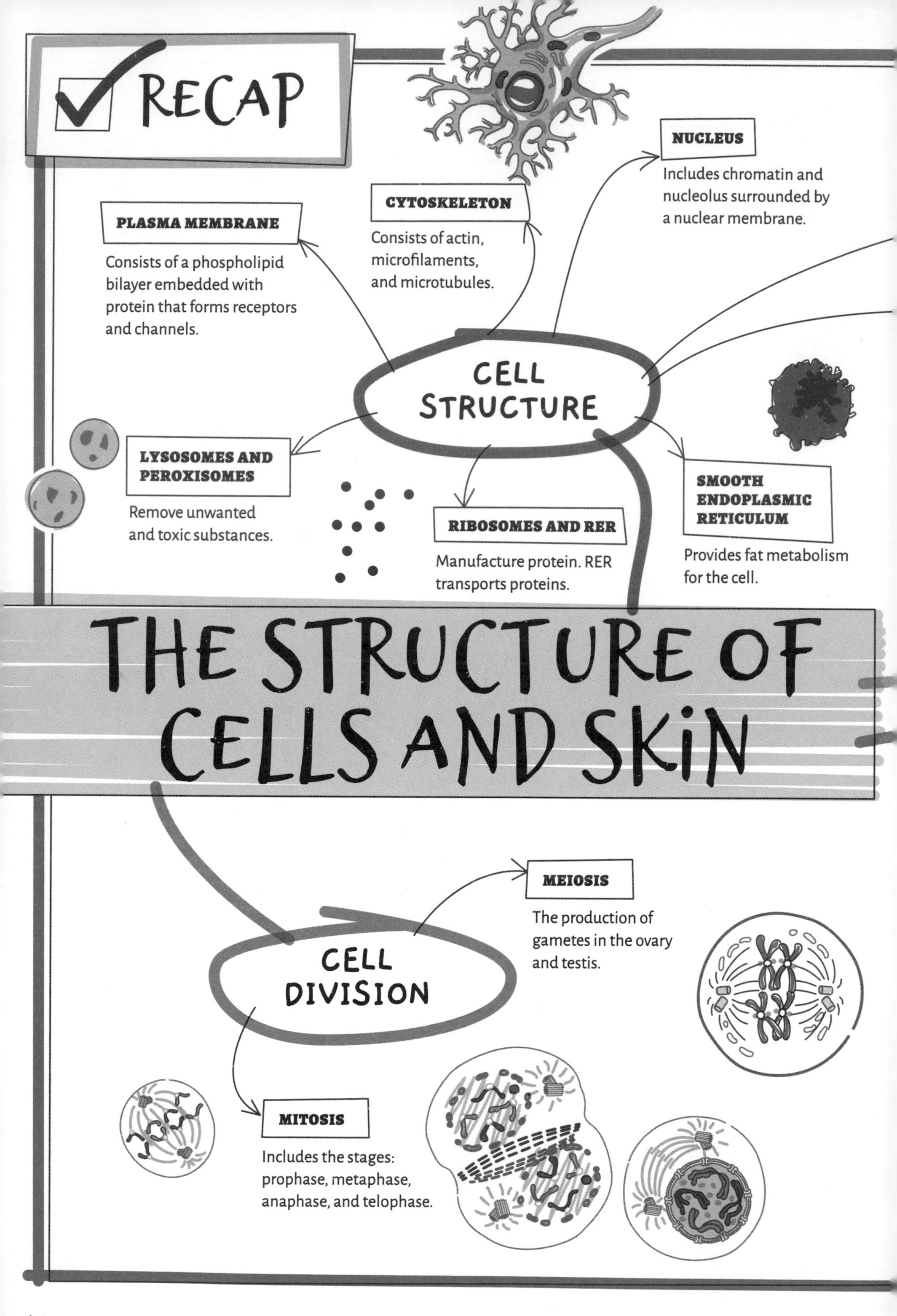
RECAP
PLASMA MEMBRANE
Consists of a phospholipid bilayer embedded with protein that forms receptors and channels.
CYTOSKELETON
Consists of actin, microfilaments, and microtubules.
NUCLEUS
Includes chromatin and nucleolus surrounded by a nuclear membrane.
CELL STRUCTURE
LYSOSOMES AND PEROXISOMES
Remove unwanted and toxic substances.
RIBOSOMES AND RER
Manufacture protein. RER transports proteins.
SMOOTH ENDOPLASMIC RETICULUM
Provides fat metabolism for the cell.
THE STRUCTURE OF CELLS AND SKIN
MEIOSIS
The production of gametes in the ovary and testis.
CELL DIVISION
MITOSIS
Includes the stages: prophase, metaphase, anaphase, and telophase.

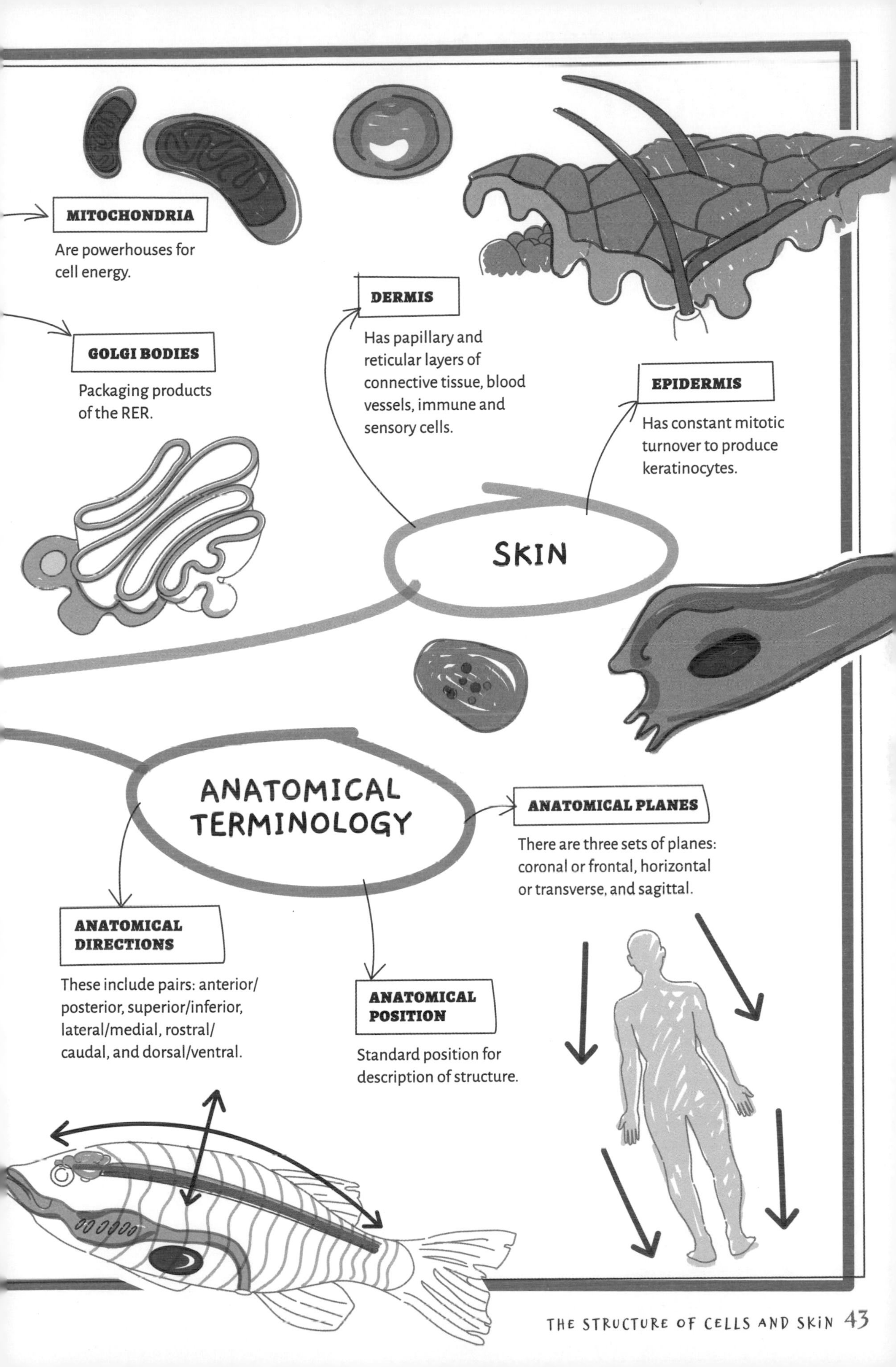
MITOCHONDRIA
Are powerhouses for cell energy.
GOLGI BODIES
Packaging products of the RER.
DERMIS
Has papillary and reticular layers of connective tissue, blood vessels, immune and sensory cells.
EPIDERMIS
Has constant mitotic turnover to produce keratinocytes.
SKIN
ANATOMICAL TERMINOLOGY
ANATOMICAL PLANES
There are three sets of planes: coronal or frontal, horizontal or transverse, and sagittal.
ANATOMICAL DIRECTIONS
These include pairs: anterior/posterior, superior/inferior, lateral/medial, rostral/caudal, and dorsal/ventral.
ANATOMICAL POSITION
Standard position for description of structure.

CHAPTER 3

SKELETON AND JOINTS

The skeletal system and its associated joints provide rigidity to the body structure as well as storing essential minerals; for example, calcium and phosphorus.

The axial skeleton of the head and trunk has a segmented arrangement derived from the body structure of the ancestral vertebrate, whereas the limbs (appendicular skeleton) develop as outgrowths of the trunk.

Bones can develop by deposition of calcium salts in cartilage models or within membrane sheets.

ORGANIZATION OF THE SKELETON

The skeleton provides a framework for body structure, protects delicate internal organs, and affords the lever arm attachments for the voluntary skeletal muscles.

The skeleton is divided into axial and appendicular parts. The **axial skeleton** (shown here in blue) runs down the midline of the body and includes the skull, hyoid bone (see pages 50, 51), vertebral column, sternum, and 12 pairs of ribs. Some bones of the lower vertebral column are fused as the sacrum.

The **appendicular skeleton** is for the limbs and their attachments by limb girdles to the axial skeleton.

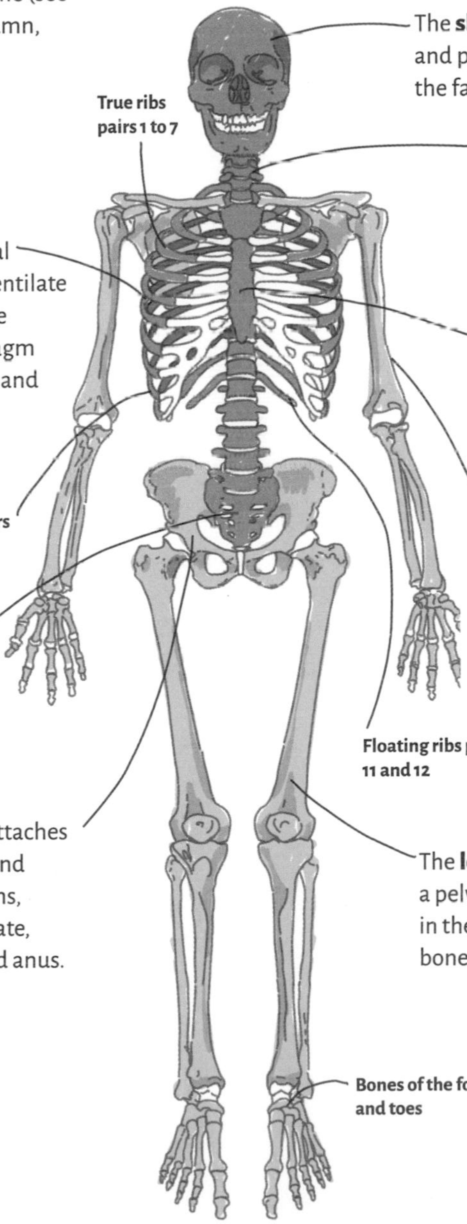

The **skull** protects the brain and provides structure for the face.

True ribs pairs 1 to 7

The **vertebral column** is the backbone of the body and combines flexibility with support. It consists of vertebrae that increase in size from top to bottom.

The **ribs** protect the lungs, heart, and upper abdominal organs. Ribs can move to ventilate the lungs. They also provide attachment for the diaphragm that separates the thoracic and abdominal cavities.

The **sternum** lies at the front of the chest and is divided into three parts, e.g., manubrium, body, and xiphoid process from superior to inferior.

False ribs pairs 8 to 10

The **upper limbs** consist of a pectoral girdle, long bones in the arm and forearm, and the bones of the wrist and hand.

The **sacrum** is part of the vertebral column and is made up of five fused vertebrae.

Floating ribs pairs 11 and 12

The **pelvic girdle** (pelvis) attaches the rest of the lower limb and also protects internal organs, e.g., urinary bladder, prostate, uterus, ovaries, rectum, and anus.

The **lower limbs** consist of a pelvic girdle, long bones in the thigh and leg, and bones of the foot and toes.

Bones of the foot and toes

JOINTS AND MOVEMENTS

Wherever bones meet there is a joint. The most familiar joints are where movement occurs, and anatomists have specific names for those actions.

When we bend at a joint (e.g., our upper limb at the elbow), we call the movement **flexion**; straightening the joint is **extension**. Moving a limb away from the body midline is **abduction**; moving toward the midline is **adduction**. Rotation can be toward the midline (**medial rotation**) or away (**lateral rotation**).

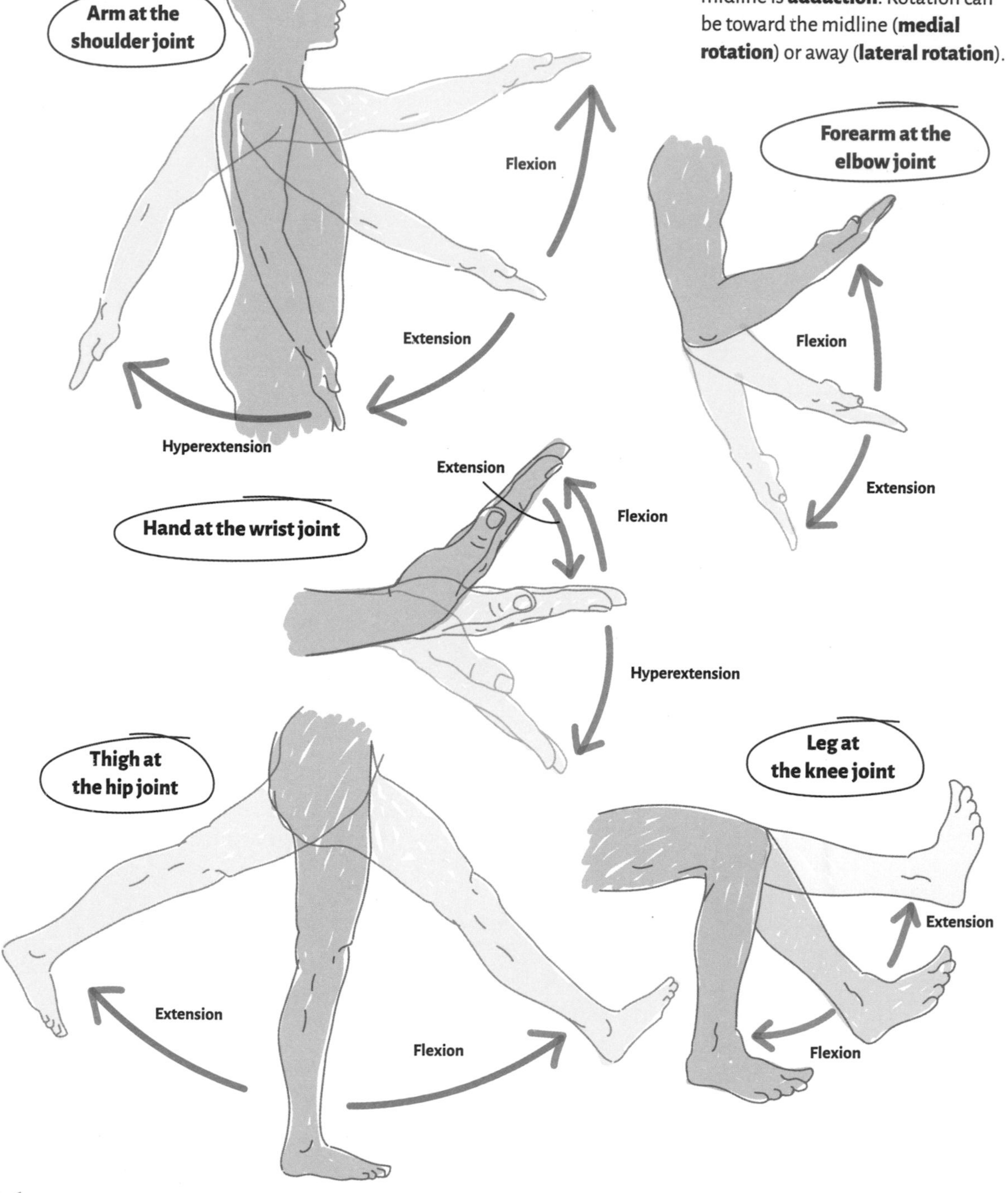

Arm at the shoulder joint
Flexion
Extension
Hyperextension
Forearm at the elbow joint
Flexion
Extension
Hand at the wrist joint
Extension
Flexion
Hyperextension
Thigh at the hip joint
Extension
Flexion
Leg at the knee joint
Extension
Flexion

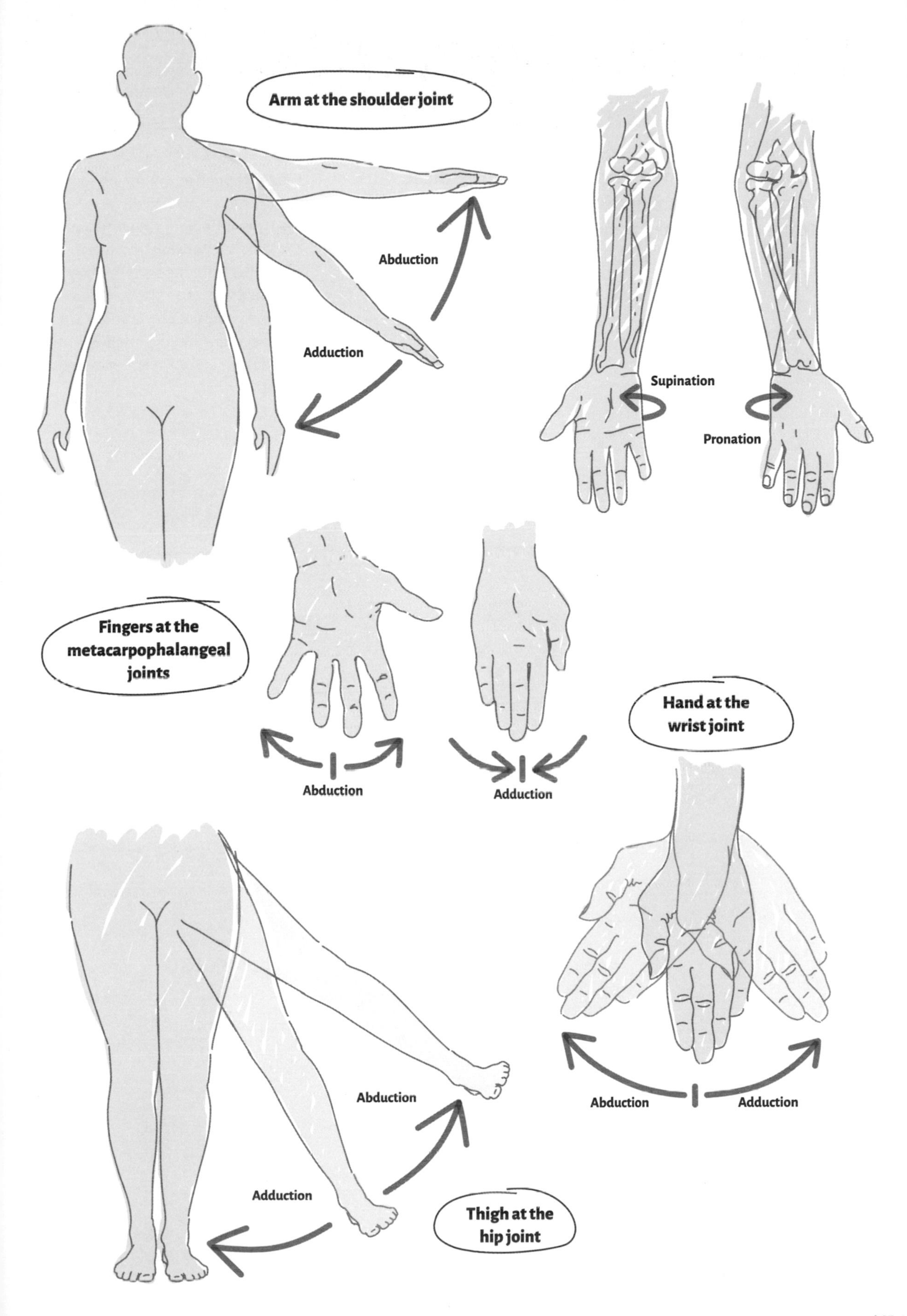
Arm at the shoulder joint
Abduction
Adduction
Supination
Pronation
Fingers at the metacarpophalangeal joints
Abduction
Adduction
Hand at the wrist joint
Abduction
Adduction
Abduction
Adduction
Thigh at the hip joint

THE STRUCTURE OF BONE

Bone is a connective tissue made of mineralized organic fibers with cellular inclusions. Although heavily mineralized, bone retains a capacity to turn over material and remodel in response to mechanical forces, mainly compressive forces and tension from muscles.

The cellular component of **bone** requires the presence of mature bone cells (osteocytes) within tiny cavities in the bone (Howship's lacunae). Bone has a good supply of blood and nerves. Damage to the blood supply of bone leads to bone death (avascular necrosis) and fracture. The external membranes of bone (periosteum) are extremely sensitive to pain, which is why cancerous deposits in bone cause so much pain through the pressure they produce.

Bone development and cells

Bone may form in a cartilage scaffold (endochondral ossification) or between membranes (intramembranous ossification). Most long bones form in a cartilage scaffold from ossification centers in the shaft and at each end.

Three types of bone cells

Osteocyte

Long processes extending through canaliculi of the bone

Osteoblast

New bone is made by **osteoblasts**, which produce the organic matrix of bone called **osteoid**. Calcium salts are deposited within the osteoid to form primitive woven bone. Once osteoblasts have finished making osteoid, they become **osteocytes**, which provide ongoing maintenance of the bone.

Another important cell type within bone is the **osteoclast**, which reabsorbs the bone. Bone formation and reabsorption are ongoing complementary processes that allow the constant reshaping of bone in response to mechanical forces. Bone is laid down along lines of compressive forces to strengthen the bone and is removed from sites where it is no longer needed.

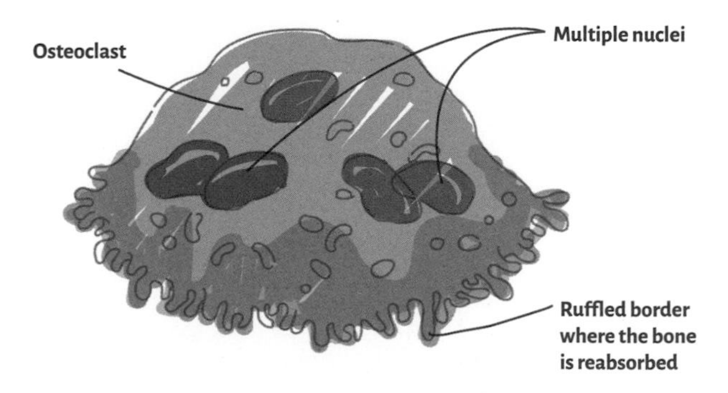
Osteoclast
Multiple nuclei
Ruffled border where the bone is reabsorbed

Structure of long bones

Typical **long bones** (e.g., the femur of the thigh) have a tubular structure to maximize strength and minimize weight. The outer wall is composed of compact bone, whereas the inner part is spongy or cancellous bone arranged in a meshwork of **trabeculae**.

Compact bone is very dense and composed of many units called **osteons** or **Haversian systems**, arranged parallel to the long axis of the bone and the usual compressive forces. **Circumferential lamellae** surround **Haversian canals** that carry blood vessels and nerves through the bone.

Perforating (Volkmann's) canals join adjacent Haversian canals to the vessels of both the **periosteum** on the exterior of the bone and the endosteal membranes lining the marrow cavity.

Spongy bone (see below) occupies the cavities of long and irregular bones and the internal space of the skull plates.

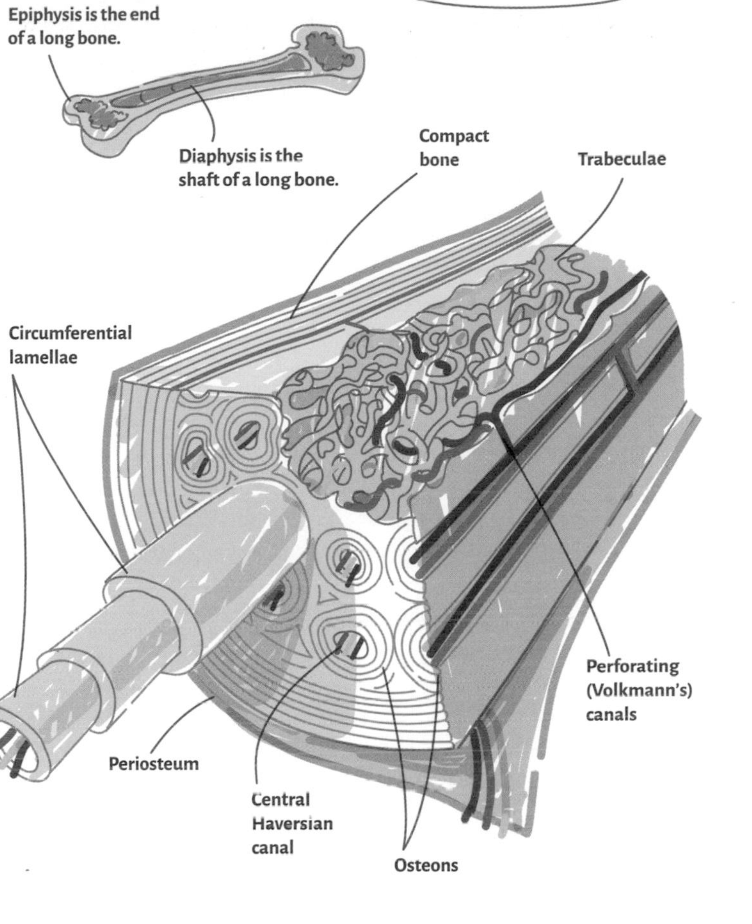

Bone structure
Epiphysis is the end of a long bone.
Diaphysis is the shaft of a long bone.
Compact bone
Trabeculae
Circumferential lamellae
Perforating (Volkmann's) canals
Periosteum
Central Haversian canal
Osteons

Bone growth

Bone lengthening occurs at the **epiphyseal plates**, which are zones of cartilage where cell division produces new osteoblasts and matrix for mineralization. The increase in bone circumference or width occurs by the laying down of bone under the periosteal membrane (appositional growth). As a bone grows in circumference, there is also removal of bone at the endosteum to optimize the strength-for-weight of the bone. Epiphyseal plates occur in the early years and disappear in puberty.

How bones develop

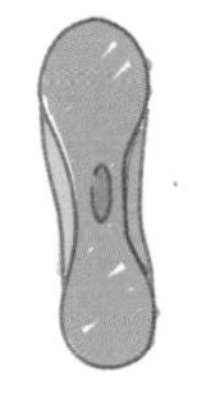

Cartilage model of a future long bone

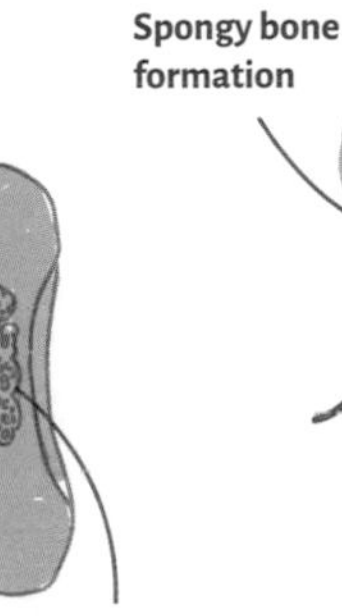

Spongy bone formation

The primary ossification center is where bone formation starts in the cartilage of the shaft.

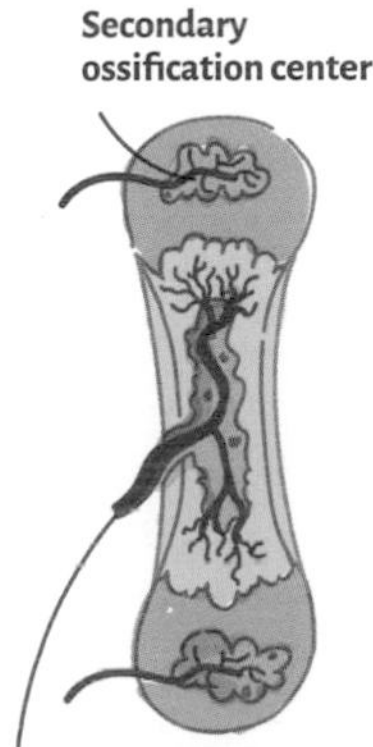

Secondary ossification center

Nutrient vessels supply the developing bone tissue.

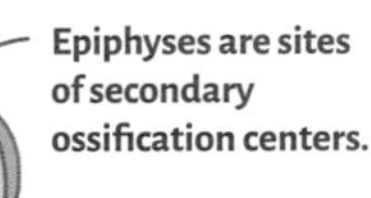

Epiphyses are sites of secondary ossification centers.
Epiphyseal plate
Diaphysis is the site of the primary ossification center that later develops a marrow cavity.

BONES OF THE AXIAL SKELETON

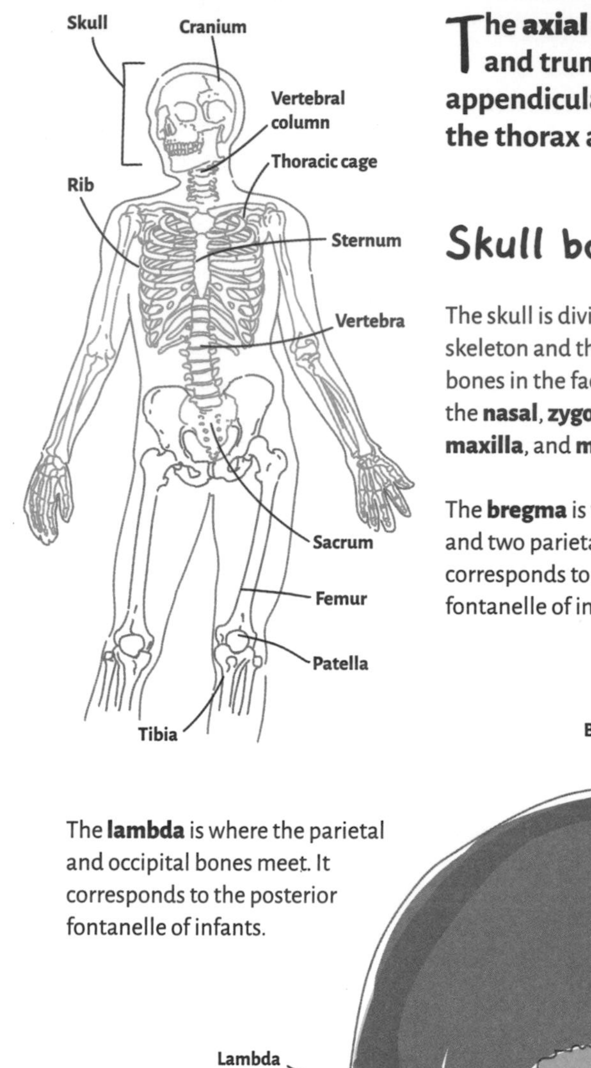

Skull
Cranium
Vertebral column
Thoracic cage
Rib
Sternum
Vertebra
Sacrum
Femur
Patella
Tibia

The axial skeleton is found within the head, neck, and trunk. It provides attachment for the appendicular skeleton and protects the viscera of the thorax and upper abdomen.

Skull bones

The skull is divided into the facial skeleton and the braincase. The bones in the facial skeleton include the **nasal**, **zygomatic**, **lacrimal**, **maxilla**, and **mandible** bones.

The **bregma** is where the frontal and two parietal bones meet. It corresponds to the anterior fontanelle of infants.

The bones in the braincase include those of the skull base (**sphenoid**, **ethmoid**, **temporal**, and **occipital**) and the flat bones of the cranial vault (frontal and parietal). A hole in the occipital bone (**foramen magnum**) allows the spinal cord to connect with the brain stem.

The skull accommodates the major sense organs— the eyes, ears, nose, and tongue—within bony cavities.

The **lambda** is where the parietal and occipital bones meet. It corresponds to the posterior fontanelle of infants.

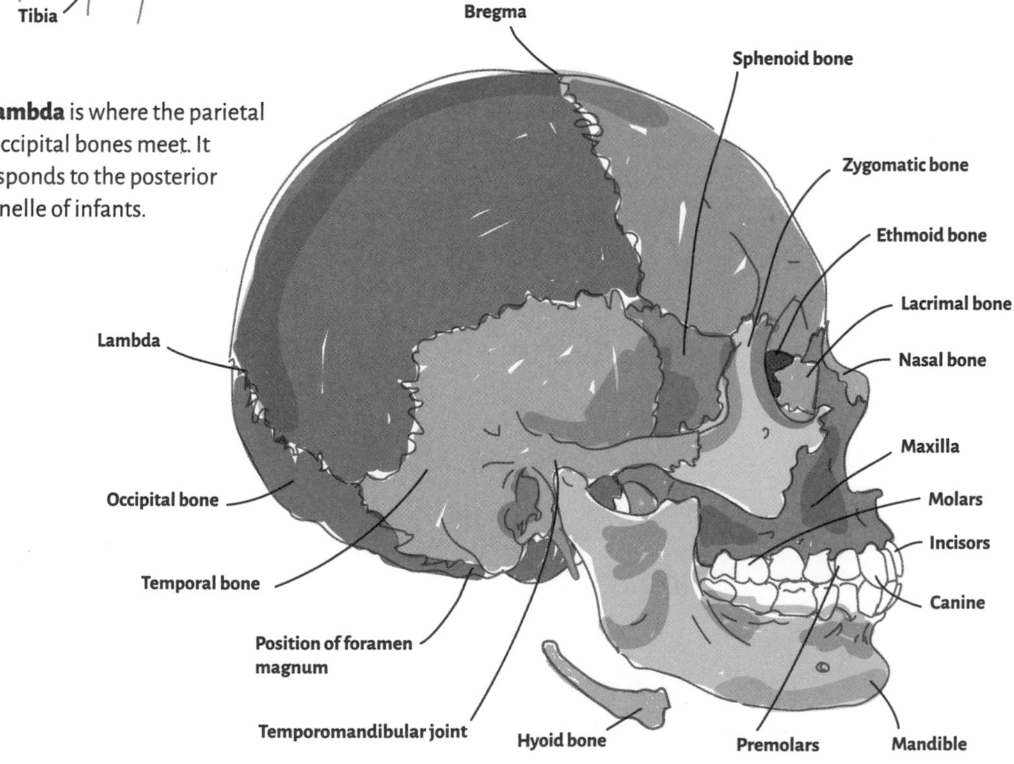

Bregma
Sphenoid bone
Zygomatic bone
Ethmoid bone
Lacrimal bone
Nasal bone
Lambda
Maxilla
Molars
Incisors
Canine
Occipital bone
Temporal bone
Position of foramen magnum
Temporomandibular joint
Hyoid bone
Premolars
Mandible

The **orbit** is a bony cavity that protects the eye and provides attachment for the extraocular muscles that move the eyeball (see the illustration on page 139).

The inner ear is sheltered deep inside the very dense petrous (petrous means rocklike) temporal bone. The high density of this bone reflects extraneous sound away from the inner ear to optimize hearing acuity.

The nasal olfactory area is located on the ethmoid bone at the roof of the nasal cavity.

Taste receptors are located mainly on the tongue, which is protected by the mandible and maxilla.

The maxilla and mandible bear the teeth (32 in a normal adult) in upper and lower dental arcades. Each quadrant of the adult mouth carries two incisors, one canine, two premolars, and three molars.

The mandible articulates with the temporal bone on each side. The **temporomandibular joint** (TMJ) can open and close, protrude and retract the jaw, and move the jaw from side to side.

Vertebral column

The **vertebral column** consists of a series of 26 irregular bones called **vertebrae**, which increase in size from the skull base to the hip bone. The vertebrae are arranged into a series of 7 in the neck (cervical), 12 in the chest (thoracic), 5 in the lower back (lumbar), and 5 at the base of the trunk (sacral vertebrae fused to form the sacrum). There are additional small bones that form the vestigial tail (coccyx) below the sacrum.

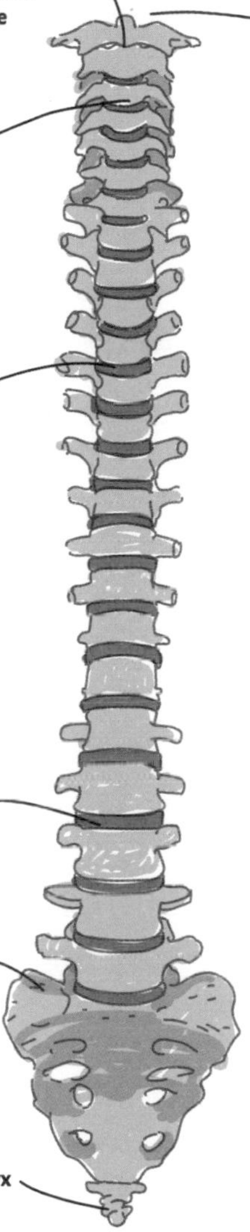

The axis is the second cervical vertebra. The atlas swivels on the axis to rotate the head.

The atlas is the first cervical vertebra and articulates with the occiput of the skull base to produce nodding movements.

7 cervical vertebrae

12 thoracic vertebrae

5 lumbar vertebrae

1 sacrum

1 coccyx

HYOID PROTECTS THE AIRWAY

The **hyoid** is an important bone in the neck that provides attachments for muscles of the tongue and throat and prevents the airway from collapsing during deep inspiration. A fractured hyoid is something forensic pathologists look for as proof of strangulation (see the illustration on page 50 for the position of the hyoid bone).

Ribs and sternum

The vital organs of the chest are protected by 12 pairs of ribs and the sternum.

The **sternum** is a flat bony structure at the front of the chest, divided into manubrium, body, and xiphoid process. The term *xiphoid* refers to the swordlike shape of this bone—the *xiphos* was the sword carried by Greek hoplites (see page 45).

True ribs pairs 1 to 7 are attached to the sternum with cartilage.

False ribs pairs 8 to 10 have an indirect attachment to the sternum, by the rib above.

Floating ribs pairs 11 and 12 have no attachment to the sternum.

BONES OF THE UPPER LIMB

The bones of the upper limb are divided into those of the shoulder or pectoral girdle and those of the rest of the limb.

Pectoral girdle

The **pectoral girdle** consists of the scapula (shoulder blade) and clavicle (collarbone).

The **clavicle** has a direct attachment to the axial skeleton through a joint with the manubrium of the sternum. It functions as a pivot, allowing the scapula and upper limb to move freely around the point of attachment of the clavicle to the sternum.

Arm and humerus

The **humerus** is the bone of the arm and articulates proximally with the scapula and distally with the radius and ulna of the forearm.

The head of the humerus can move freely on the glenoid cavity to allow the arm to be raised above the head.

The articulation of the distal humerus with the ulna is a hinge-like joint that allows only flexion and extension of the elbow.

The joint between the humerus and radius allows pivoting of the radius around its long axis for rotation of the forearm.

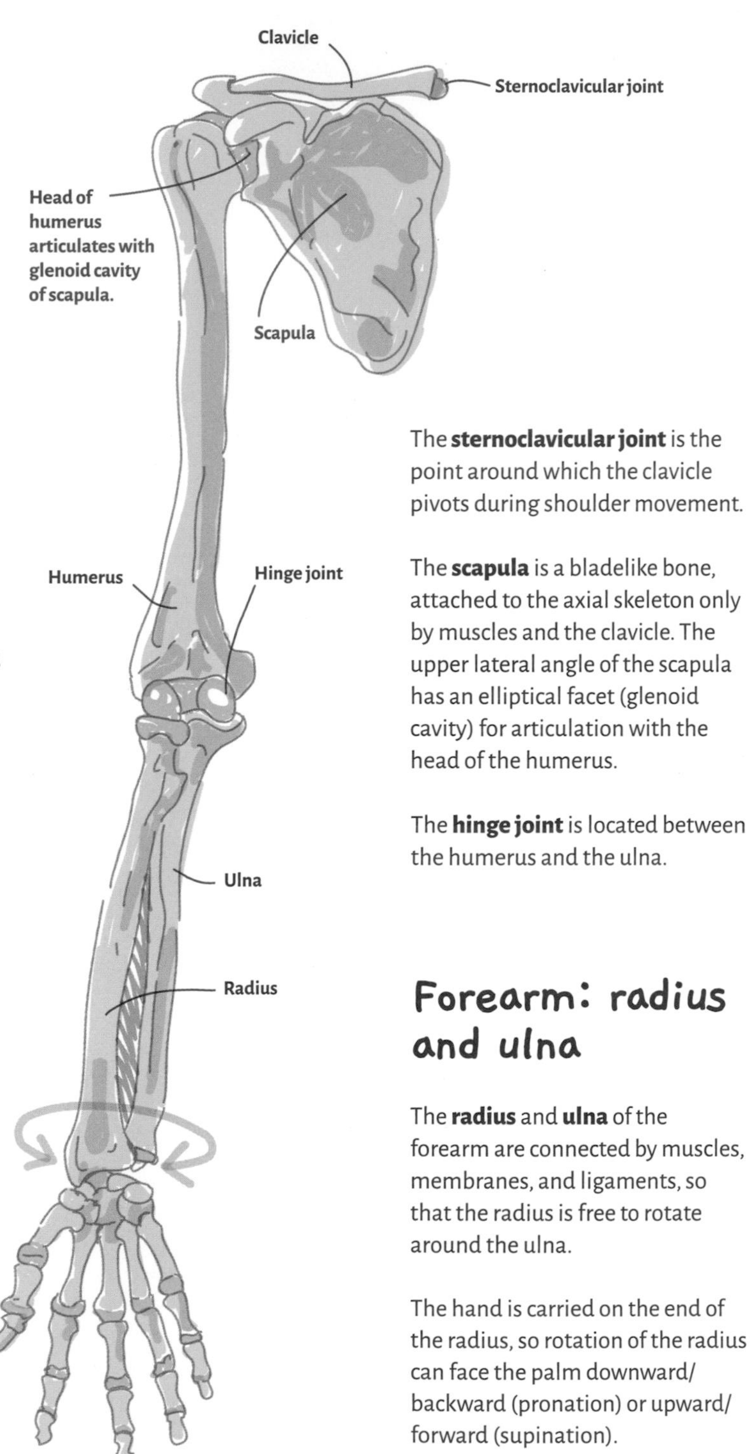

Clavicle
Sternoclavicular joint
Head of humerus articulates with glenoid cavity of scapula.
Scapula
Humerus
Hinge joint
Ulna
Radius

The **sternoclavicular joint** is the point around which the clavicle pivots during shoulder movement.

The **scapula** is a bladelike bone, attached to the axial skeleton only by muscles and the clavicle. The upper lateral angle of the scapula has an elliptical facet (glenoid cavity) for articulation with the head of the humerus.

The **hinge joint** is located between the humerus and the ulna.

Forearm: radius and ulna

The **radius** and **ulna** of the forearm are connected by muscles, membranes, and ligaments, so that the radius is free to rotate around the ulna.

The hand is carried on the end of the radius, so rotation of the radius can face the palm downward/backward (pronation) or upward/forward (supination).

Wrist and hand

Carpals are wrist bones arranged in two rows of four. In the proximal row are **scaphoid**, **lunate**, **triquetrum**, and **pisiform** bones. In the distal row are **trapezium**, **trapezoid**, **capitate**, and **hamate** bones. The **elliptical wrist joint** is between the distal radius and the scaphoid and lunate. It allows flexion and extension of the wrist and movement of the hand toward or away from the body midline (adduction/abduction).

Fingers

The bones of the fingers (digits) are called **phalanges**.

Digit 1 (the thumb) has two phalanges: proximal and distal.

Digits 2 to 5 have three phalanges: proximal, middle, and distal.

Joints between phalanges, called **interphalangeal joints**, are hinge joints. Those between the proximal phalanges and **metacarpals (metacarpophalangeal)** are condylar joints that allow flexion/extension and some sideways movement (abduction/adduction).

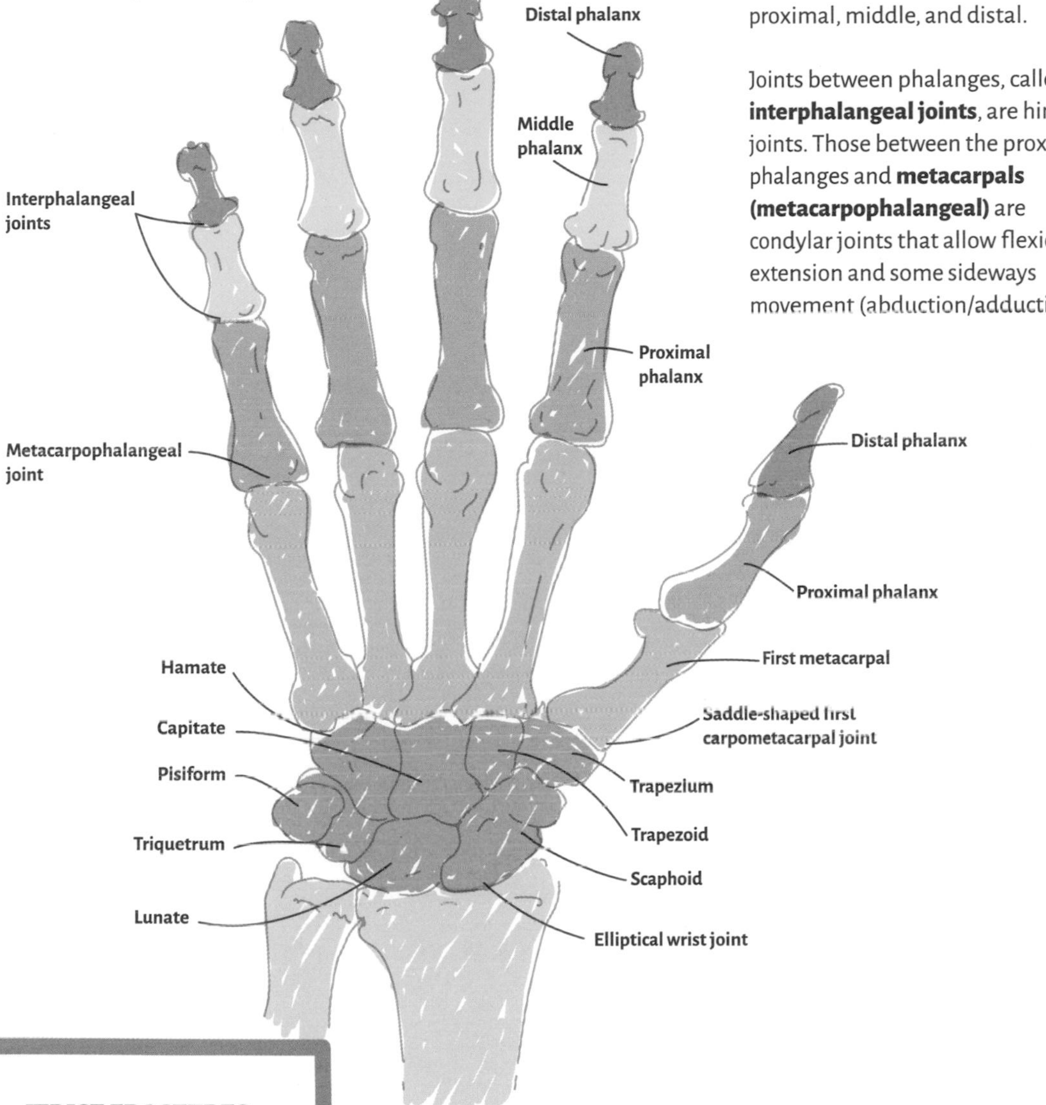

Distal phalanx
Middle phalanx
Interphalangeal joints
Proximal phalanx
Distal phalanx
Metacarpophalangeal joint
Proximal phalanx
First metacarpal
Hamate
Saddle-shaped first carpometacarpal joint
Capitate
Pisiform
Trapezium
Trapezoid
Triquetrum
Scaphoid
Lunate
Elliptical wrist joint

WRIST FRACTURES

Fractures of the distal radius and/or scaphoid are common due to falls onto the outstretched hand. The elongated scaphoid can also be fractured across its waist, leading to avascular necrosis (death) of one half.

Palm

The **palm** of the hand is formed of five metacarpals, which articulate with the distal row of carpal bones. The **first carpometacarpal** is the joint between the first metacarpal of the thumb and the trapezium.

The trapezium is particularly important for human hand function. It is rotated out of the plane of the palm and has a saddle shape that allows the thumb to sweep across the palm to perform opposition movements with all the other fingers (bringing the pad of thumb and fingers together).

BONES OF THE LOWER LIMB

Bones of the lower limb consist of the pelvic girdle and the rest of the limb. The two hip bones and sacrum form a highly stable ring, which transmits body weight and protects the soft pelvic organs.

Pelvic girdle

The **pelvic girdle** has a single hip bone (os coxa), which forms from the fusion of three bones (**pubis**, **ilium**, and **ischium**) during fetal and neonatal life.

Pelvis is Latin for "bucket" or "basin" and refers to the basin shape of the paired os coxa and sacrum together.

Three bones meet at the cuplike **acetabulum** (hip socket) for articulation with the head of the femur.

The pubis of one side articulates with its pair through the **pubic symphysis**.

The **sacrum** articulates with the os coxa on each side.

The ilium articulates with the sacrum at the **sacroiliac joint**. Dense ligaments stabilize the sacroiliac joint, which is usually only mobile during late pregnancy, when the birth canal must be as wide and flexible as possible.

Prior to fusion of the three hip bones, the acetabulum is occupied by a triradiate cartilage, which is shaped like the Mercedes logo.

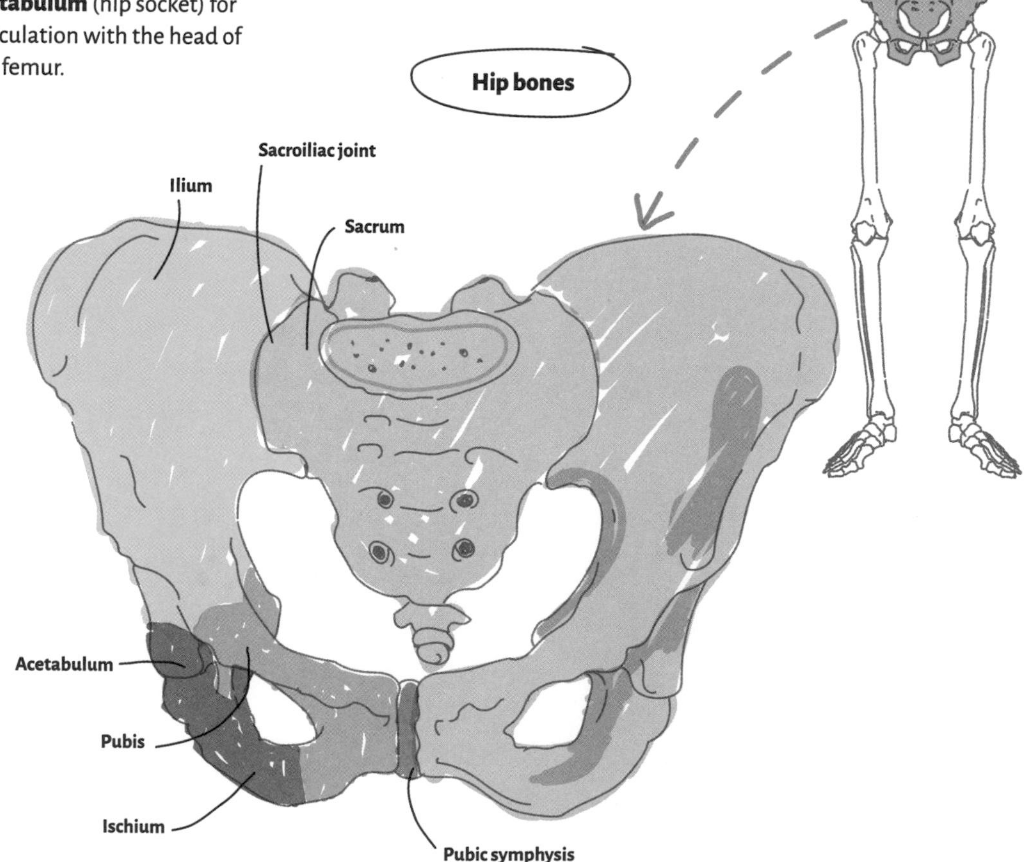
Hip bones
Sacroiliac joint
Ilium
Sacrum
Acetabulum
Pubis
Ischium
Pubic symphysis

Thigh and leg bones

The bone of the thigh is the **femur**.

The **patella** is a sesamoid bone embedded in the tendon of the quadriceps femoris muscle of the anterior thigh.

The **tibia** is the weight-bearing bone of the leg. The proximal tibia has paired and flattened articular surfaces for the femoral condyles, separated by an **intercondylar region** for attachment of ligaments. The distal tibia has a bump called the **medial malleolus**, which is felt at the inner ankle.

The fibula and tibia together form the upper surface of the **talocrural joint** with the talus of the foot.

The **head of femur**'s proximal end articulates in a ball-and-socket joint with the acetabulum. This hip joint is less mobile than the shoulder joint but allows some movement in three axes, i.e., flexion/extension, abduction/adduction, and rotation.

The distal femur has paired **condyles** that articulate with one of the bones of the leg (the tibia) and the patella at the knee joint.

The **fibula** lies on the lateral side of the leg and is mainly for muscle attachments. The fibula articulates with the tibia both proximally and distally. The distal end of the fibula forms the **lateral malleolus**, which is felt at the outside of the ankle.

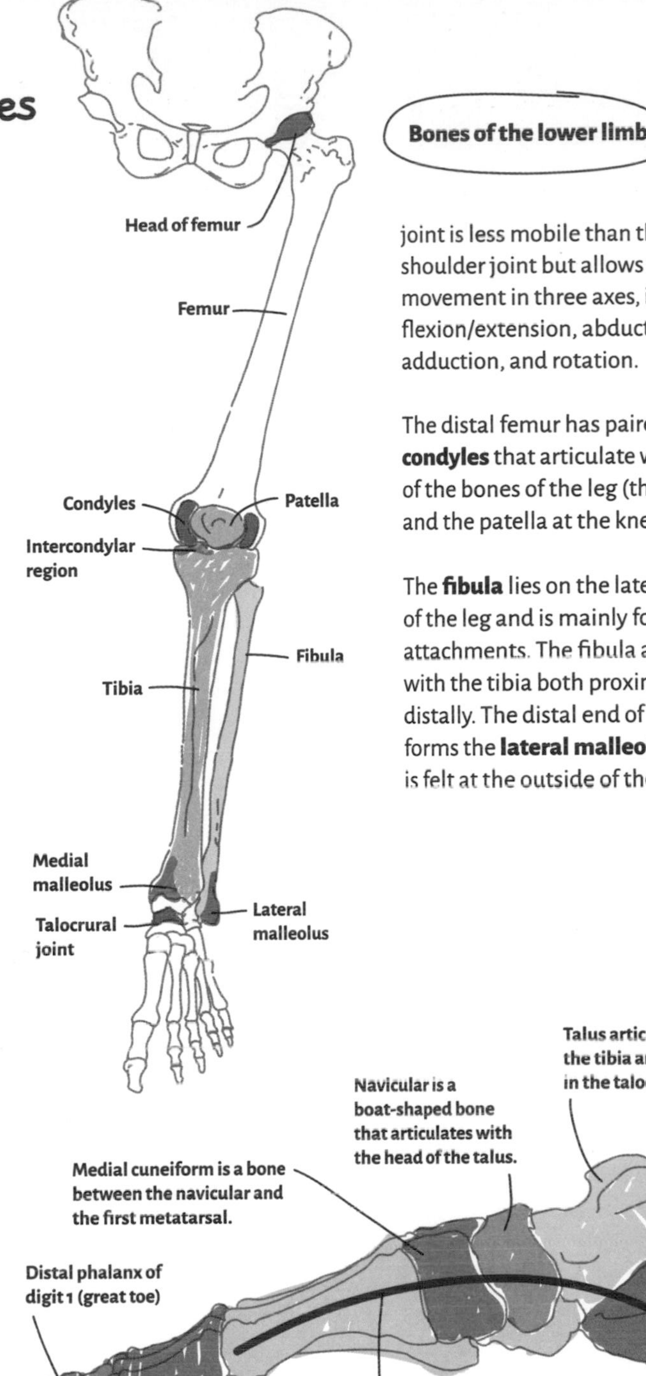
Bones of the lower limb
Head of femur
Femur
Condyles
Patella
Intercondylar region
Fibula
Tibia
Medial malleolus
Lateral malleolus
Talocrural joint
Talus articulates with the tibia and fibula in the talocrural joint.
Navicular is a boat-shaped bone that articulates with the head of the talus.
Medial cuneiform is a bone between the navicular and the first metatarsal.
Distal phalanx of digit 1 (great toe)
Proximal phalanx of digit 1 (great toe)
Medial longitudinal arch of the foot
Calcaneus forms the heel and supports the talus.

Foot and toes

The proximal foot contains seven tarsal bones: **talus**, **calcaneus**, **navicular**, and cuboid, and intermediate, lateral, and **medial cuneiforms**. The more distal part of the foot is made up of five metatarsals. The weight of the body is transmitted through the talus and calcaneus to the ground, but also forward through the navicular, cuneiforms, and metatarsals to the base of the big toe.

The toes or digits are composed of phalanges, much like the fingers. The great toe (digit 1) has two phalanges: **proximal** and **distal**. Digits 2 to 5 have three phalanges in each toe: proximal, middle, and distal.

Most weight bearing passes to the ground at the tuberosity (rounded prominence) of the calcaneus, which forms our bony heel. The first metatarsal, medial cuneiform, navicular, talus, and calcaneus form a medial longitudinal arch of the foot, which may collapse to give flat feet (pes planus).

JOINTS OF THE BODY

The place where two or more bones meet is called a joint. Body joints may be fibrous, cartilaginous, or synovial, referring to the substances between the two bones (fibrous tissue, cartilage, or synovial fluid).

Fibrous joints

Fibrous joints are usually immobile or poorly mobile. They are found between the bones of the skull (**sutures**) and between the teeth and the jaw (**gomphoses**).

The **gomphosis** is formed by **periodontal ligaments** around the teeth, which anchor the teeth firmly into the maxilla or mandible.

Dense fibrous tissue in the sutures, as well as the interlocking pattern of the bone edges, lock the skull bones together.

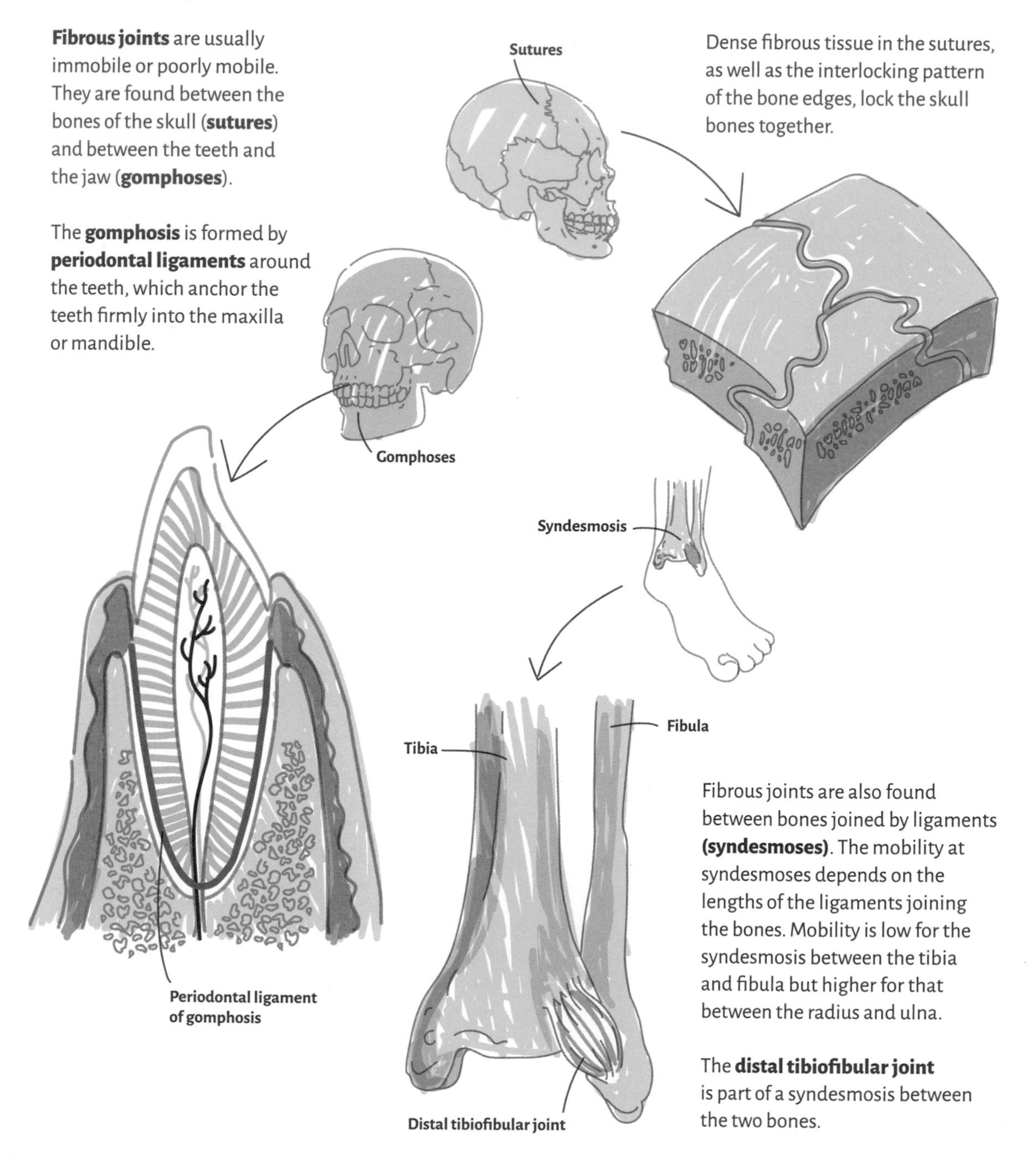

Sutures
Gomphoses
Syndesmosis
Periodontal ligament of gomphosis
Tibia
Fibula
Distal tibiofibular joint

Fibrous joints are also found between bones joined by ligaments **(syndesmoses)**. The mobility at syndesmoses depends on the lengths of the ligaments joining the bones. Mobility is low for the syndesmosis between the tibia and fibula but higher for that between the radius and ulna.

The **distal tibiofibular joint** is part of a syndesmosis between the two bones.

Cartilaginous joints

In **cartilaginous joints**, the articulating bones are joined by cartilage.

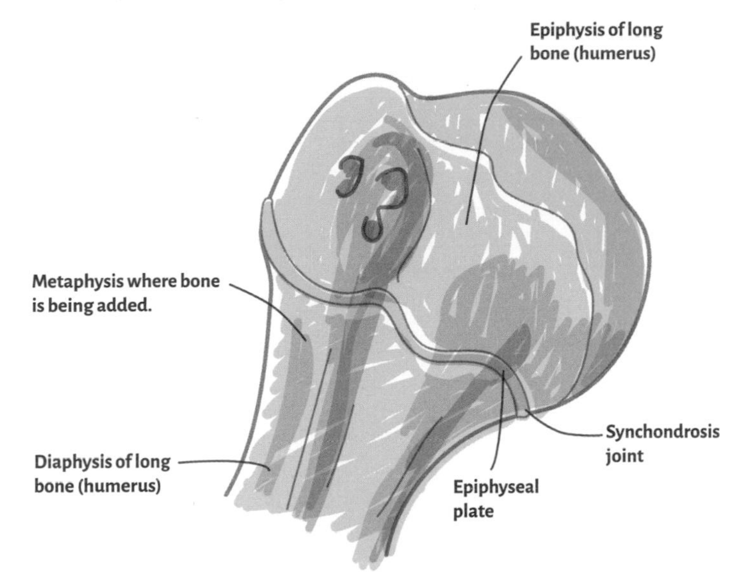

Epiphysis of long bone (humerus)
Metaphysis where bone is being added.
Diaphysis of long bone (humerus)
Epiphyseal plate
Synchondrosis joint

Synchondroses are joints where two bones (or ossification centers) are joined by hyaline (glassy) cartilage. The best examples are the **epiphyseal plates**. These are growth zones and will disappear when the individual has completed adolescence.

Symphyses are joints (see **symphyseal joint** opposite) where fibrocartilage joins the two bones. They are found between the bodies of vertebrae (intervertebral disks) and between the two pubic bones of the pelvis, called the **pubic symphysis**.

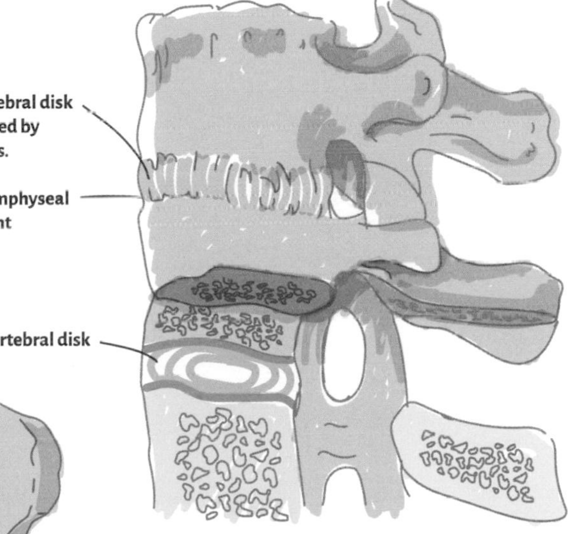

Intervertebral disk surrounded by ligaments.
Symphyseal joint
Intervertebral disk

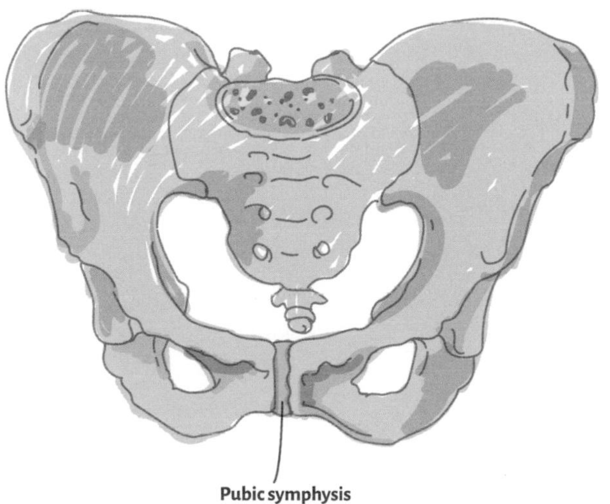

Pubic symphysis

Symphyses are slightly mobile joints that combine strength with some flexibility. The flexibility of the pubic symphysis can be increased during late pregnancy by the hormone relaxin, to help widen the birth canal.

Synovial joints

Synovial joints are often classified by their shape and free range of movement. The stability of synovial joints depends on the conformity of joint surfaces (how well they fit), the ligaments strengthening the joint, and the tone of the muscles that cross the joint. This is why good muscle strength is important for avoiding joint instability and the wear and tear of osteoarthritis.

FEATURES OF SYNOVIAL JOINTS

★Smooth **articular cartilage** of the hyaline (glassy) type at the joint surfaces to absorb compressive forces
★A joint or synovial cavity filled with synovial fluid to minimize friction
★An **articular capsule** that provides stability to the joint through its fibrous tissues (fibrous capsule)
★A **synovial membrane** lining the interior of the articular capsule and producing the synovial fluid that fills the joint
★Viscous synovial fluid itself, which resembles raw egg white in consistency, and lubricates the articular joint surfaces
★Reinforcing ligaments usually outside the joint capsule (extracapsular) but sometimes inside the capsule (intracapsular)
★Fluid-filled bursae alongside the joint, which may connect with the joint cavity
★Rich nerve supply to detect joint pain and monitor joint stretching

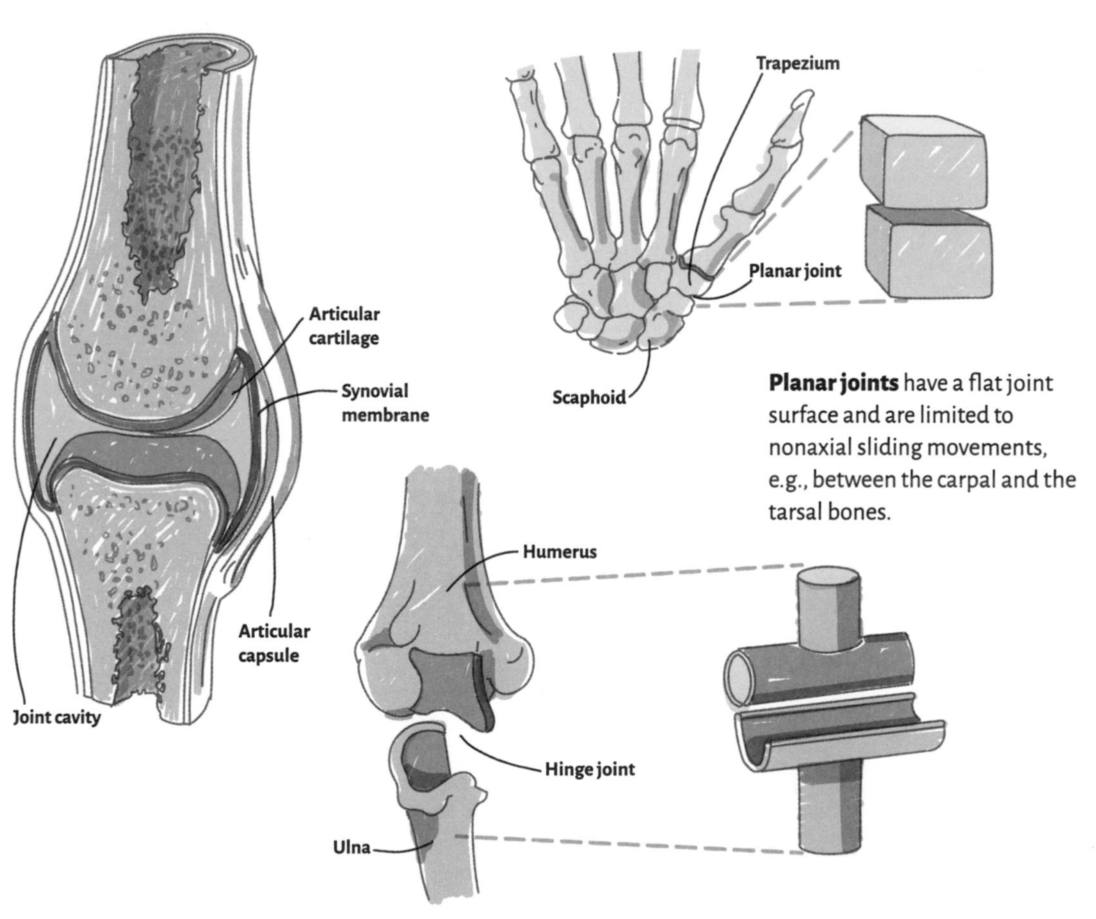
Synovial joint
Articular cartilage
Synovial membrane
Articular capsule
Joint cavity
Trapezium
Planar joint
Scaphoid
Humerus
Hinge joint
Ulna

Planar joints have a flat joint surface and are limited to nonaxial sliding movements, e.g., between the carpal and the tarsal bones.

Hinge joints allow only uniaxial flexion and extension, e.g., between the humerus and the ulna, or between the phalanges of the digits.

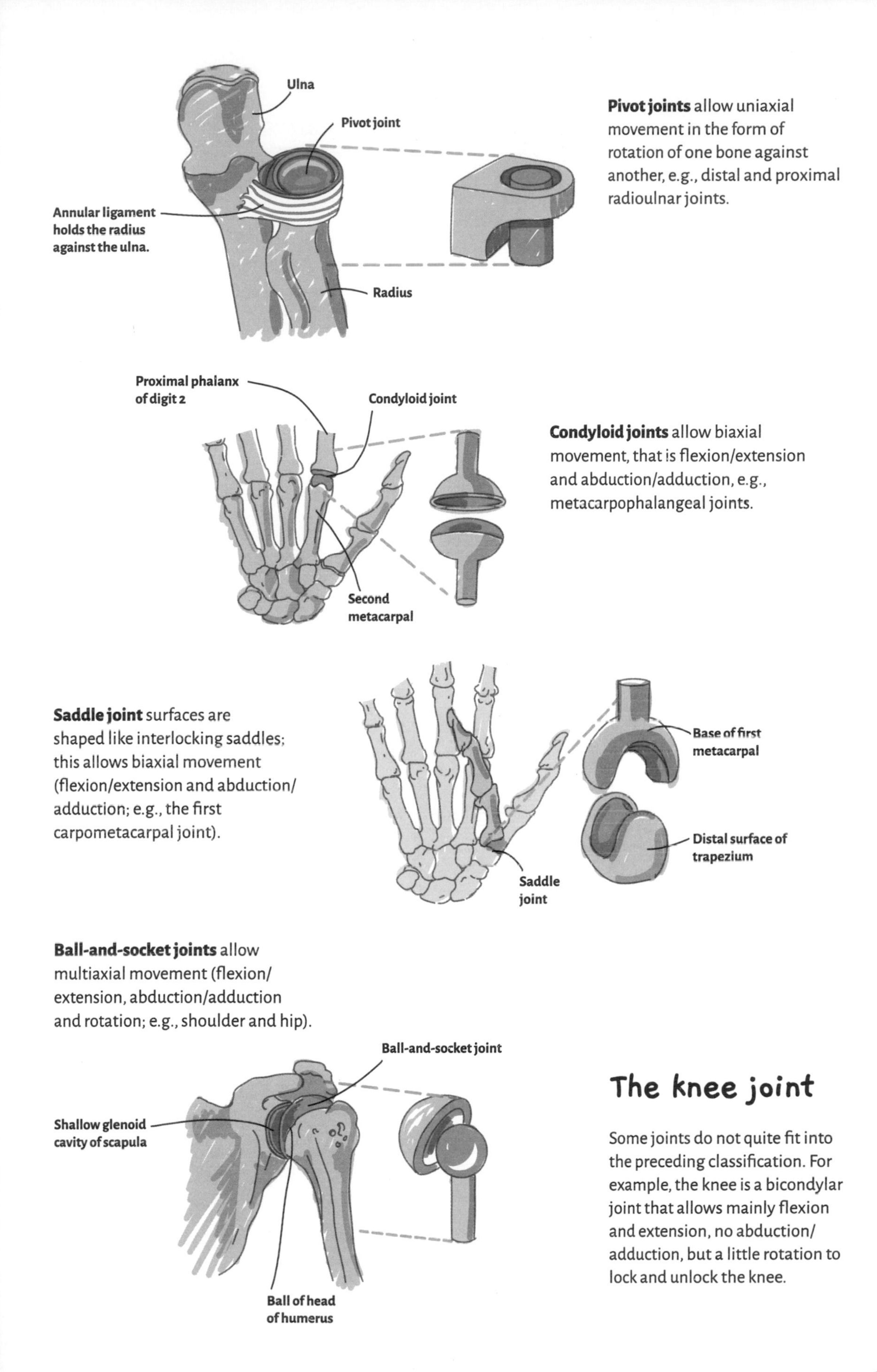
Ulna
Pivot joint
Annular ligament holds the radius against the ulna.
Radius
Proximal phalanx of digit 2
Condyloid joint
Second metacarpal
Base of first metacarpal
Distal surface of trapezium
Saddle joint
Ball-and-socket joint
Shallow glenoid cavity of scapula
Ball of head of humerus

Pivot joints allow uniaxial movement in the form of rotation of one bone against another, e.g., distal and proximal radioulnar joints.

Condyloid joints allow biaxial movement, that is flexion/extension and abduction/adduction, e.g., metacarpophalangeal joints.

Saddle joint surfaces are shaped like interlocking saddles; this allows biaxial movement (flexion/extension and abduction/adduction; e.g., the first carpometacarpal joint).

Ball-and-socket joints allow multiaxial movement (flexion/extension, abduction/adduction and rotation; e.g., shoulder and hip).

The knee joint

Some joints do not quite fit into the preceding classification. For example, the knee is a bicondylar joint that allows mainly flexion and extension, no abduction/adduction, but a little rotation to lock and unlock the knee.

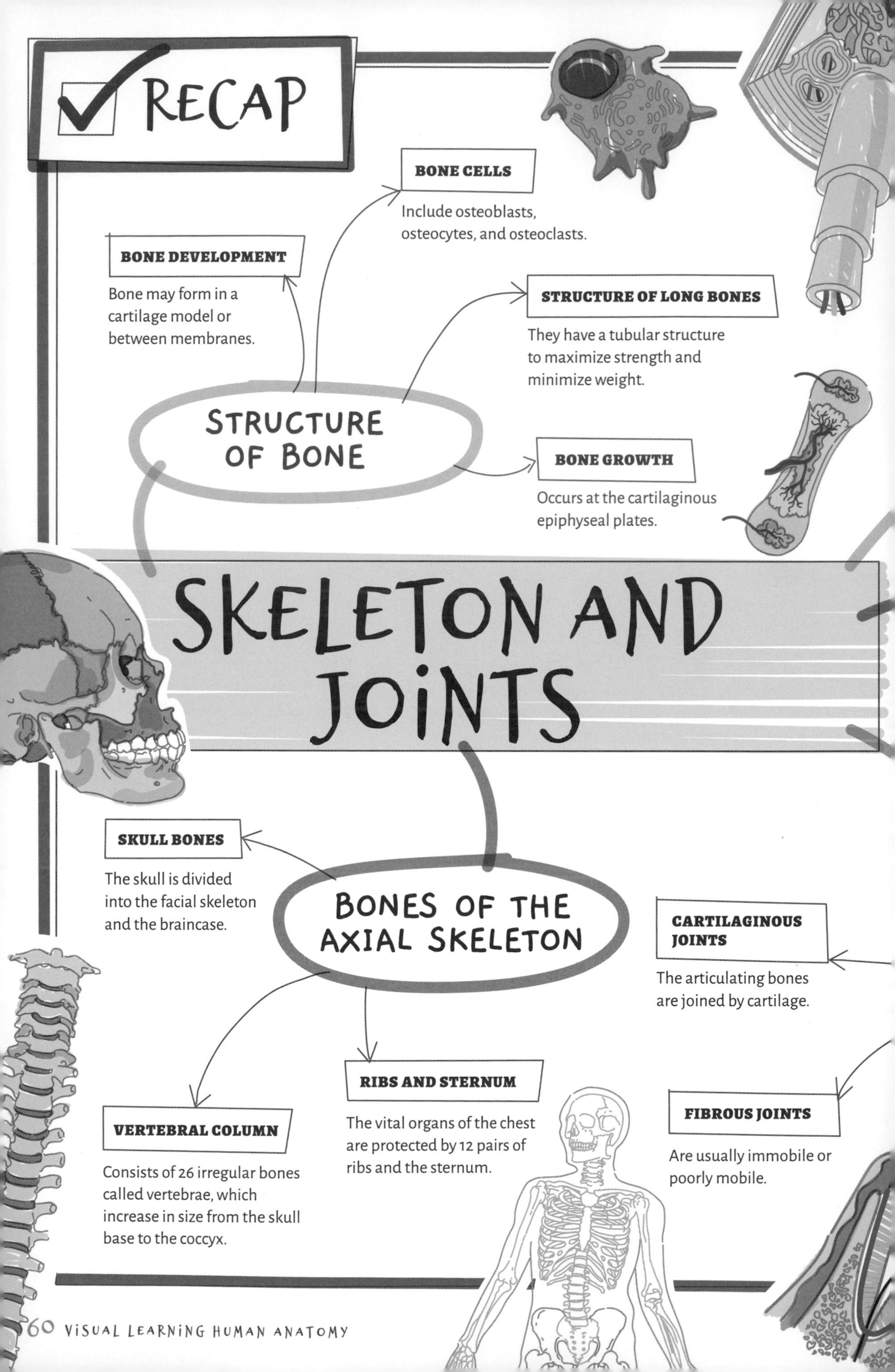
RECAP
SKELETON AND JOINTS
STRUCTURE OF BONE
BONE DEVELOPMENT
Bone may form in a cartilage model or between membranes.
BONE CELLS
Include osteoblasts, osteocytes, and osteoclasts.
STRUCTURE OF LONG BONES
They have a tubular structure to maximize strength and minimize weight.
BONE GROWTH
Occurs at the cartilaginous epiphyseal plates.
BONES OF THE AXIAL SKELETON
SKULL BONES
The skull is divided into the facial skeleton and the braincase.
VERTEBRAL COLUMN
Consists of 26 irregular bones called vertebrae, which increase in size from the skull base to the coccyx.
RIBS AND STERNUM
The vital organs of the chest are protected by 12 pairs of ribs and the sternum.
CARTILAGINOUS JOINTS
The articulating bones are joined by cartilage.
FIBROUS JOINTS
Are usually immobile or poorly mobile.

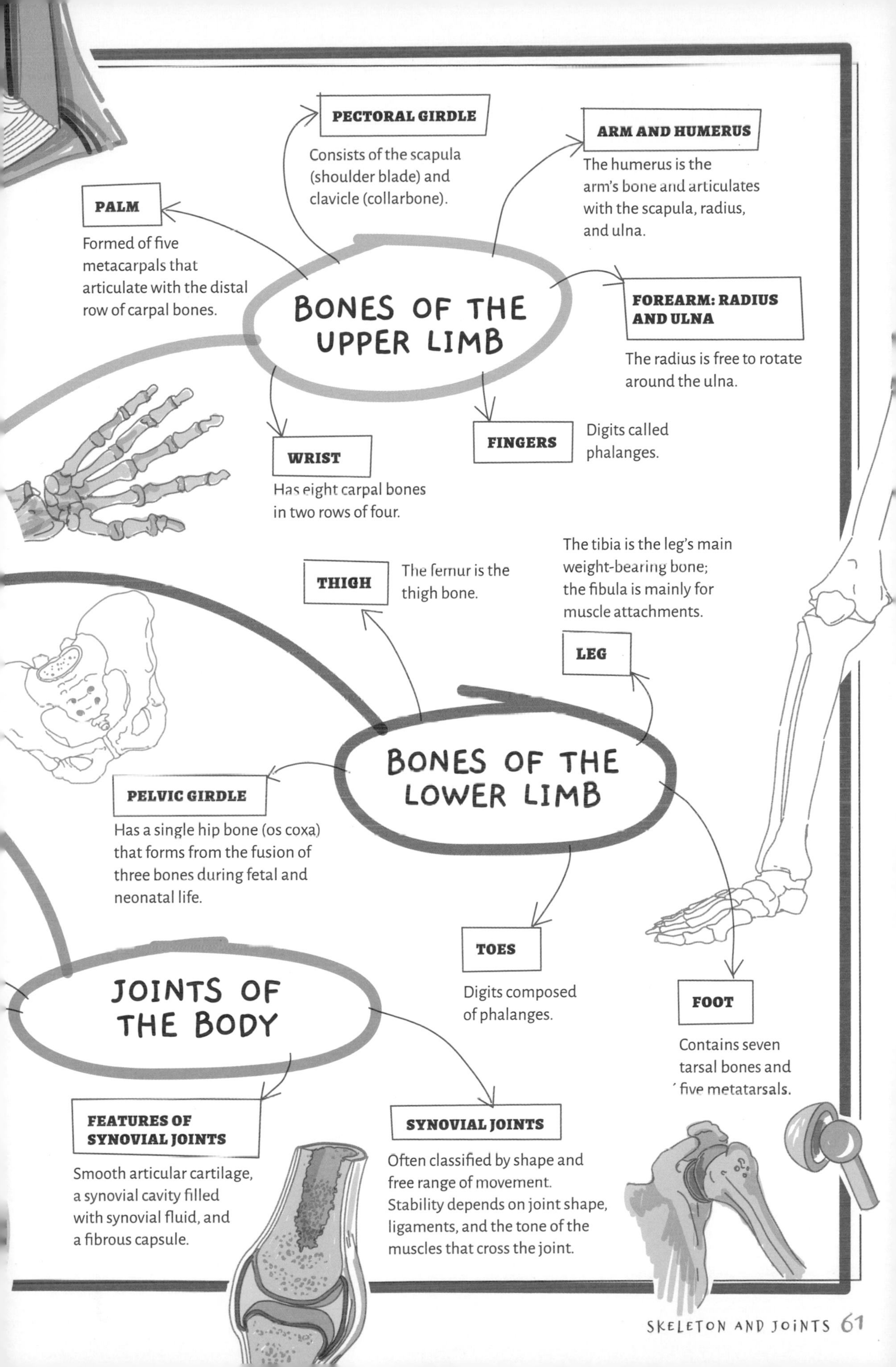

BONES OF THE UPPER LIMB
PECTORAL GIRDLE
Consists of the scapula (shoulder blade) and clavicle (collarbone).
ARM AND HUMERUS
The humerus is the arm's bone and articulates with the scapula, radius, and ulna.
PALM
Formed of five metacarpals that articulate with the distal row of carpal bones.
FOREARM: RADIUS AND ULNA
The radius is free to rotate around the ulna.
WRIST
Has eight carpal bones in two rows of four.
FINGERS
Digits called phalanges.
BONES OF THE LOWER LIMB
THIGH
The femur is the thigh bone.
LEG
The tibia is the leg's main weight-bearing bone; the fibula is mainly for muscle attachments.
PELVIC GIRDLE
Has a single hip bone (os coxa) that forms from the fusion of three bones during fetal and neonatal life.
TOES
Digits composed of phalanges.
FOOT
Contains seven tarsal bones and five metatarsals.
JOINTS OF THE BODY
FEATURES OF SYNOVIAL JOINTS
Smooth articular cartilage, a synovial cavity filled with synovial fluid, and a fibrous capsule.
SYNOVIAL JOINTS
Often classified by shape and free range of movement. Stability depends on joint shape, ligaments, and the tone of the muscles that cross the joint.

CHAPTER 4

MUSCULAR SYSTEM

The muscular system produces movement of the body, thanks to the attachment of the striated (striped) skeletal or voluntary muscles to the skeleton. Trunk muscles may contract isometrically (i.e., keep the same length) to maintain posture while standing or sitting.

The muscles of the body wall protect the delicate internal organs of the thoracic and abdominal cavities and assist vital internal functions, such as lung ventilation, urination (micturition), and defecation.

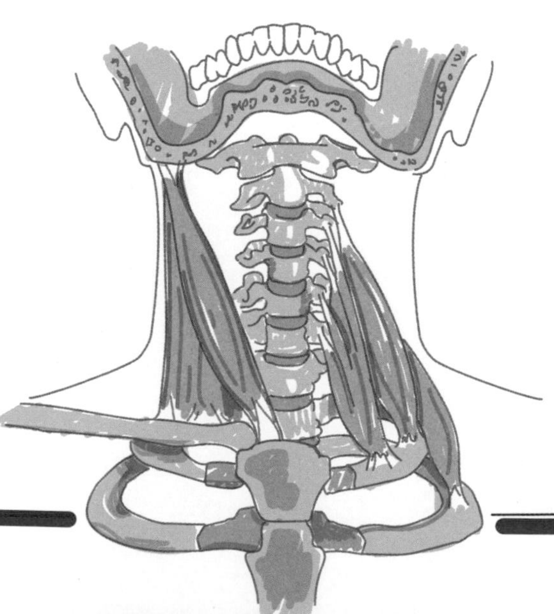

HOW TENDONS ATTACH MUSCLES TO BONES

Skeletal muscles consist of one or more muscle bellies, with at least two tendinous attachments to bone. The medial or proximal attachment of a muscle is called the origin, while the more distal or lateral attachment is called the insertion.

All three types of mechanical levers can be found in the body: the first class lever, the second class lever, and the third class lever.

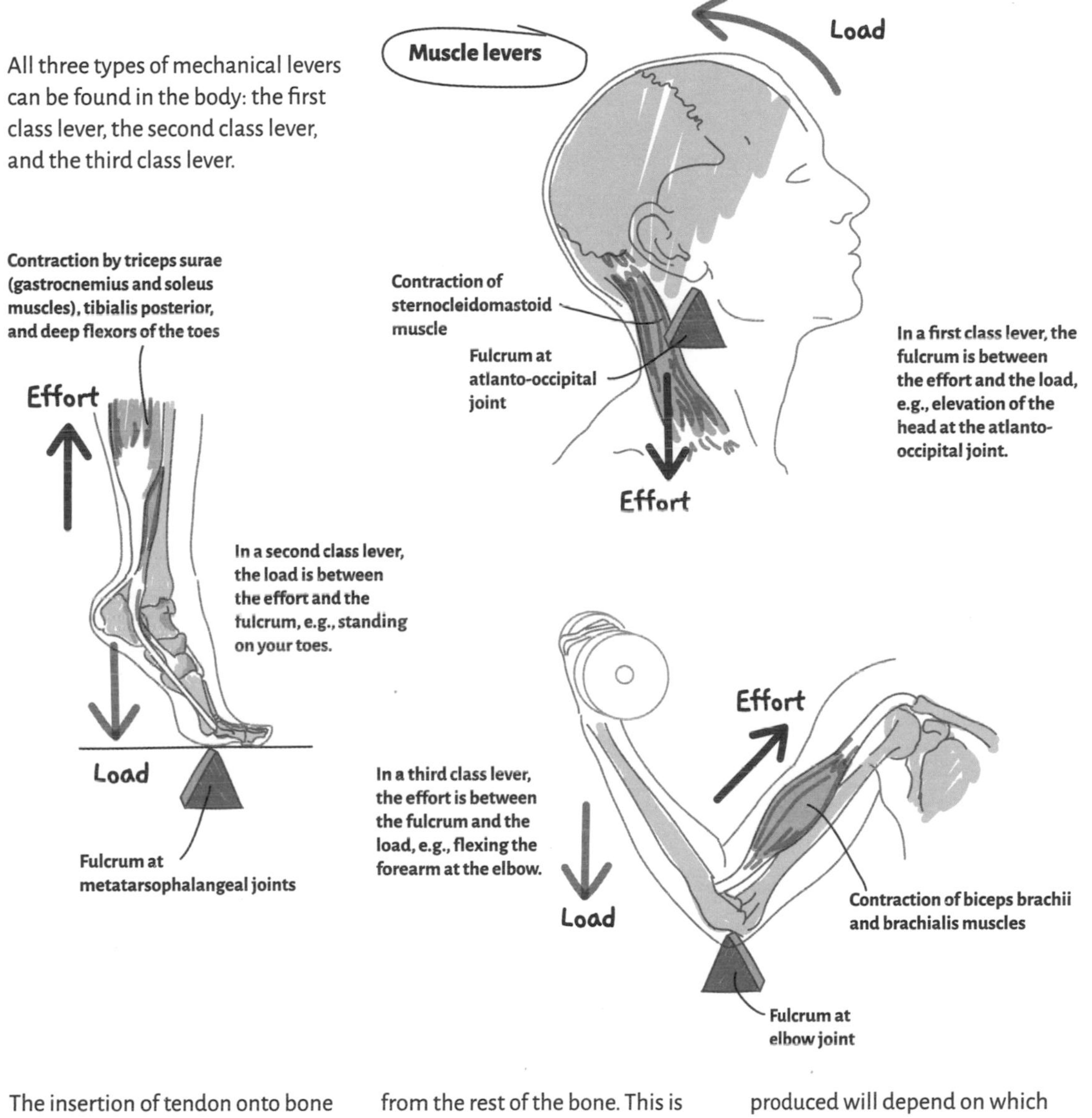

Muscle levers
Load
Contraction of sternocleidomastoid muscle
Fulcrum at atlanto-occipital joint
Effort
In a first class lever, the fulcrum is between the effort and the load, e.g., elevation of the head at the atlanto-occipital joint.
Contraction by triceps surae (gastrocnemius and soleus muscles), tibialis posterior, and deep flexors of the toes
Effort
In a second class lever, the load is between the effort and the fulcrum, e.g., standing on your toes.
Load
Fulcrum at metatarsophalangeal joints
In a third class lever, the effort is between the fulcrum and the load, e.g., flexing the forearm at the elbow.
Effort
Load
Contraction of biceps brachii and brachialis muscles
Fulcrum at elbow joint

The insertion of tendon onto bone is a specialized region that has a tensile strength close to, or even greater than, the bone itself. The tendon may be so strong that sudden muscle tension will pull the bone of the insertion site away from the rest of the bone. This is called an **avulsion fracture**.

Muscle function usually involves contraction of the belly to bring the origin and insertion closer together. The actual movement produced will depend on which end is fixed. So the triceps brachii of the arm can raise the hand above the head when the upper limb is free or elevate the trunk during push-ups when the hand is on the floor.

HEAD AND FACIAL MUSCLES

The muscles of the head and neck include the facial muscles, the muscles of mastication (chewing), extraocular muscles, and the muscles of the soft palate, pharynx, and larynx.

How facial muscles move facial skin

The **facial muscles** are also known as the muscles of facial expression. They have at least one attachment to the dermis of the facial skin, in order that facial expressions can be changed. Facial muscles are controlled by the facial nerve (cranial nerve CN7) from the brain stem. Facial muscles can be circular, surrounding openings in the face.

Some facial muscles are sheetlike, such as the **frontalis** on the forehead and the **platysma** on the neck. Other small facial muscles are tiny straps, such as **zygomaticus major** and **minor** and **levator labii superioris**.

The **buccinator** is a deeper facial muscle in the front cheek that helps move food between the teeth when we chew.

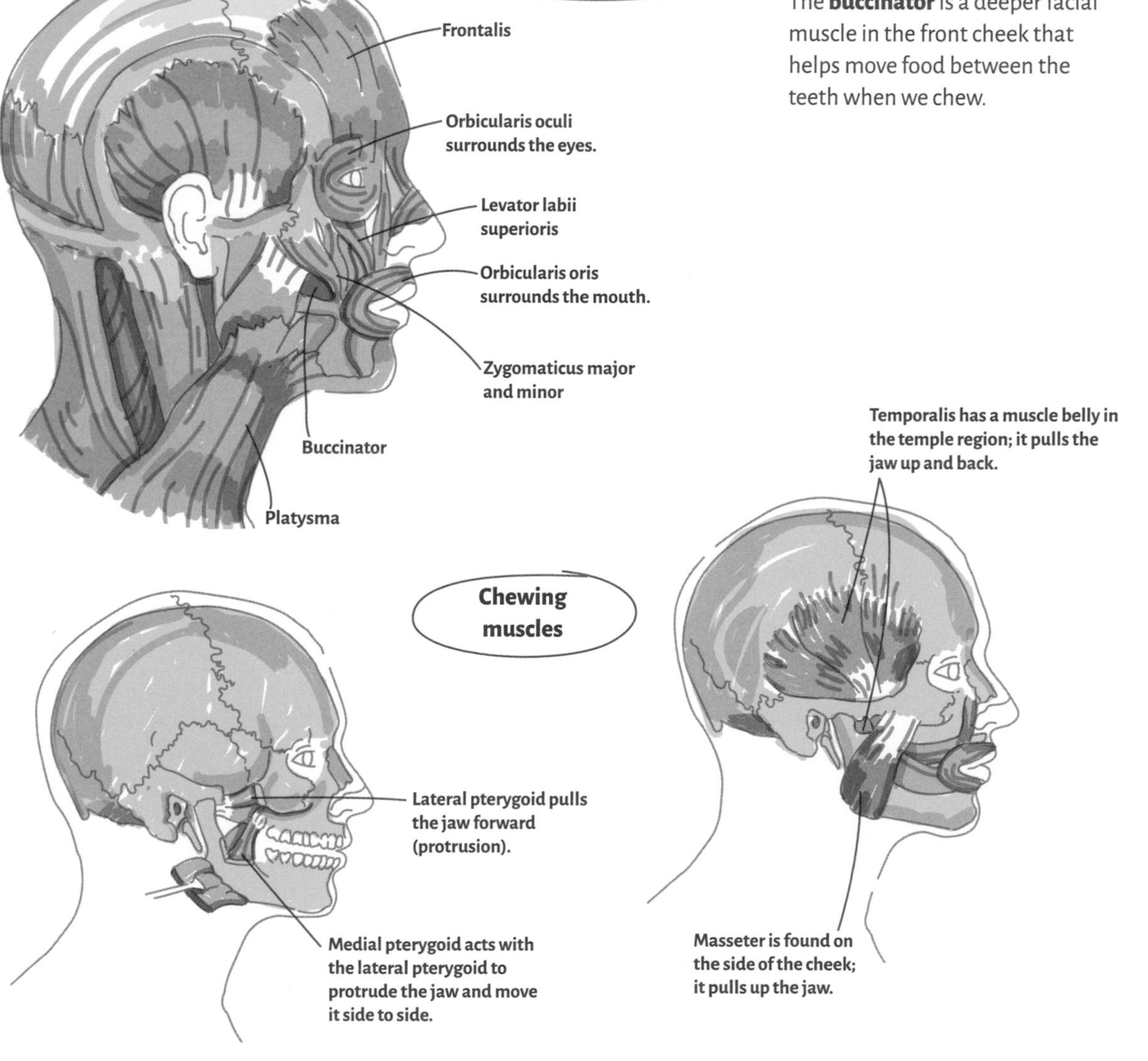
Head muscles
Frontalis
Orbicularis oculi surrounds the eyes.
Levator labii superioris
Orbicularis oris surrounds the mouth.
Zygomaticus major and minor
Buccinator
Platysma
Chewing muscles
Lateral pterygoid pulls the jaw forward (protrusion).
Medial pterygoid acts with the lateral pterygoid to protrude the jaw and move it side to side.
Temporalis has a muscle belly in the temple region; it pulls the jaw up and back.
Masseter is found on the side of the cheek; it pulls up the jaw.

Extraocular muscles

Extraocular muscles move the eyeball. Four recti—**superior rectus**, **medial rectus**, **inferior rectus**, and **lateral rectus**—are arranged at 90 degrees around the eye and move the eye up, medially, down, and laterally, respectively.

The levator palpebrae superioris elevates the eyelid.

Two oblique muscles, **superior oblique** and **inferior oblique**, can turn the eye down or up respectively, when the eye is turned nasally.

Extraocular muscles are supplied by the oculomotor, trochlear, and abducens nerves from the brain stem.

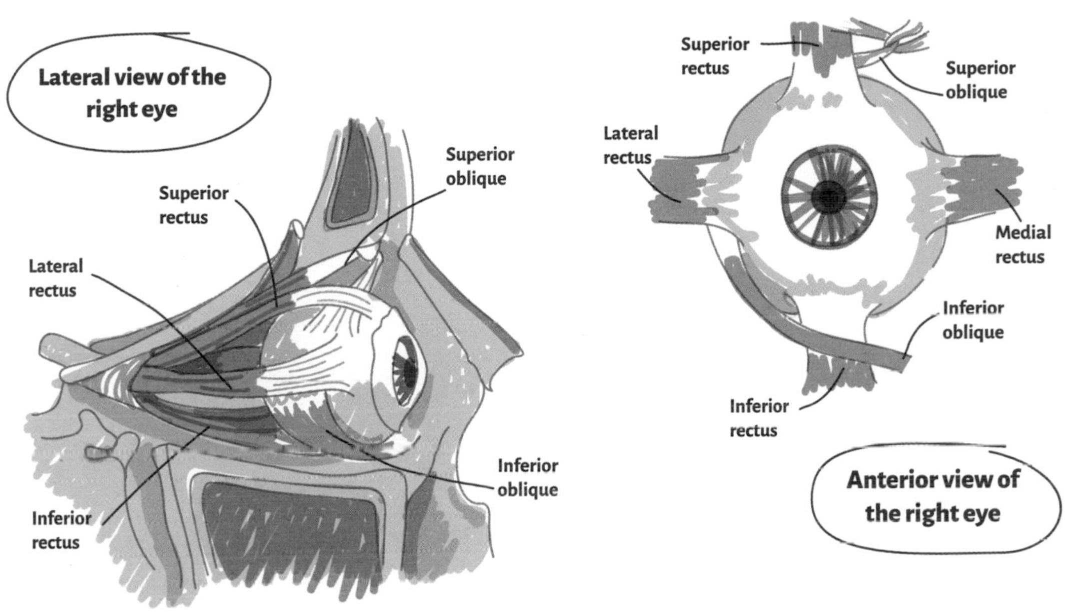
Lateral view of the right eye
Superior rectus
Superior oblique
Lateral rectus
Inferior rectus
Inferior oblique
Superior rectus
Superior oblique
Lateral rectus
Medial rectus
Inferior oblique
Inferior rectus
Anterior view of the right eye

Tongue, palate, and pharynx

The **tongue** is a muscular organ. **Intrinsic muscles** change tongue shape and are arranged in three planes. **Extrinsic muscles** change tongue position.

The tongue is important in swallowing, speech, and moving food during mastication. Most tongue muscles are supplied by the hypoglossal nerve from the brain stem.

The **palate** is a bony and muscular shelf separating the nasal and oral cavities. Palatal muscles are mainly controlled by the vagus nerve and close the nose during swallowing.

The **pharynx** is a muscular tube suspended from the skull base. Contraction of the pharynx under control of the vagus nerve squeezes swallowed food down to the **esophagus.**

Bony palate
Palate
Soft (muscular) palate
Intrinsic muscles
Pharynx
Tongue
Sagittal section of the oral cavity and pharynx
Extrinsic muscles
Esophagus

NECK AND TRUNK MUSCLES

The muscles of the trunk maintain posture, bend and rotate the head and torso, support the abdominal and pelvic organs, and play a vital role in lung ventilation. Muscles of the back extend down the vertebral column.

Neck muscles

Muscles of the neck are arranged anterior and posterior to the vertebral column.

Muscles anterior to the vertebral column include the **sternocleidomastoid**, which stabilizes and rotates the head, and the **scalene muscles**, which elevate the ribs.

Sternocleidomastoid

Clavicle

Manubrium of sternum

Muscles posterior to the vertebral column extend the neck and head and elevate the scapula and shoulders.

Scalene muscles elevate upper ribs.

Rib 1

Rib 2

Diaphragm

The **diaphragm** is the most important inspiratory muscle. It is a double-domed fibromuscular membrane that separates the thoracic and abdominal cavities.

Opening in diaphragm for inferior vena cava

Opening in diaphragm for esophagus

Fibrous central part of diaphragm

Muscular part of diaphragm

Contraction lowers the domes and increases the height of the thoracic cavity, drawing air into the lungs. The diaphragm also assists in raising intra-abdominal pressure during heavy lifting (to support back muscles), coughing, vomiting, defecation, and childbirth.

Intercostal muscles

Intercostal muscles fill the space between adjacent ribs. They can elevate or depress ribs to help in lung ventilation (inspiration and expiration, respectively). Mainly, they prevent in-drawing of the intercostal space when the diaphragm contracts.

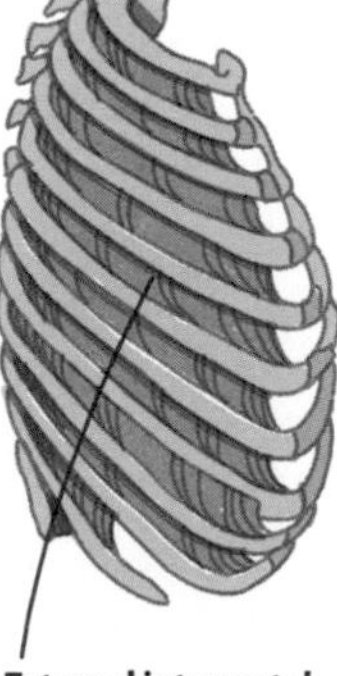

External intercostal muscles run inferiorly and anteriorly.

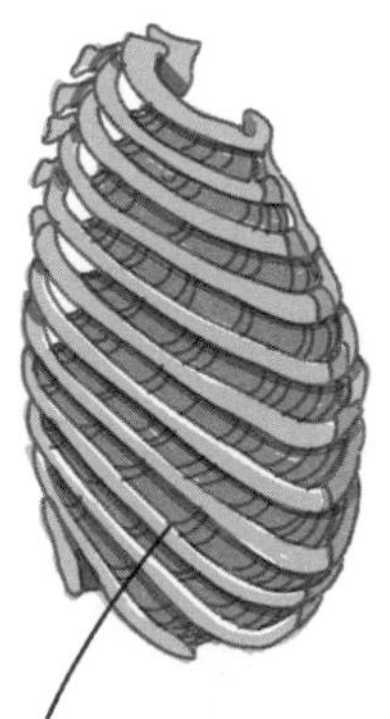

Internal intercostal muscles run superiorly and anteriorly.

Anterior abdominal muscles

The **abdominal wall muscles** protect and support the internal organs. They are important **expiratory muscles** that produce exhalation by compressing the abdomen and indirectly raising the diaphragm. They also contract rapidly during sneezing and coughing. Abdominal muscle contraction raises pressure to expel the feces during defecation, stomach contents during vomiting, and the fetus during childbirth.

The abdominal wall muscles consist of three layers of lateral muscles: **external oblique**, **internal oblique**, and **transversus abdominis**, from outside to in.

At the front of the abdominal wall is the **rectus abdominis**, which is a trunk flexor. Contracting the two obliques on one side flexes the trunk to that side (lateral flexion).

Contracting the external oblique on one side and the internal oblique on the other turns the upper trunk to the side of the active internal oblique.

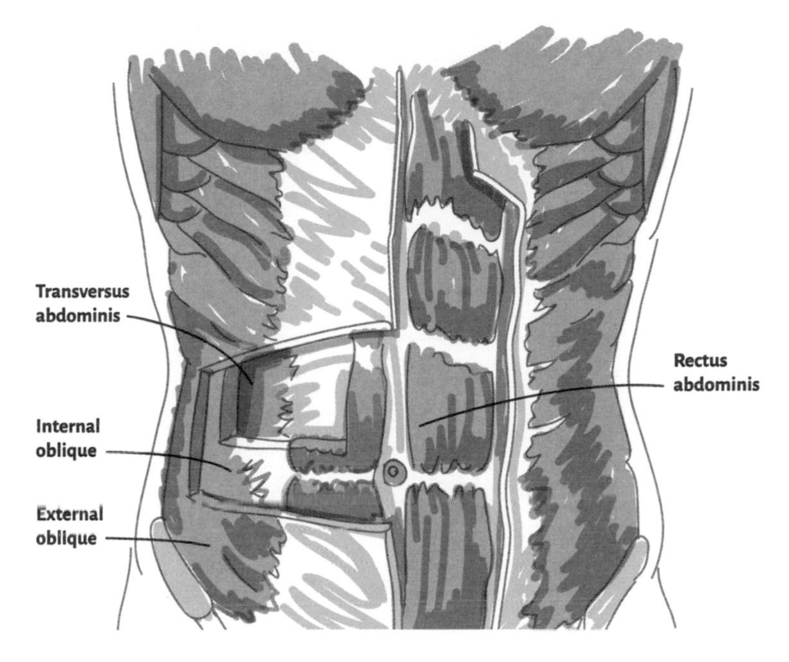
Transversus abdominis
Internal oblique
External oblique
Rectus abdominis

Posterior abdominal wall muscles

The **quadratus lumborum** attaches to the 12th rib, lumbar vertebrae, and iliac crest. It is an important postural muscle and assists the lateral flexion of the trunk.

The **psoas major** arises from lumbar vertebrae and inserts onto the femur. It flexes the thigh at the hip.

The **iliacus** arises from the iliac fossa and inserts (by common tendon with psoas major) onto the femur. It flexes the thigh at the hip.

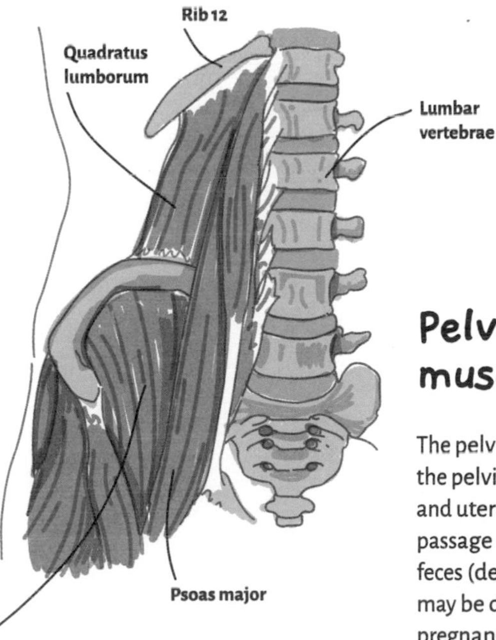
Rib 12
Quadratus lumborum
Lumbar vertebrae
Psoas major
Iliacus

Pelvic floor muscles

The pelvic floor muscles support the pelvic organs (urinary bladder and uterus) and control the passage of urine (micturition) and feces (defecation). The pelvic floor may be damaged by multiple pregnancies and vaginal delivery; this may lead to urinary and fecal incontinence in some women.

UPPER LIMB MUSCLES

The muscles of the upper limb consist of the shoulder muscles that move the upper limb around its base, elbow flexors and extensors, the muscles of the forearm, and the intrinsic muscles of the hand.

Shoulder muscles

Shoulder muscles can be divided into the deltoid, an anterior pectoralis group, a posteromedial group—**latissimus dorsi** (see page 69), serratus anterior, and **trapezius**—and the rotator cuff group.

The rounded contour of the shoulder is formed by the **deltoid**. Its anterior fibers flex the humerus, superior fibers abduct the humerus, and posterior fibers extend the humerus.

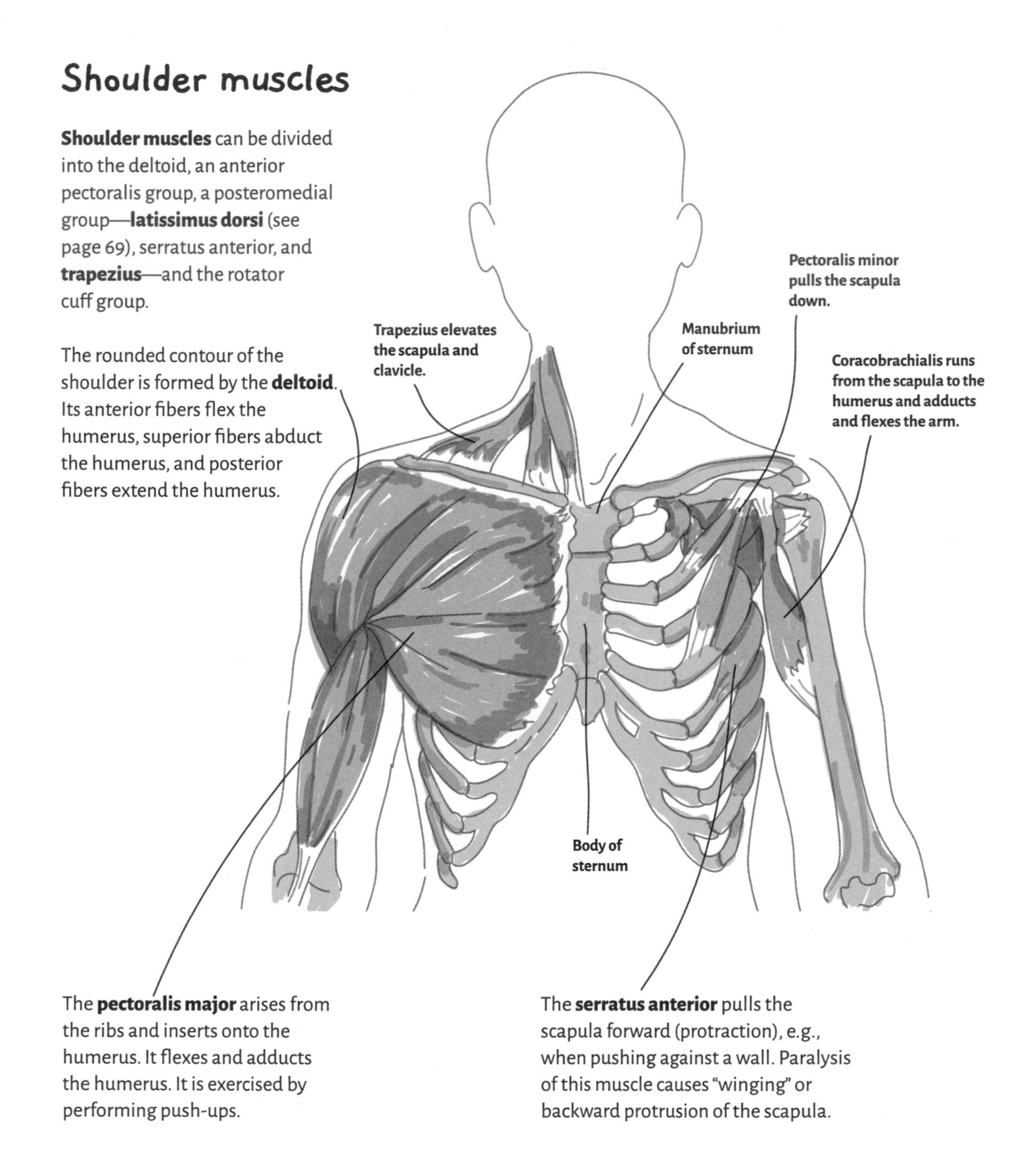

Trapezius elevates the scapula and clavicle.
Manubrium of sternum
Pectoralis minor pulls the scapula down.
Coracobrachialis runs from the scapula to the humerus and adducts and flexes the arm.
Body of sternum

The **pectoralis major** arises from the ribs and inserts onto the humerus. It flexes and adducts the humerus. It is exercised by performing push-ups.

The **serratus anterior** pulls the scapula forward (protraction), e.g., when pushing against a wall. Paralysis of this muscle causes "winging" or backward protrusion of the scapula.

Rotator cuff muscles attach around the shoulder joint capsule, which consists of **supraspinatus** (abducts the arm), **subscapularis** (medially rotates the arm), and **teres minor** and **infraspinatus** (laterally rotate the arm).

The **teres major** is not part of the rotator cuff. It runs from the scapula to the shaft of the humerus and extends and medially rotates the humerus.

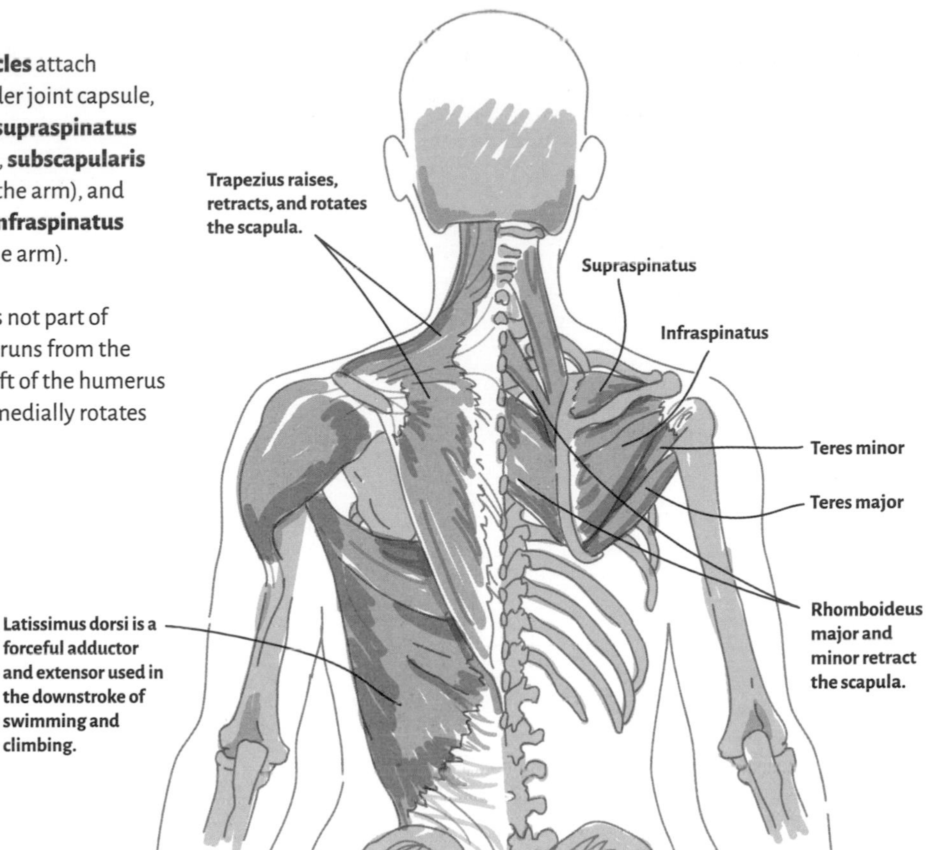
Trapezius raises, retracts, and rotates the scapula.
Supraspinatus
Infraspinatus
Teres minor
Teres major
Rhomboideus major and minor retract the scapula.
Latissimus dorsi is a forceful adductor and extensor used in the downstroke of swimming and climbing.

Elbow flexors and extensors

Muscles on the anterior arm (**biceps brachii** and **brachialis**) flex the elbow. Biceps brachii also crosses the shoulder joint and can flex the arm. Because its tendon wraps around the radius, biceps brachii is a powerful supinator of the forearm (faces palm anterior or superior).

The muscle on the posterior arm (**triceps brachii**) is a powerful extensor of the elbow. It has three heads, one attaching to the scapula (long head) and two to the humerus (lateral and medial heads). The muscle inserts onto the olecranon (point of elbow) of the ulna.

Biceps brachii

Brachialis

Triceps brachii

Anterior view of arm muscles

Posterior view of arm muscles

Forearm and hand muscles

The muscles of the forearm are grouped into two compartments: a wrist and digit flexor group on the anterior forearm and a wrist and digit extensor group on the posterior forearm.

The intrinsic muscles of the hand come in three groups: hypothenar muscles, thenar muscles, and palmar muscles.

The flexor group includes the following muscles: flexor digitorum superficialis, flexor digitorum profundus, flexor carpi radialis, flexor carpi ulnaris, and flexor pollicis longus.

Hypothenar muscles: Found at the base of the little finger, they flex and abduct the little finger and bring it into opposition with the thumb.

Flexor digitorum superficialis: Flexes the proximal interphalangeal joint.

Flexor carpi radialis: Lies on the radial side of the forearm; it flexes and abducts the wrist.

Pronator teres: Round, superficial muscle that pronates the forearm (rotates hand to face palm inferior or posterior).

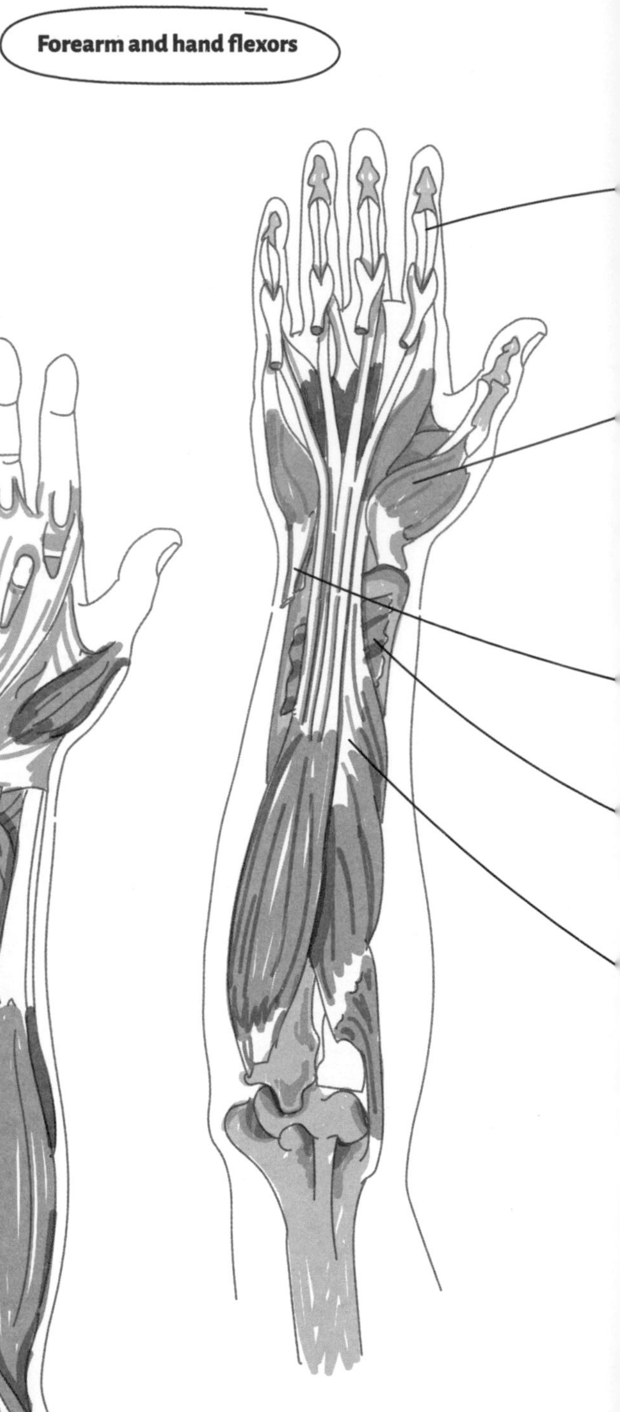
Forearm and hand flexors

Anterior view of the right forearm and hand showing deep muscles

Anterior view of the right forearm and hand showing superficial muscles

Forearm and hand extensors

Flexor digitorum profundus: Flexes the distal interphalangeal joint, a type of hinge joint.

Thenar muscles: Found at the base of the thumb; they flex and abduct the thumb and bring it into opposition with other digits. Opposition is a functionally important movement that allows the pad of the thumb to be brought into contact with the pads of the other fingers.

Flexor carpi ulnaris: Found on the ulnar side of the forearm; it flexes and adducts the wrist.

Pronator quadratus: Deep four-sided muscle that pronates the forearm, facing the palm inferior or posterior.

Flexor pollicis longus: Flexes the interphalangeal joint of the thumb. It is an important muscle in power grip.

Palmar muscles: Located inside the palm and between the metacarpals; they abduct and adduct the fingers. Interossei extend the fingers at the interphalangeal joints and flex the metacarpophalangeal joints. This position of the hand is important for fine finger movements in the precision grip, e.g., when threading a needle.

Extensor indicis: Independent extension of the index finger (digit 2) to point. This is an important muscle for pointing (extending) the index finger while the other fingers are flexed.

Extensor pollicis longus: Extends the interphalangeal joint of the thumb.

Extensor pollicis brevis: Extends and abducts the proximal phalanx at the metacarpophalangeal joint of the thumb.

Extensor digiti minimi: Independent extension of the little finger (digit 5).

Extensor digitorum: Extends the fingers (digits 2 to 5) at the metacarpophalangeal joints.

Extensor carpi radialis longus and **brevis**: Extends and abducts the wrist.

Brachioradialis: Flexes the elbow when the forearm is partially pronated (rotated).

Posterior view of the left forearm and the hand showing extensor muscles

LOWER LIMB MUSCLES

The muscles of the lower limb include those in the gluteal (buttock) region, thigh, leg, and foot. These muscles are primarily used for standing, climbing, and walking, and do not have the precise control of hand muscles.

Gluteal muscles

Muscles of the gluteal region are in three layers.

Gluteus medius and **gluteus minimus**: Abduct the thigh at the hip joint and support the pelvis during the stance phase of walking.

Gluteus maximus: The rump muscle extends the thigh at the hip joint during stair climbing.

Deep **gluteal muscles** laterally rotate the thigh at the hip joint:
Piriformis
Obturator internus and **externus**
gemelli
Quadratus femoris.

Thigh muscles

The muscles of the thigh are in three compartments: anterior group, medial group, and posterior group.

Anterior group

Quadriceps femoris (rectus femoris, vastus medialis, vastus intermedius, vastus lateralis): Extend the knee.

Rectus femoris: The most anterior of the quadriceps group, it crosses the hip, so it also flexes the thigh.

Sartorius: Crosses both the hip and knee joints and flexes both to sit cross-legged.

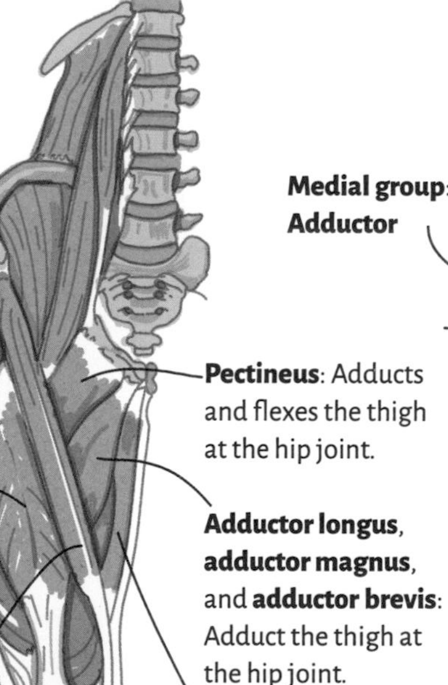

Medial group: Adductor

Pectineus: Adducts and flexes the thigh at the hip joint.

Adductor longus, **adductor magnus**, and **adductor brevis**: Adduct the thigh at the hip joint.

Gracilis: Adducts and medially rotates the thigh.

Posterior group: Hamstrings

Semimembranosus, **semitendinosus**, **biceps femoris**: All cross the hip and knee joints. They extend the thigh at the hip joint and flex the knee.

Semimembranosus

Semitendinosus

Biceps femoris

Leg and foot muscles

Muscles of the leg are in three groups: anterior, posterior, and lateral.

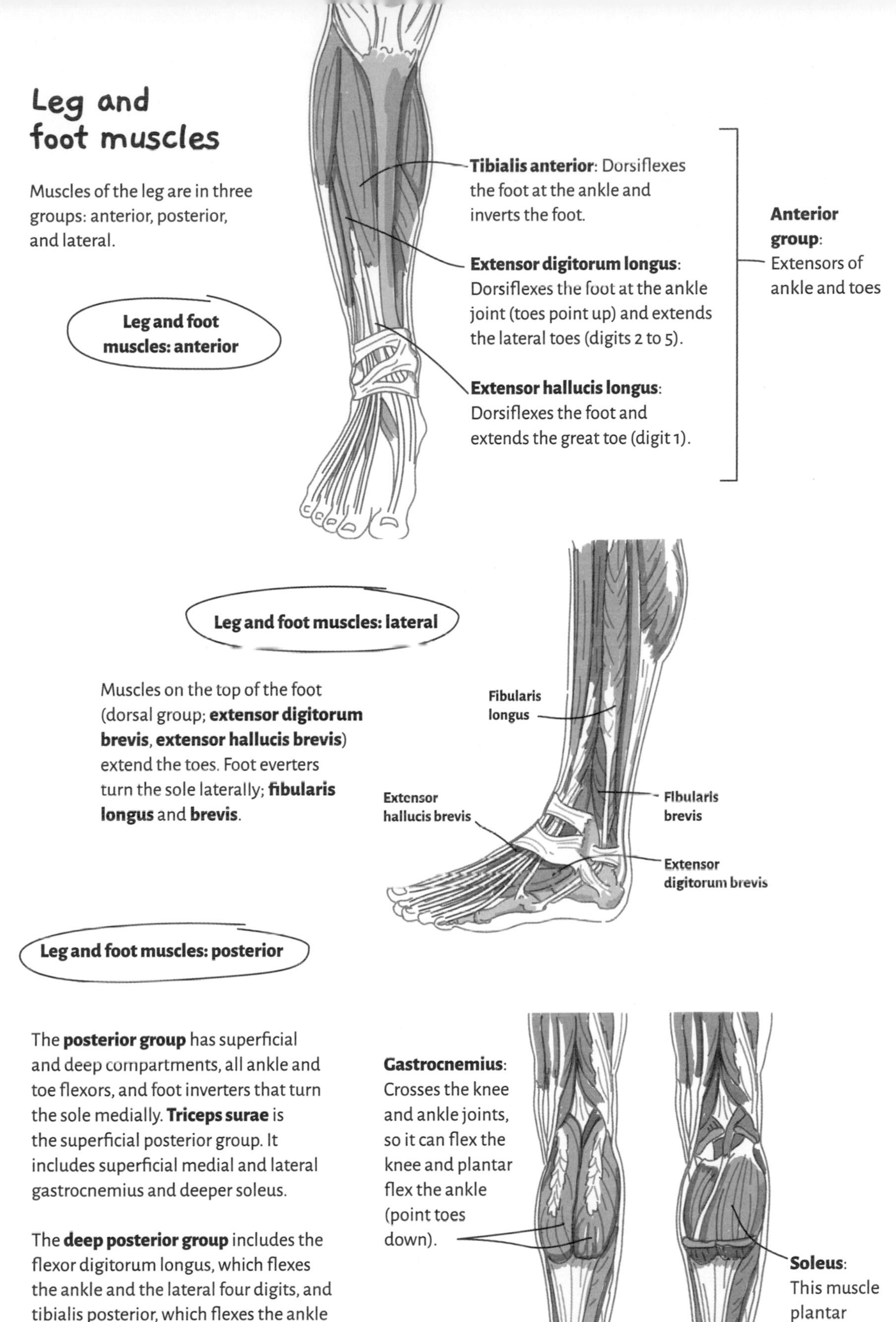

Fibularis longus
Fibularis brevis
Extensor hallucis brevis
Extensor digitorum brevis

Leg and foot muscles: anterior

Tibialis anterior: Dorsiflexes the foot at the ankle and inverts the foot.

Extensor digitorum longus: Dorsiflexes the foot at the ankle joint (toes point up) and extends the lateral toes (digits 2 to 5).

Extensor hallucis longus: Dorsiflexes the foot and extends the great toe (digit 1).

Anterior group: Extensors of ankle and toes

Leg and foot muscles: lateral

Muscles on the top of the foot (dorsal group; **extensor digitorum brevis**, **extensor hallucis brevis**) extend the toes. Foot everters turn the sole laterally; **fibularis longus** and **brevis**.

Leg and foot muscles: posterior

The **posterior group** has superficial and deep compartments, all ankle and toe flexors, and foot inverters that turn the sole medially. **Triceps surae** is the superficial posterior group. It includes superficial medial and lateral gastrocnemius and deeper soleus.

The **deep posterior group** includes the flexor digitorum longus, which flexes the ankle and the lateral four digits, and tibialis posterior, which flexes the ankle and inverts the foot (not shown here).

Flexor hallucis longus flexes the ankle and flexes the great toe.

Gastrocnemius: Crosses the knee and ankle joints, so it can flex the knee and plantar flex the ankle (point toes down).

Soleus: This muscle plantar flexes the ankle.

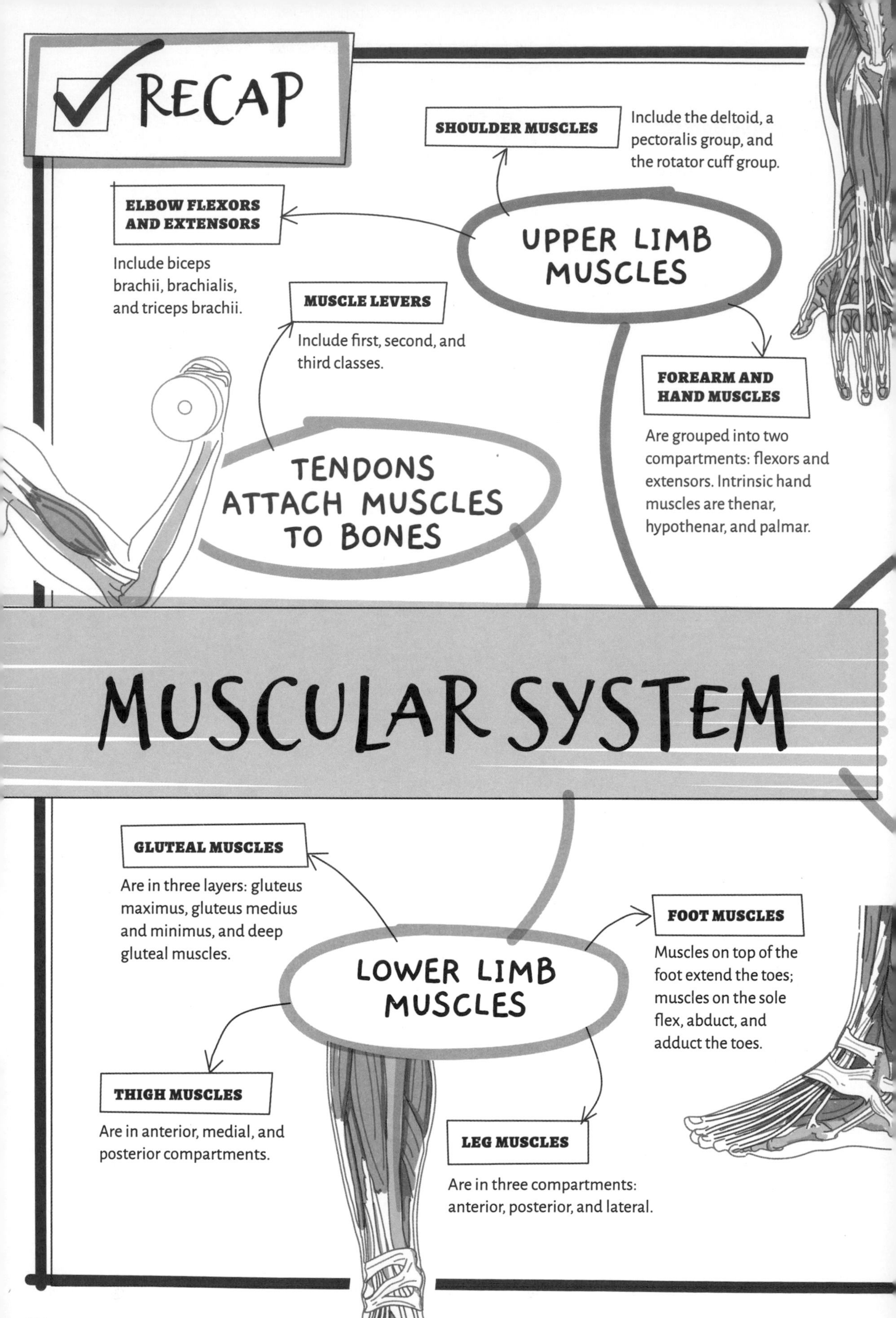
RECAP
SHOULDER MUSCLES
Include the deltoid, a pectoralis group, and the rotator cuff group.
ELBOW FLEXORS AND EXTENSORS
Include biceps brachii, brachialis, and triceps brachii.
UPPER LIMB MUSCLES
MUSCLE LEVERS
Include first, second, and third classes.
FOREARM AND HAND MUSCLES
Are grouped into two compartments: flexors and extensors. Intrinsic hand muscles are thenar, hypothenar, and palmar.
TENDONS ATTACH MUSCLES TO BONES
MUSCULAR SYSTEM
GLUTEAL MUSCLES
Are in three layers: gluteus maximus, gluteus medius and minimus, and deep gluteal muscles.
FOOT MUSCLES
Muscles on top of the foot extend the toes; muscles on the sole flex, abduct, and adduct the toes.
LOWER LIMB MUSCLES
THIGH MUSCLES
Are in anterior, medial, and posterior compartments.
LEG MUSCLES
Are in three compartments: anterior, posterior, and lateral.

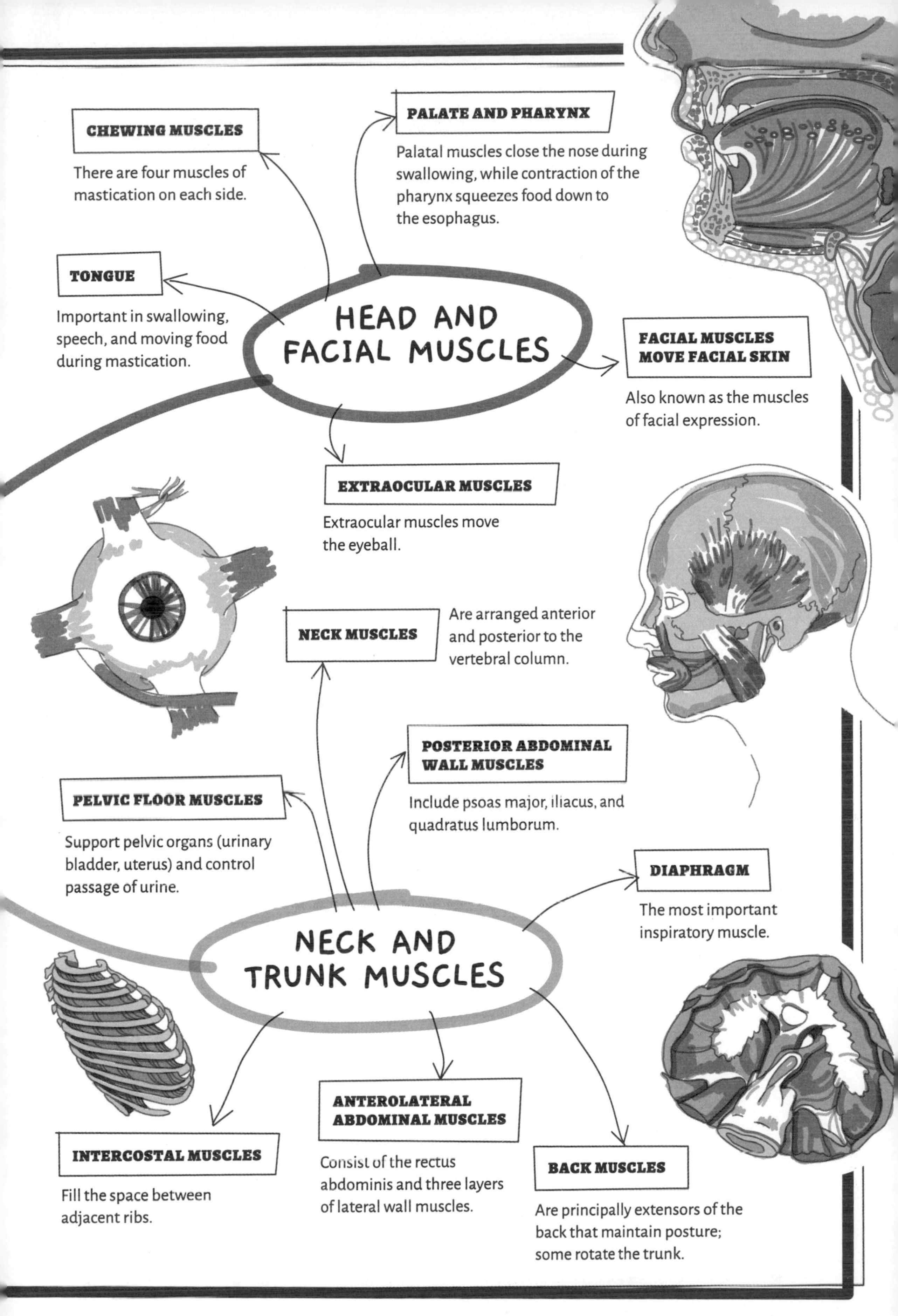
HEAD AND FACIAL MUSCLES
CHEWING MUSCLES
There are four muscles of mastication on each side.
PALATE AND PHARYNX
Palatal muscles close the nose during swallowing, while contraction of the pharynx squeezes food down to the esophagus.
TONGUE
Important in swallowing, speech, and moving food during mastication.
FACIAL MUSCLES MOVE FACIAL SKIN
Also known as the muscles of facial expression.
EXTRAOCULAR MUSCLES
Extraocular muscles move the eyeball.
NECK AND TRUNK MUSCLES
NECK MUSCLES
Are arranged anterior and posterior to the vertebral column.
POSTERIOR ABDOMINAL WALL MUSCLES
Include psoas major, iliacus, and quadratus lumborum.
PELVIC FLOOR MUSCLES
Support pelvic organs (urinary bladder, uterus) and control passage of urine.
DIAPHRAGM
The most important inspiratory muscle.
INTERCOSTAL MUSCLES
Fill the space between adjacent ribs.
ANTEROLATERAL ABDOMINAL MUSCLES
Consist of the rectus abdominis and three layers of lateral wall muscles.
BACK MUSCLES
Are principally extensors of the back that maintain posture; some rotate the trunk.

CHAPTER 5

NERVOUS SYSTEM AND SENSES

The nervous system has central and peripheral components. The central component (brain and spinal cord) is protected within the dorsal body cavity formed by the skull and vertebral column. Sensory information reaches the brain by cranial and spinal nerves. Some responses to sensation are reflexive (e.g., knee jerk, flexion withdrawal from heat, or pupillary constriction) but most require central processing at the spinal cord or brain level. These responses are then sent through motor pathways to muscles and glands of the body.

NEURON STRUCTURE

The typical neuron has specialized structures for the input and output of information. The dendrites are the input pathway, and the axon is the output pathway. There are about 80,000,000 neurons in the average brain.

The structure of a neuron

Cell body
Nucleus
Dendrites
The axon hillock is where the action potential starts.
Axon
Action potential
Schwann cell
Terminal branches

Neuronal **cell bodies** have abundant rough endoplasmic reticulum (Nissl substance), which makes structural proteins and other essential chemicals.

Dendrites are structures that form a treelike structure (*dendron* is Greek for "tree"). They receive chemical signals across the synapses (gaps) from the axons of other neurons.

The **action potential** or nerve impulse is a wave of electrical activity that runs down the axon from the axon hillock to the axon terminal.

The **terminal branches** of the axon end in synaptic boutons (buttons) that make contact with either other nerve cells or muscle fibers.

The **nucleus** commands the activities of the neuron and contains all the genetic information that the neuron needs.

The **axon** is the nerve fiber that carries the output from the neuron. Axons are usually coated with a fatty sheath called myelin, which speeds up conduction by as much as 100 times. Axons carry a wave of electrical activity called an action potential, which starts at the axon hillock and spreads down to the axon terminal.

Most neurons have contacts with other neurons across chemical **synapses**. These are tiny structures about 1/1000 mm (1/25,000 inch) wide, where neurotransmitter chemicals are released from the axon terminal of one cell to affect the electrical activity of the dendrite of another cell.

In the peripheral nervous system, **Schwann cells** make the myelin sheath that speeds up the rate of conduction of the action potential or nerve impulse as much as 100 times.

FUNCTIONAL ORGANIZATION OF THE NERVOUS SYSTEM

The three basic functions of the nervous system are sensation (input), central processing (integration), and motor (output). The motor and sensory functions cross the boundary between the peripheral and the central nervous system, so there are sensory and motor elements in both PNS and CNS.

Sensory function

Sensory receptors detect the external or internal environment and code the information for transmission to the CNS. The neurons that carry sensory information are called **afferents** (from Latin *ad fero*, "carry toward"). Sensory neurons are found in the cranial and spinal nerves.

Integrative function

Integration involves storing and analyzing sensory information and making behavioral decisions based on that information.

Retinal ganglion cells: Located in the retina of the eye are sensory neurons that transmit to the brain by the optic nerve.

Multiple integrative neurons: Located inside the brain, they process sensory information and make decisions about actions.

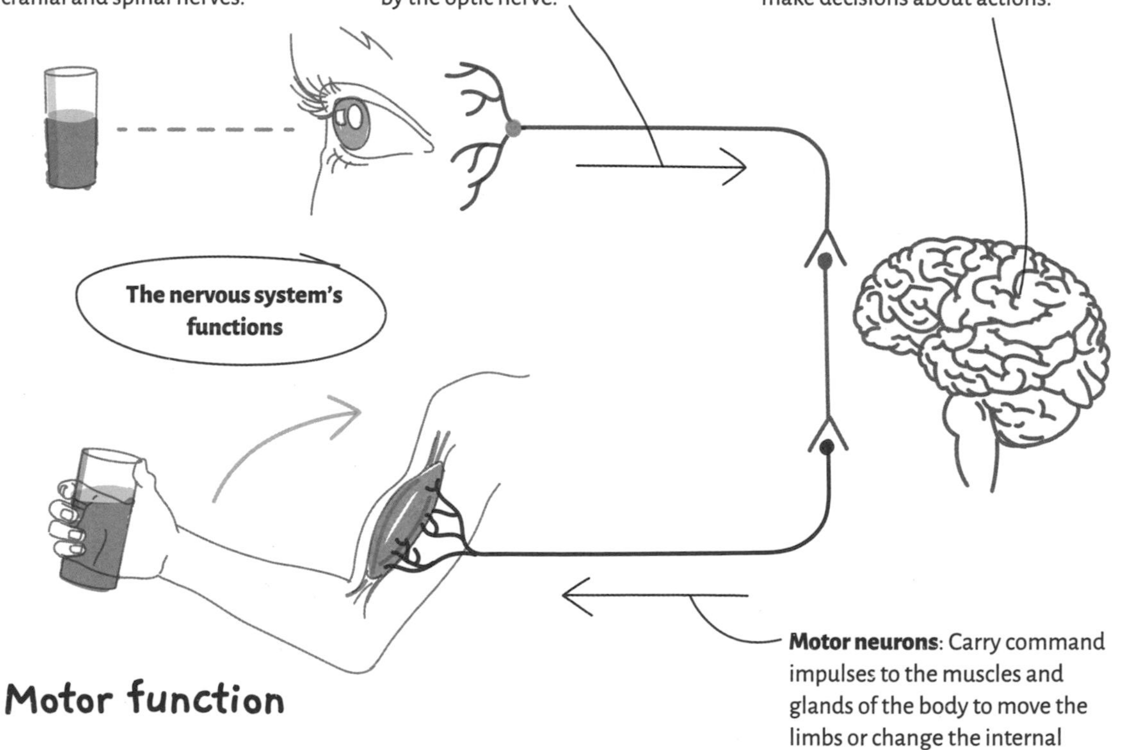
The nervous system's functions

Motor neurons: Carry command impulses to the muscles and glands of the body to move the limbs or change the internal body state.

Motor function

The nervous system responds to the world through motor function. Motor neurons send information out of the CNS and are called **efferents** (from Latin *e fero*, "carry away"). Efferents may act on smooth, cardiac, or skeletal muscles and glands of the internal organs or skin.

Somatic nervous system

The somatic nervous system is that part of the peripheral nervous system concerned with the body surface, bones, joints, and skeletal muscles. **Somatic sensory neurons** carry information from the skin, muscle spindles, joint stretch receptors, and special senses (e.g., eye and ear). **Somatic motor neurons** carry information from the CNS to skeletal (voluntary) muscles only.

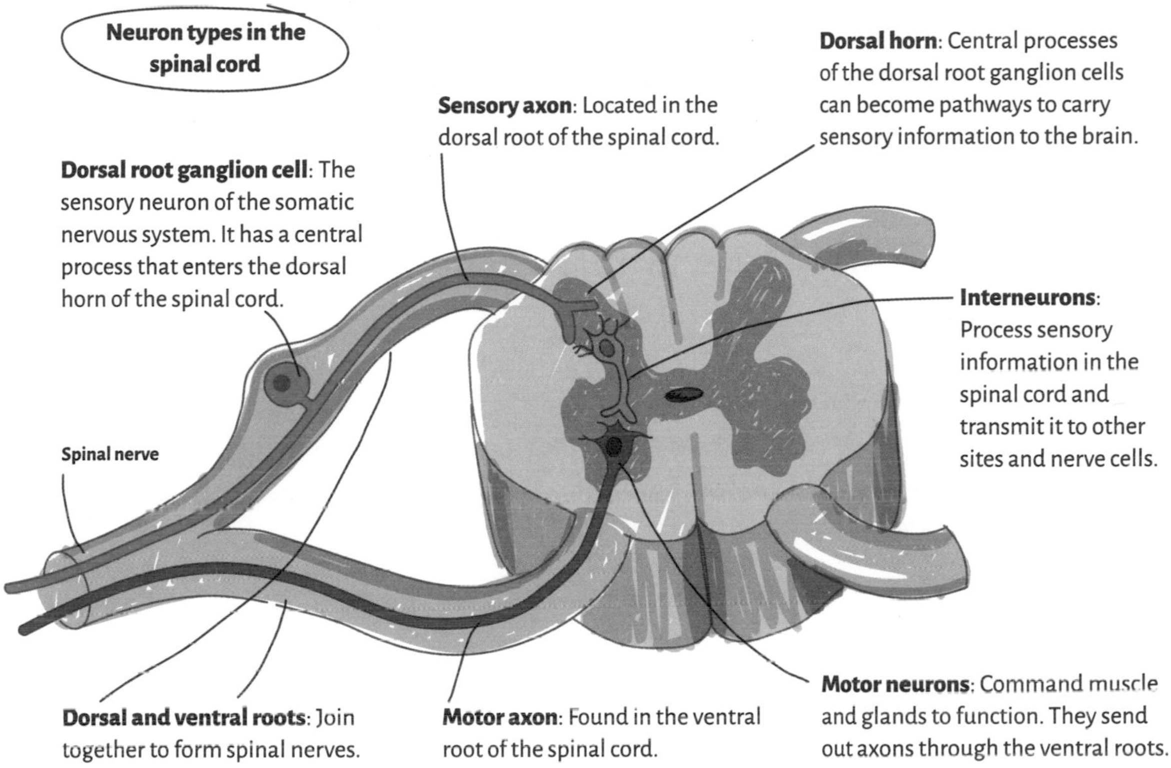

Neuron types in the spinal cord

Dorsal root ganglion cell: The sensory neuron of the somatic nervous system. It has a central process that enters the dorsal horn of the spinal cord.

Sensory axon: Located in the dorsal root of the spinal cord.

Dorsal horn: Central processes of the dorsal root ganglion cells can become pathways to carry sensory information to the brain.

Interneurons: Process sensory information in the spinal cord and transmit it to other sites and nerve cells.

Spinal nerve

Dorsal and ventral roots: Join together to form spinal nerves.

Motor axon: Found in the ventral root of the spinal cord.

Motor neurons: Command muscle and glands to function. They send out axons through the ventral roots.

Autonomic nervous system

The autonomic nervous system (ANS) is concerned with the semiautomatic control of internal organs. The ANS has **visceral sensory neurons** that carry information from internal organs to the CNS, and **visceral motor (autonomic) neurons**, that control smooth muscle and the glands of visceral organs.

Visceral sensory neurons also convey pain from the digestive tract (gut) wall, which can arise from excessive gut traction or distension (bowel obstruction), or when the gut epithelium is eroded.

The ANS is traditionally divided into the sympathetic and parasympathetic nervous systems. The **sympathetic nervous system** is for the emergency expenditure of energy. Output is from thoracic and upper lumbar levels of the spinal cord (T1 to L1).

The **parasympathetic nervous system** is for "resting and digesting" and replenishing energy reserves. Output is from the oculomotor, facial, glossopharyngeal, and vagus cranial nerves (CN 3, 7, 9, 10) and sacral spinal segments S2 to S4.

Enteric nervous system

The enteric nervous system (ENS) has sensory, motor, and integrative neurons. Like the ANS, the ENS is entirely involuntary.

Sensory ENS neurons monitor chemistry within the digestive tract and stretching of the wall.

Motor ENS neurons control gut smooth muscle to produce **peristalsis** (the rhythmic contraction that moves food slowly down the digestive tract). They also control gastric acid and intestinal gland secretion.

BRAIN STRUCTURE AND FUNCTION

The brain is divided into the forebrain and the brain stem, with the cerebellum (little cerebrum) attached to the brain stem. The human forebrain is greatly expanded as the cerebral cortex, but also important are the deeper forebrain structures (diencephalon and striatum).

Brain development

The brain develops from an embryonic tube, as the nervous tissue shares its origins with the skin and starts as a flat tadpole-shaped surface (**neural plate**) on the dorsum of the embryo. This flat surface rolls into a tube to make the primitive brain (**neural tube**). The rostral end develops three bulges (**primary brain vesicles**) for the fore-, mid-, and hindbrain. The caudal end becomes the spinal cord. The forebrain vesicle expands to form the **telencephalon** (adult cerebral cortex and basal ganglia) and **diencephalon** (adult thalamus). The midbrain vesicle becomes the **mesencephalon** (midbrain). The hindbrain vesicle develops into a **metencephalon** (pons) and **myelencephalon** (medulla oblongata).

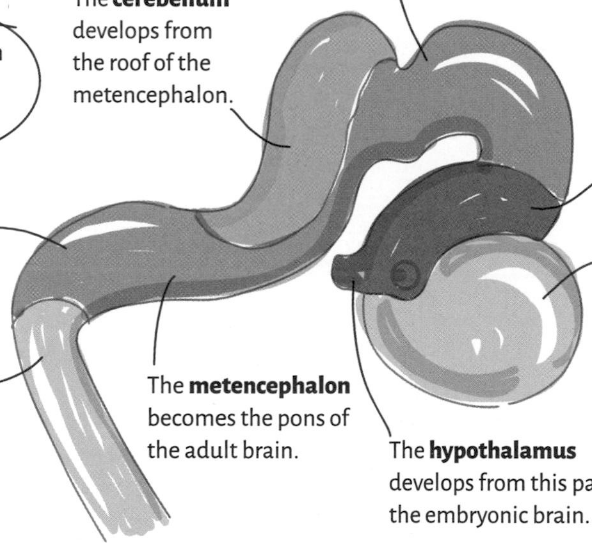

The developing human brain at five weeks after fertilization

The **cerebellum** develops from the roof of the metencephalon.

The **mesencephalon** becomes the adult midbrain.

The **diencephalon** becomes the adult thalamus, prethalamus, and pretectum.

The **myelencephalon** becomes the medulla oblongata of the adult brain.

The **telencephalon** grows rapidly to form the cerebral hemispheres of the brain that overlie all other brain parts.

Embryonic spinal cord is the most caudal part of the neural tube.

The **metencephalon** becomes the pons of the adult brain.

The **hypothalamus** develops from this part of the embryonic brain.

Brain stem

All vertebrates' brain stems have a similar structure, because their functions have been the same for 500 million years. Thus, the brain stem is an ancient part of the brain.

Key components of the brain stem include the cranial nerve nuclei (for sensory and motor function of the head, neck, and internal organs), ascending and descending pathways connecting the brain with the spinal cord, and the reticular formation for many automatic functions.

The brain stem consists of the midbrain, pons, and medulla oblongata. The midbrain connects with the pretectum rostrally, and the medulla oblongata connects with the spinal cord caudally through the foramen magnum of the skull base.

The **cerebellum** is an outgrowth of the brain stem. It develops as an offshoot from the metencephalon, called the **rhombic lip**. These paired lips meet in the midline during the middle trimester and form the body of the cerebellum. The cerebellum has inputs from the inner ear, spinal cord, and pons and coordinates motor activity throughout the body.

Deep parts of the forebrain

The deep parts of the forebrain develop from the embryonic diencephalon and deep telencephalon. The **diencephalon** forms the pretectum, thalamus, and prethalamus. Parts of the deep telencephalon include the striatum, pallidum, and septum. Modern anatomical understanding now identifies the hypothalamus with the telencephalon.

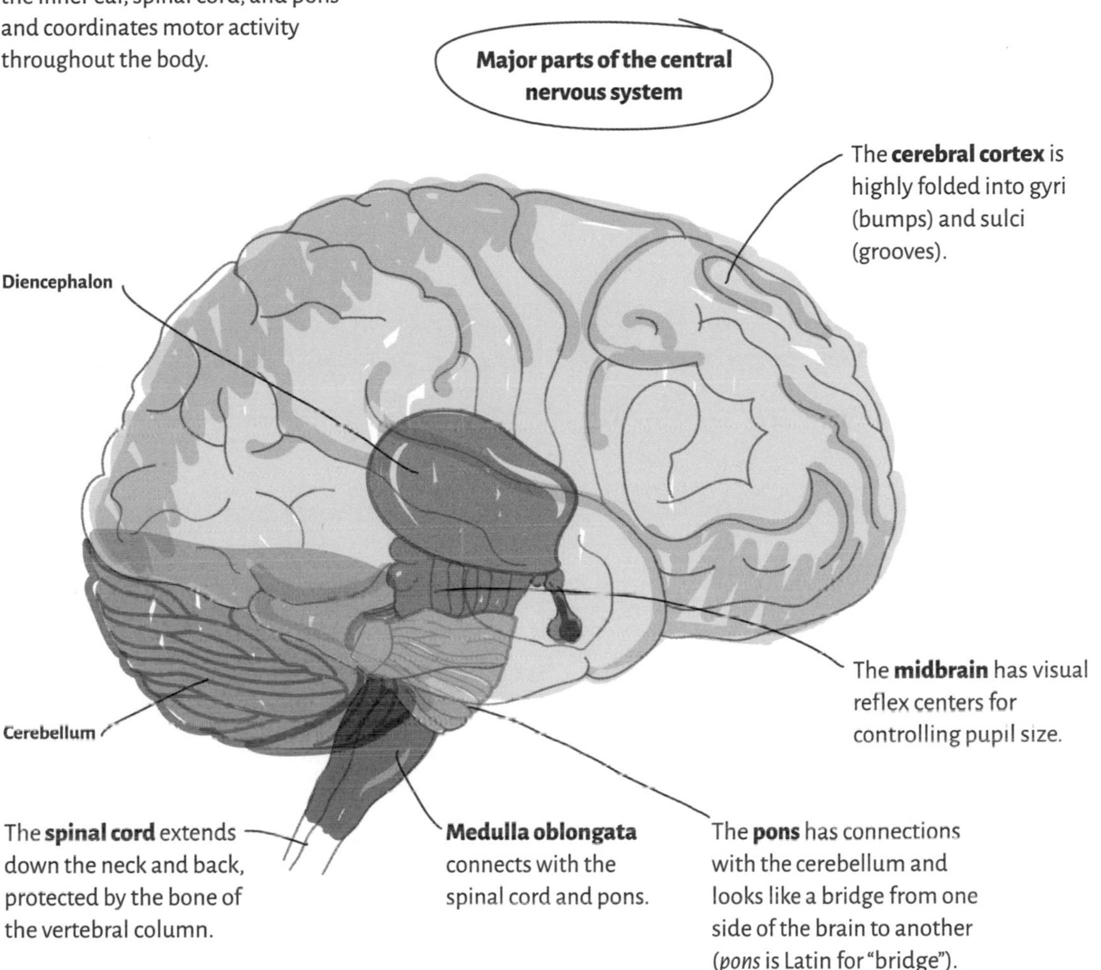

Major parts of the central nervous system
The cerebral cortex is highly folded into gyri (bumps) and sulci (grooves).
Diencephalon
The midbrain has visual reflex centers for controlling pupil size.
Cerebellum
The spinal cord extends down the neck and back, protected by the bone of the vertebral column.
Medulla oblongata connects with the spinal cord and pons.
The pons has connections with the cerebellum and looks like a bridge from one side of the brain to another (pons is Latin for "bridge").

Folded cortical surface

The outer surface of the **cerebrum** of the forebrain is highly folded to form the cerebral cortex in two hemispheres connected by the corpus callosum. The surface has an area of 0.12 square meter (1.3 square feet). The elevations are called **gyri** (sing. gyrus), and the intervening grooves are called **sulci** (sing. sulcus).

The **cerebral cortex** is composed of gray matter containing about 15 billion neurons in a six-layered structure. Some parts of the cortex are partially hidden, e.g., the hippocampus. An **olfactory bulb** also extends from the underside of the forebrain (see pages 102 and 103).

CORTICAL FUNCTIONAL REGIONS

The surface of the brain is called the cerebral cortex and is formed into folds and grooves (gyri and sulci). Discrete areas of the cortical surface serve different functions: movement, sensation, language, judgment, and planning.

Lobes, fissures, and sulci

The **cortical surface** is divided into four surface lobes (frontal, parietal, temporal, and occipital), each roughly beneath the bone of the same name. The **lateral fissure** separates the frontal and temporal lobes.

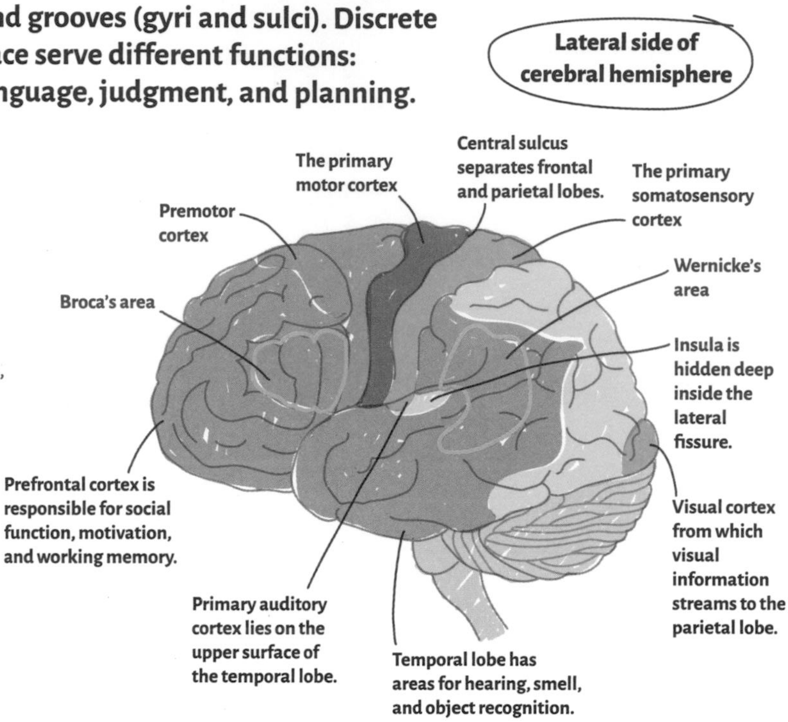

Lateral side of cerebral hemisphere

Premotor cortex

The primary motor cortex

Central sulcus separates frontal and parietal lobes.

The primary somatosensory cortex

Wernicke's area

Broca's area

Insula is hidden deep inside the lateral fissure.

Prefrontal cortex is responsible for social function, motivation, and working memory.

Visual cortex from which visual information streams to the parietal lobe.

Primary auditory cortex lies on the upper surface of the temporal lobe.

Temporal lobe has areas for hearing, smell, and object recognition.

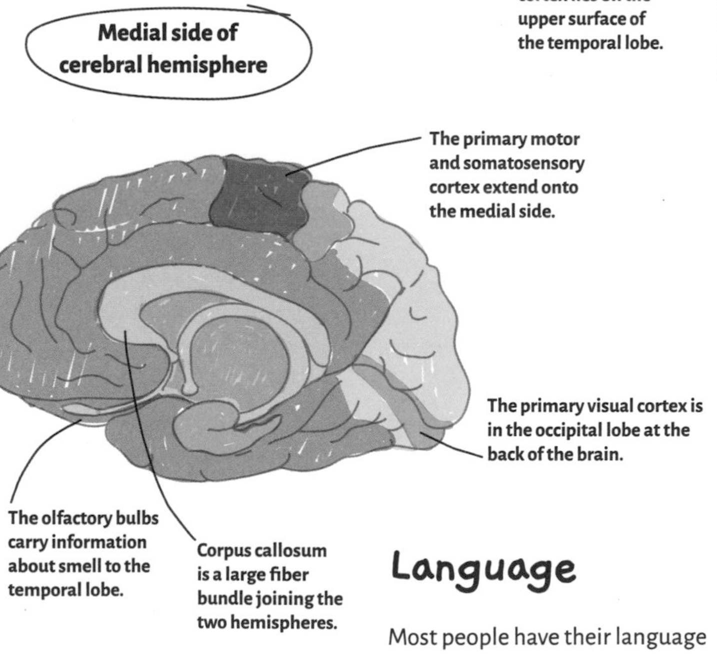

Medial side of cerebral hemisphere

The primary motor and somatosensory cortex extend onto the medial side.

The primary visual cortex is in the occipital lobe at the back of the brain.

The olfactory bulbs carry information about smell to the temporal lobe.

Corpus callosum is a large fiber bundle joining the two hemispheres.

Language

Most people have their language areas in the left hemisphere. Two key areas are **Broca's** (for word formulation) and **Wernicke's** (for receptive language, word choice, and sentence construction).

Touch

Touch, pain, temperature, joint position, and vibration are represented in a functional region of the cortex called the **primary somatosensory cortex**, which is located on a bulge called the **postcentral gyrus**.

Different body parts are represented on different cortical regions (a feature called **somatotopic organization**), with the head lowest and lateral, then the upper limb, trunk, lower limb, and genitals from lateral to medial. The face and hands have large representations because they are behaviorally important and need more nervous tissue to process information.

Vision

Visual space is represented on the **primary visual cortex** of the occipital lobe. The visual cortex has a map on it's surface of the visual world (visuotopic organization), with a large area of the cortex devoted to central detailed vision.

Visual information streams to the parietal lobe (dorsal visual stream) for analysis of the position of objects in visual space. Visual information also streams to the temporal lobe (ventral visual stream) for the identification of objects.

Working memory

The **dorsolateral prefrontal cortex** performs working memory, which is the ability to store sequences of actions for immediate activities (e.g., following a recipe or entering a phone number).

Hearing

Hearing is represented on the **primary auditory cortex** at the superior surface of the temporal lobe. This region extends into the depths of the lateral fissure.

Different frequencies of sound (pitches) are represented in different parts of the cortex (tonotopic organization). The auditory cortex streams to the receptive language area (Wernicke's).

Motor and somatosensory maps on the cerebral cortex

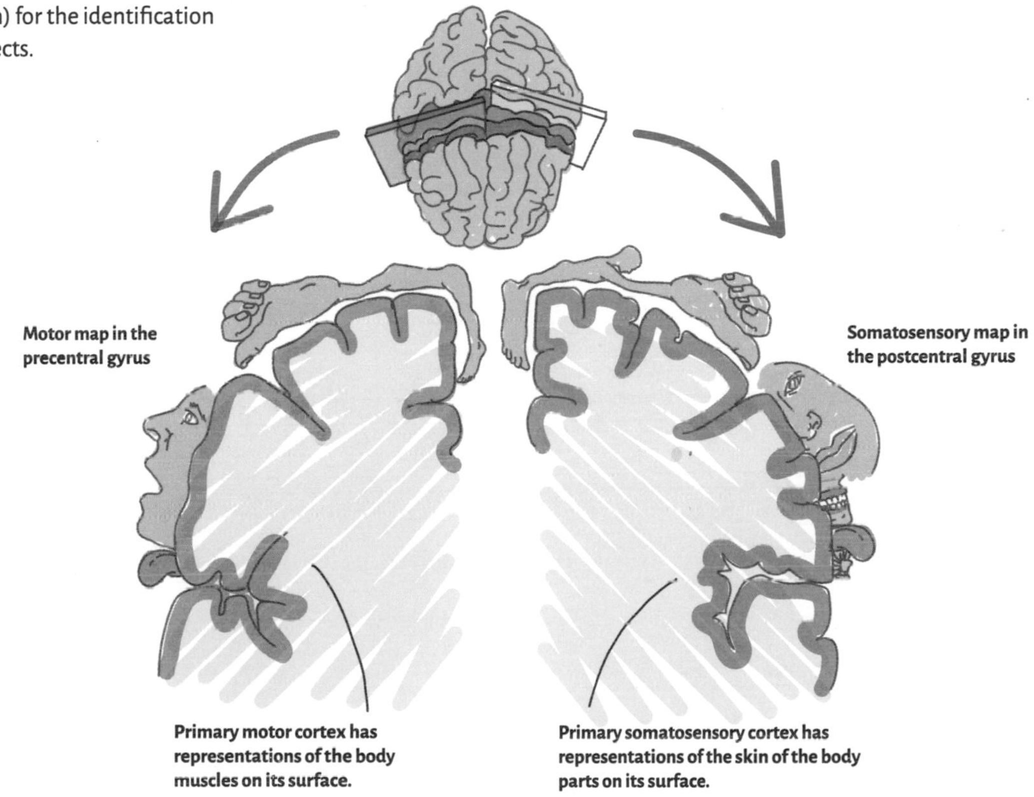

Motor map in the precentral gyrus

Somatosensory map in the postcentral gyrus

Primary motor cortex has representations of the body muscles on its surface.

Primary somatosensory cortex has representations of the skin of the body parts on its surface.

Smell and taste

Information about odors is processed on a small area on the medial side of the temporal lobe. Taste is processed in the frontal cortex inside the lateral fissure and in a deep brain region called the **insula**.

Planning and judgment

The **prefrontal cortex** anterior to the motor areas is responsible for social function, planning, and judgment.

Movement

Cortical regions in the frontal lobe control movement. The **premotor cortex** initiates commands that are sent to the **primary motor cortex**, located on the **precentral gyrus**. Muscles of the body parts are mapped onto the cortical surface (musculotopic organization) with large areas devoted to the facial and hand muscles.

BRAIN STEM AND CEREBELLUM

The brain stem attaches the spinal cord to the fore- and midbrain. It carries the ascending and descending pathways for the spinal cord and also serves other important automatic functions: breathing, control of blood pressure and heart rate, and control of digestive activity.

Cranial nerves (CN)

Cranial nerve 2 (**optic nerve**) attaches above the brain stem. Cranial nerves 3 to 12 attach to the brain stem. Cranial nerves 3 and 4 (**oculomotor** and **trochlear**) attach to the midbrain; cranial nerve 5 (**trigeminal**) attaches to the pons; nerves 6, 7, and 8 (**abducens**, **facial**, and **vestibulocochlear**) attach along the junction between the pons and medulla; and nerves 9, 10, 11, and 12 (**glossopharyngeal**, **vagus**, **accessory**, and **hypoglossal**) attach to the medulla.

Reticular formation

Reticular formation consists of a group of neurons from the brain stem that are interconnected for the purposes of breathing and blood circulation. It includes centers for control of the respiratory rhythm; blood pressure; and force, rate, and speed of heart muscle contraction.

The pons and medulla contain separate centers for control of ventilation. These centers set the respiratory rhythm (cycles of inspiration and expiration) in response to the levels of oxygen and carbon dioxide in the blood.

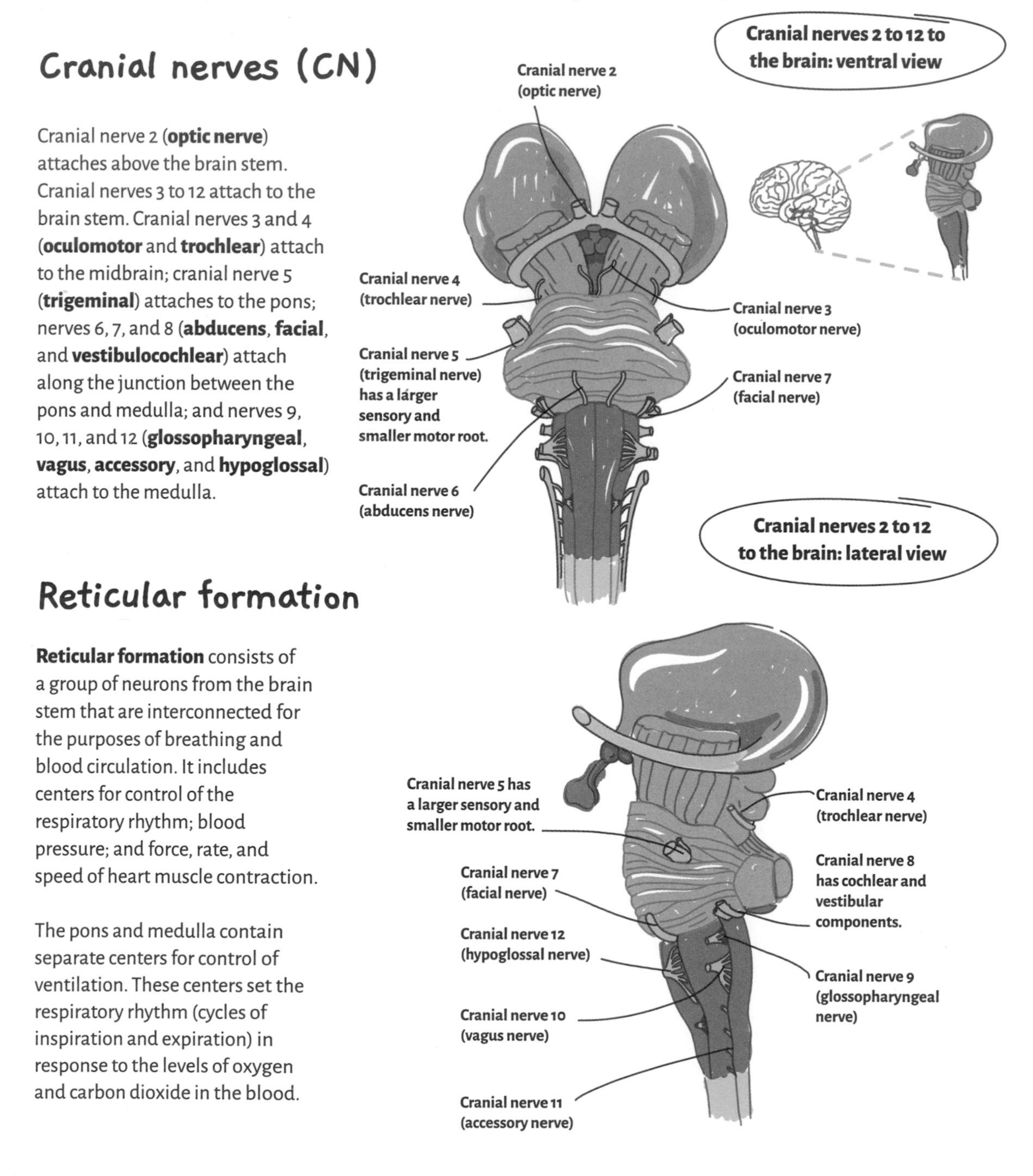

Cranial nerves 2 to 12 to the brain: ventral view
Cranial nerve 2 (optic nerve)
Cranial nerve 4 (trochlear nerve)
Cranial nerve 3 (oculomotor nerve)
Cranial nerve 5 (trigeminal nerve) has a larger sensory and smaller motor root.
Cranial nerve 7 (facial nerve)
Cranial nerve 6 (abducens nerve)
Cranial nerves 2 to 12 to the brain: lateral view
Cranial nerve 5 has a larger sensory and smaller motor root.
Cranial nerve 4 (trochlear nerve)
Cranial nerve 8 has cochlear and vestibular components.
Cranial nerve 7 (facial nerve)
Cranial nerve 12 (hypoglossal nerve)
Cranial nerve 9 (glossopharyngeal nerve)
Cranial nerve 10 (vagus nerve)
Cranial nerve 11 (accessory nerve)

The respiratory centers send commands to the phrenic nucleus in the cervical spinal cord to drive the diaphragm and to the thoracic spinal cord to drive the intercostal muscles.

Other centers in the pons and medulla control heart rate and blood pressure. They send commands to the autonomic neurons in the vagal nucleus of the medulla and the sympathetic neurons of the spinal cord.

The neuron groups that use serotonin, norepinephrine (noradrenaline), and dopamine as neurotransmitters are also located in the reticular formation. They project to diverse areas of the brain to control sensory attention, sleep, mood, and emotional responses.

Brain stem and cerebellum

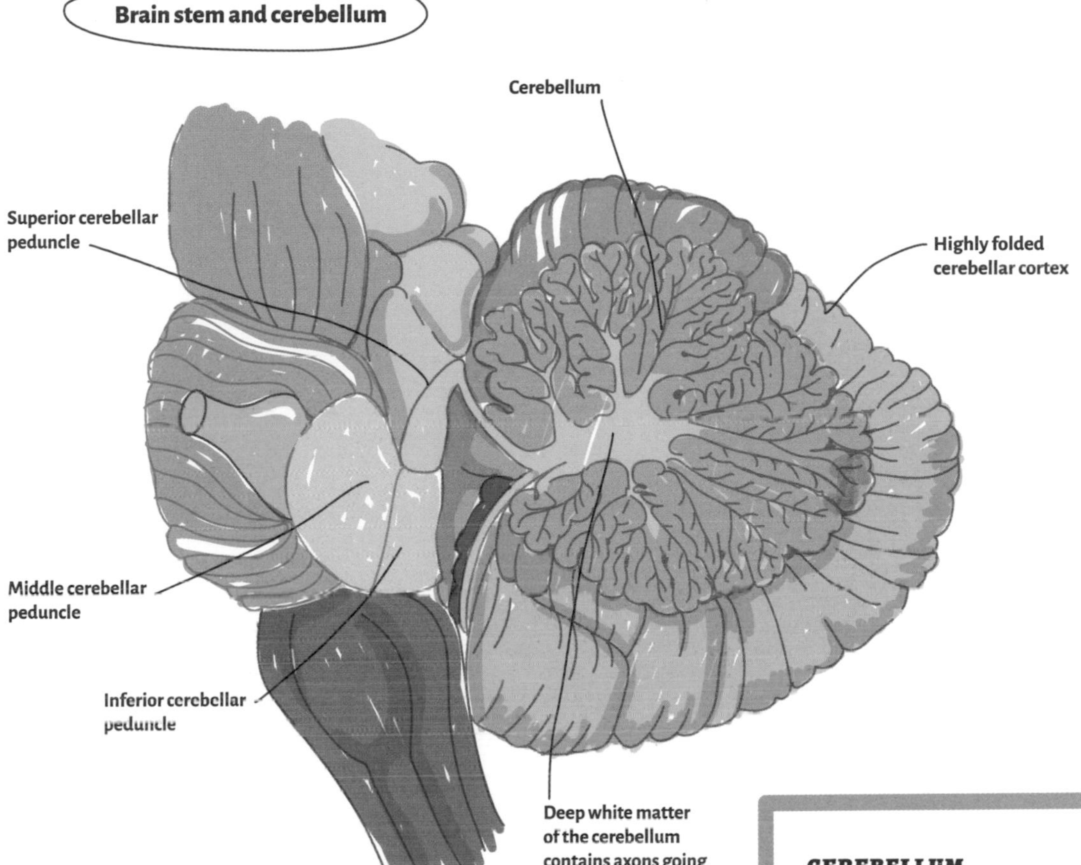
Cerebellum
Superior cerebellar peduncle
Highly folded cerebellar cortex
Middle cerebellar peduncle
Inferior cerebellar peduncle
Deep white matter of the cerebellum contains axons going to and from the cerebellar cortex.

Cerebellum

The **cerebellum** ("little brain") gets its name from its resemblance to the cerebrum. The cerebellum plays a key role in motor coordination, ensuring that movements are carried out with smooth and sequential muscle activation. It carries stored motor routines in its surface layer (**cerebellar cortex**) that can be activated on the instructions from the motor cortex of the cerebrum (see above).

The cerebellum is joined to the brain stem by three **cerebellar peduncles** (superior, middle, and inferior). It consists of a highly folded cerebellar cortex around a large core of nerve fibers (**deep white matter**). Embedded inside the cerebellum are the deep cerebellar nuclei (neuron groups); these nerve cells send out information from the cerebellum.

CEREBELLUM FUNCTIONS

The cerebellum has diverse functions that include:

★ The coordination of eye and head movements using information about balance and rotation of the head
★ The regulation of muscle tone through pathways to the spinal cord
★ The activation of the appropriate muscles in sequence during pre-programmed motor tasks

SPINAL CORD STRUCTURE AND FUNCTION

The spinal cord is about 45 cm (18 inches) long and extends from the base of the skull to the middle of the back just below the rib cage. It contains ascending and descending pathways and neurons to process sensory and motor function.

Basic structure of the spinal cord

The spinal cord has a central core of **gray matter** (neurons and their dendrites) surrounded by a shell of **white matter** (ascending and descending nerve pathways). The gray matter is divided into dorsal horn, intermediate zone, and ventral horn. A remnant of the neural tube (**central canal**) lies in the center of the spinal cord.

The **dorsal horn** processes sensory input, such as temperature and pain superficially, and complex touch, vibration, and muscle stretch in deeper areas.

The **intermediate zone** processes sensory input from internal organs and contains neurons to control internal organs.

The **ventral horn** contains motor neurons that drive skeletal muscles.

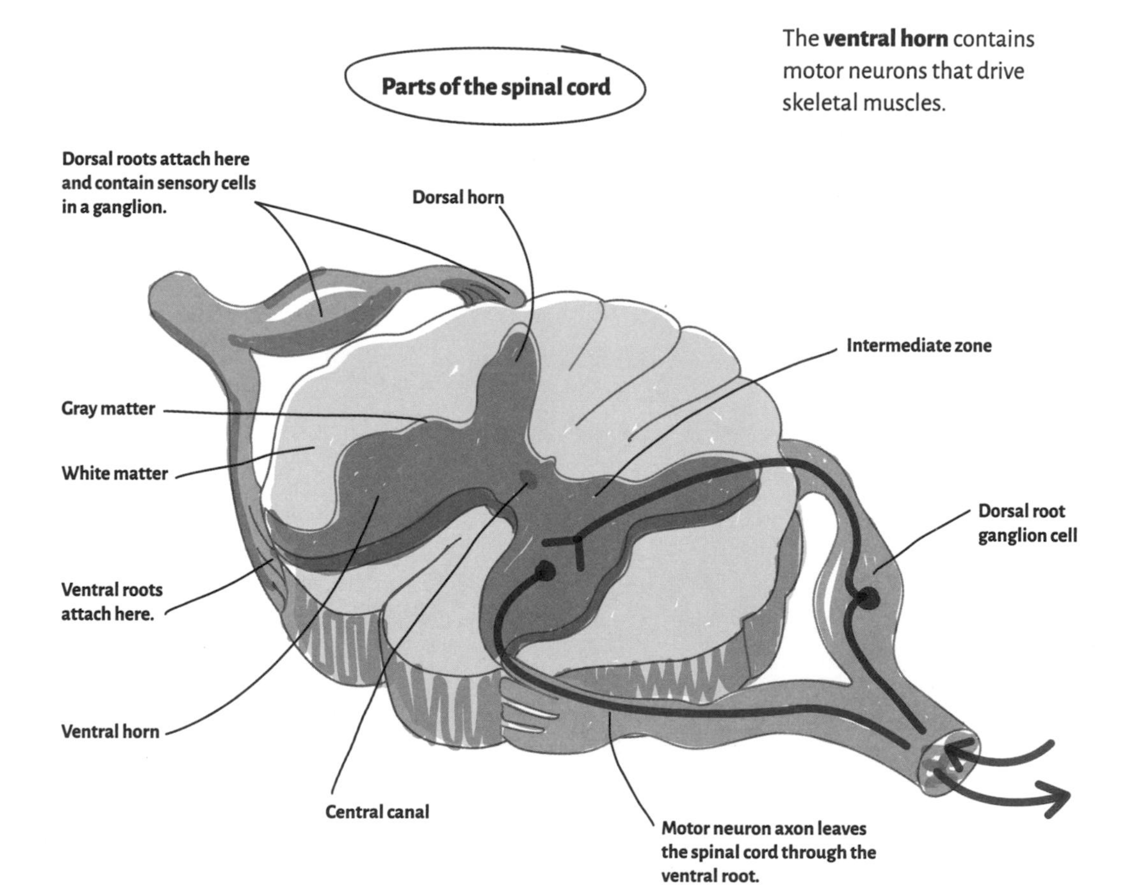

Parts of the spinal cord
Dorsal roots attach here and contain sensory cells in a ganglion.
Dorsal horn
Intermediate zone
Gray matter
White matter
Dorsal root ganglion cell
Ventral roots attach here.
Ventral horn
Central canal
Motor neuron axon leaves the spinal cord through the ventral root.

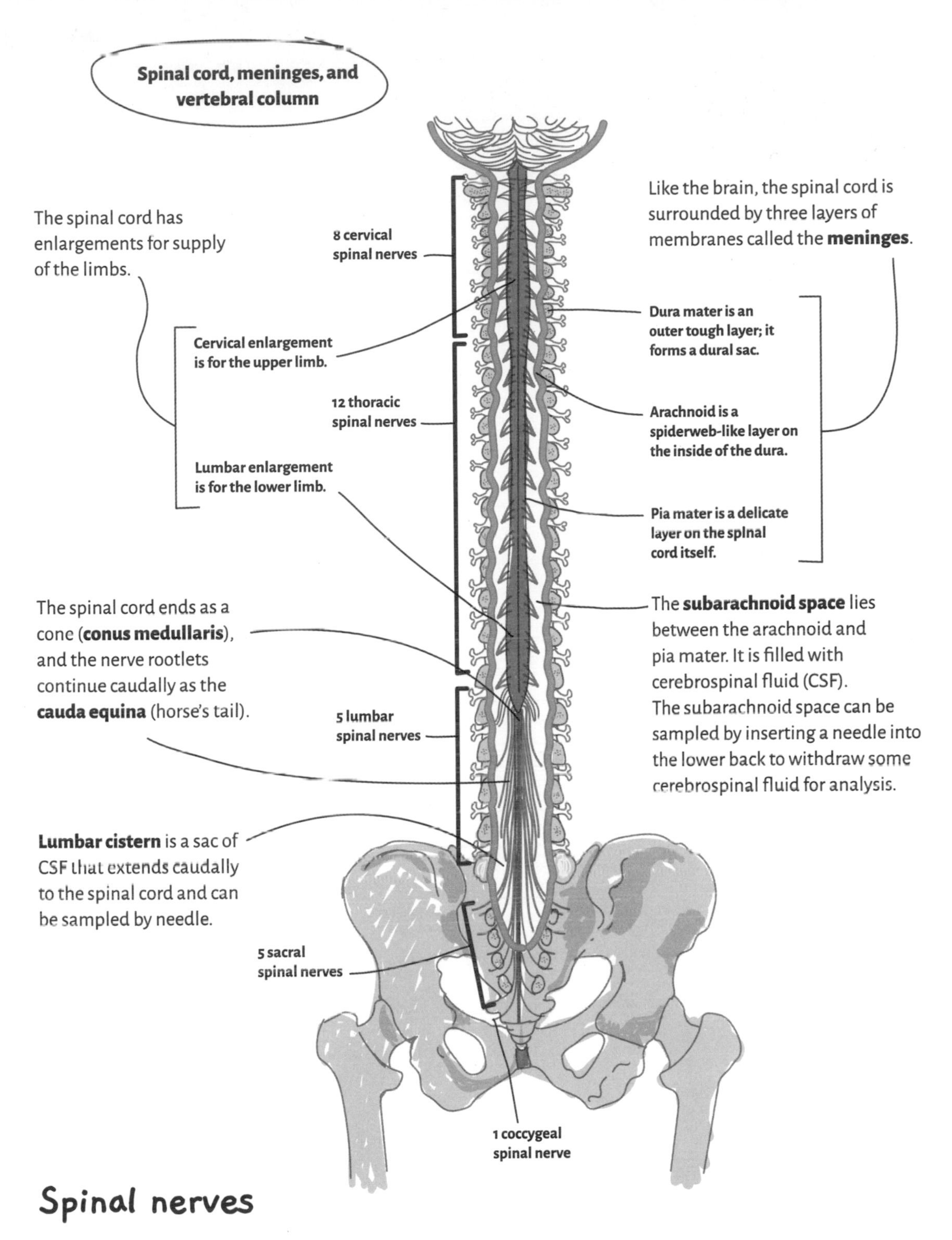

Spinal cord, meninges, and vertebral column
The spinal cord has enlargements for supply of the limbs.
8 cervical spinal nerves
Cervical enlargement is for the upper limb.
12 thoracic spinal nerves
Lumbar enlargement is for the lower limb.
Like the brain, the spinal cord is surrounded by three layers of membranes called the **meninges**.
Dura mater is an outer tough layer; it forms a dural sac.
Arachnoid is a spiderweb-like layer on the inside of the dura.
Pia mater is a delicate layer on the spinal cord itself.
The spinal cord ends as a cone (**conus medullaris**), and the nerve rootlets continue caudally as the **cauda equina** (horse's tail).
The **subarachnoid space** lies between the arachnoid and pia mater. It is filled with cerebrospinal fluid (CSF). The subarachnoid space can be sampled by inserting a needle into the lower back to withdraw some cerebrospinal fluid for analysis.
5 lumbar spinal nerves
Lumbar cistern is a sac of CSF that extends caudally to the spinal cord and can be sampled by needle.
5 sacral spinal nerves
1 coccygeal spinal nerve

Spinal nerves

There are 31 pairs of spinal nerves attached to the spinal cord, numbered according to region. They are formed from the junction of dorsal and ventral roots.

Cervical: C1 to C8
Thoracic: T1 to T12
Lumbar: L1 to L5
Sacral: S1 to S5
Coccygeal: Co1

Special groups of spinal nerves form plexuses for the limbs: for the brachial plexus nerves C5 to T1 and for the lumbosacral plexus nerves L2 to S3.

Each spinal nerve carries somatic sensory and motor nerve fibers. Those in the thoracolumbar outflow (T1 to L1) and in the sacral outflow (S2 to S4) segments also carry autonomic nerve fibers (viscerosensory and visceromotor).

Ascending pathways

Ascending pathways carry information rostrally (up) and are sensory.

Dorsal columns: Carry complex (two-point discrimination or fine) touch, vibration, and proprioceptive information to the brain stem. The dorsal columns include the fasciculus cuneatus and gracilis.

Fasciculus gracilis: Carries fine touch, vibration, and conscious proprioception (joint position sense) from the lower part of the body.

Spinocerebellar tracts: Carry information about muscle tension and joint position to the cerebellum, so it can control movement.

Fasciculus cuneatus: Carries fine touch, vibration, and conscious proprioception (joint position sense) from the upper part of the body.

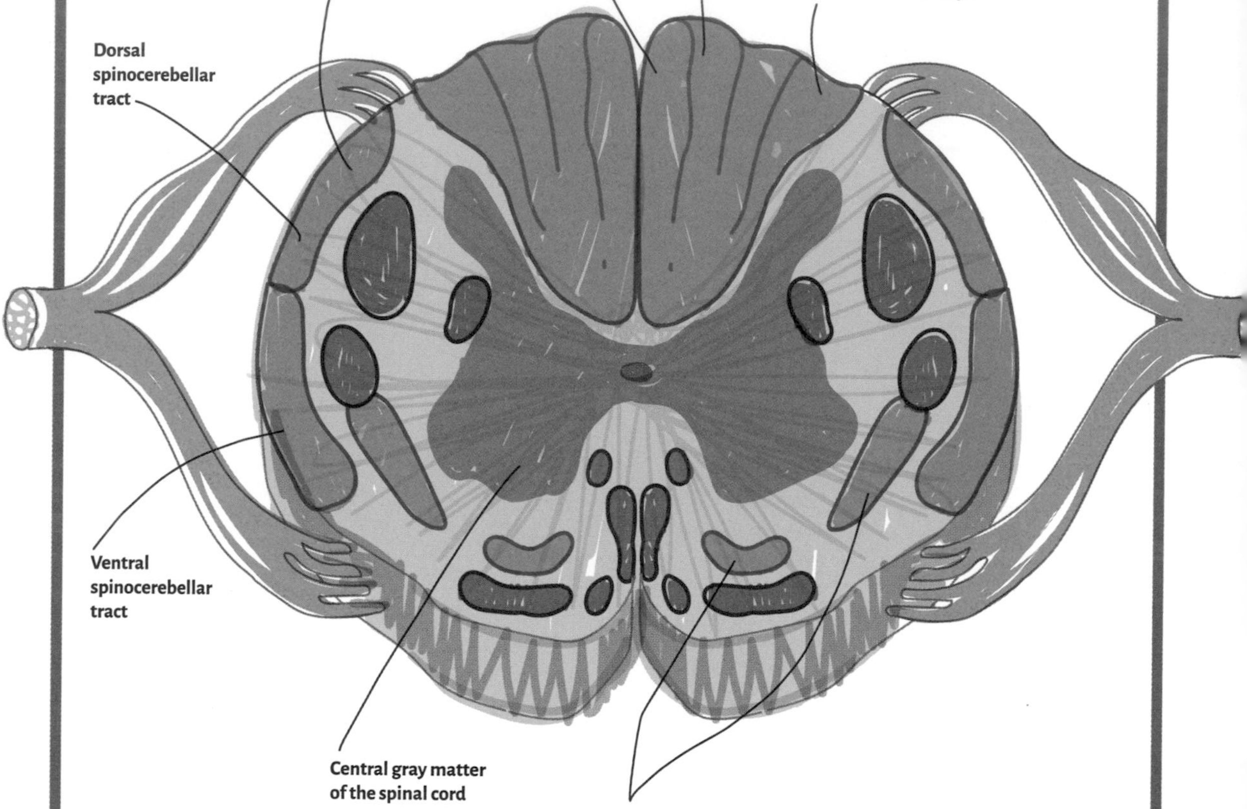

Dorsal spinocerebellar tract

Ventral spinocerebellar tract

Central gray matter of the spinal cord

Spinothalamic tracts: Carry pain, temperature, and simple touch to the thalamus. The spinothalamic tract includes ventral and lateral parts.

Descending pathways

Descending pathways carry information caudally (down) the spinal cord and are mostly motor, although some can modify sensory inputs.

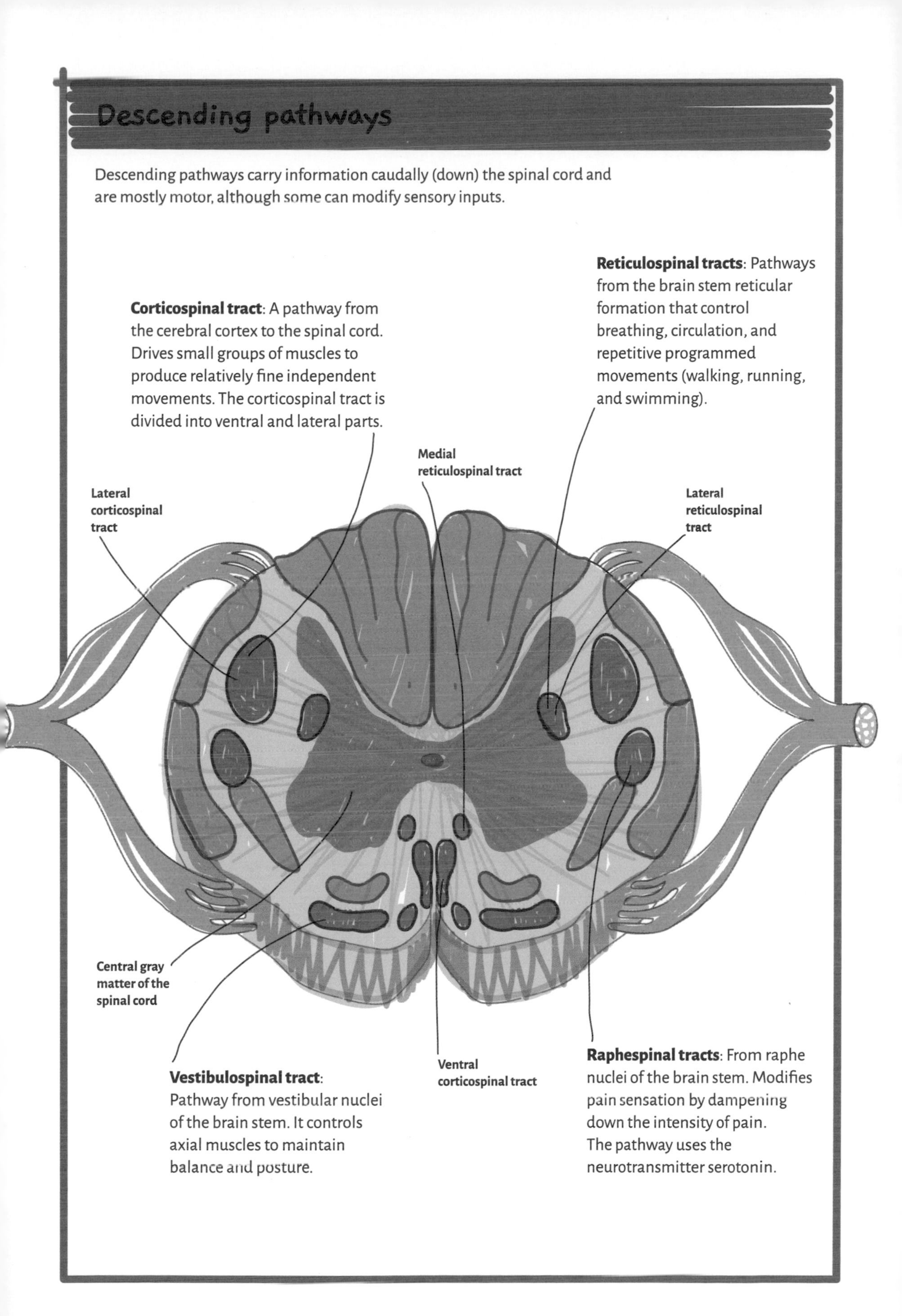

Reticulospinal tracts: Pathways from the brain stem reticular formation that control breathing, circulation, and repetitive programmed movements (walking, running, and swimming).
Corticospinal tract: A pathway from the cerebral cortex to the spinal cord. Drives small groups of muscles to produce relatively fine independent movements. The corticospinal tract is divided into ventral and lateral parts.
Medial reticulospinal tract
Lateral corticospinal tract
Lateral reticulospinal tract
Central gray matter of the spinal cord
Vestibulospinal tract: Pathway from vestibular nuclei of the brain stem. It controls axial muscles to maintain balance and posture.
Ventral corticospinal tract
Raphespinal tracts: From raphe nuclei of the brain stem. Modifies pain sensation by dampening down the intensity of pain. The pathway uses the neurotransmitter serotonin.

NERVES OF THE HEAD AND NECK

The nerves of the head and neck include the 12 cranial nerves attached to the brain and the branches of the upper cervical spinal cord. Two cranial nerves (olfactory and optic) attach to the forebrain, and the rest of the cranial nerves to the brain stem.

How many cranial nerves are there?

There are 12 pairs of cranial nerves in humans, and embryonic humans have one more (CN0). This is the nervus terminalis from the nose to the forebrain, which is lost during development.

Olfactory nerve (CN1) is only a sensory nerve composed of multiple fine nerve fibers that reach from the olfactory epithelium to the olfactory bulb. The processing of smell information occurs in the bulb and the olfactory cortex of the forebrain.

Optic nerve (CN2) is a purely sensory nerve that contains about 1 to 1.5 million nerve fibers and starts at the back of the eyeball. Each nerve fiber is the axon of one retinal ganglion cell (see page 97). Nerve fibers from the nasal half of each retina cross at the optic chiasm below the hypothalamus. Temporal retina fibers stay on the same side, but nasal retinal fibers cross. Fibers carrying information about the left visual field cross to the right side of the brain and vice versa.

Extraocular muscle nerves are **oculomotor (CN3)**, **trochlear (CN4)**, and **abducens (CN6)**, which are exclusively motor nerves to drive the extraocular muscles that move the eyeball.

Oculomotor nerve (CN3) drives the superior rectus, the medial rectus, the inferior rectus, and the inferior oblique muscles. It also contains parasympathetic fibers that drive smooth muscle constricting the pupil and controlling lens shape for focusing.

Trigeminal nerve (CN5) is a mixed sensory and motor nerve. It supplies sensation to the face, such as touch, pain, and temperature. It also controls the muscles of mastication. It has three branches (hence the name), called **ophthalmic**, **maxillary**, and **mandibular** that supply the forehead, cheek, and jaw, respectively.

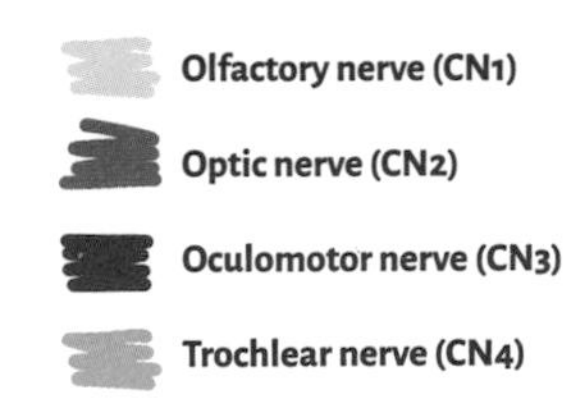

Olfactory nerve (CN1)
Optic nerve (CN2)
Oculomotor nerve (CN3)
Trochlear nerve (CN4)

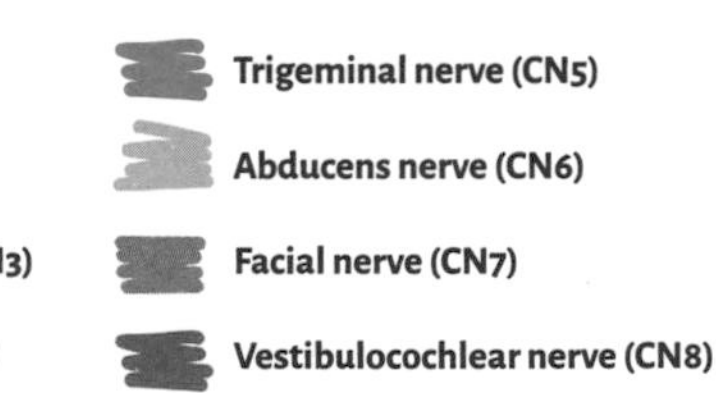

Trigeminal nerve (CN5)
Abducens nerve (CN6)
Facial nerve (CN7)
Vestibulocochlear nerve (CN8)

Facial nerve (CN7) is another mixed sensory and motor nerve. It supplies the muscles of facial expression and the lacrimal, submandibular, and sublingual salivary glands. It serves taste from the anterior two-thirds of the tongue.

Vestibulocochlear nerve (CN8) provides sensation from the inner ear for hearing (cochlear division) and vestibular function (head, balance, and acceleration; vestibular division).

Glossopharyngeal nerve (CN9) is a mixed sensory and motor nerve. It supplies one muscle in the pharynx and the parotid gland. It serves sensation from the palate and the pharynx.

Vagus nerve (CN10) is also a mixed sensory and motor nerve. It is so-called because it wanders through the neck, chest, and upper abdomen. It supplies the muscles of the pharynx and the larynx, most muscles of the soft palate, the esophagus, the stomach glands, the smooth muscle of the stomach, the small intestine, and the upper large intestine. It also provides sensation to the larynx, airways, lungs, and alimentary canal.

Accessory nerve (CN11) is another purely motor nerve. It supplies the sternocleidomastoid and the upper trapezius muscles.

Hypoglossal nerve (CN12) is also a purely motor nerve. It supplies the intrinsic and extrinsic muscles of the tongue.

Brain stem and cranial nerves

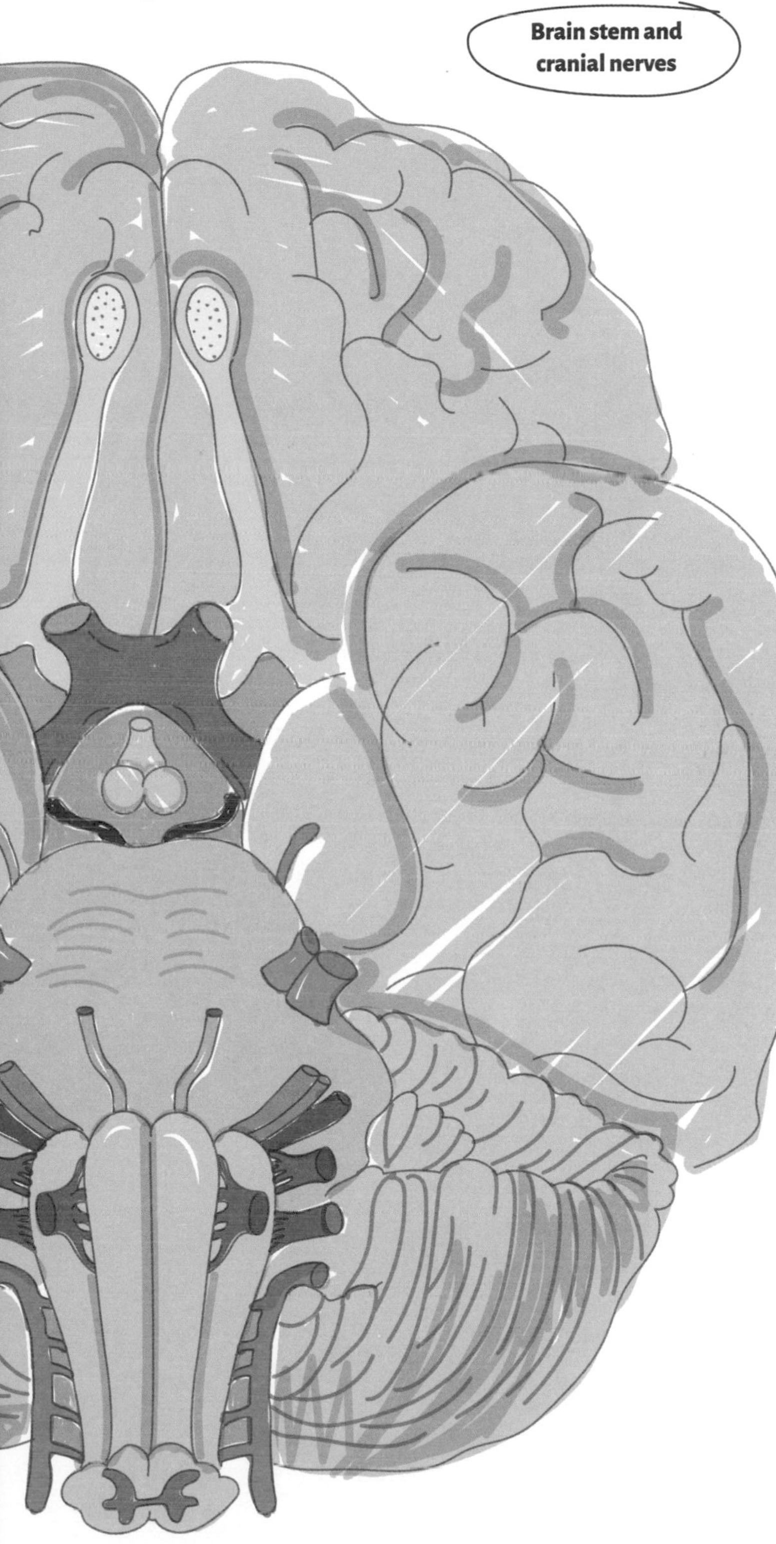

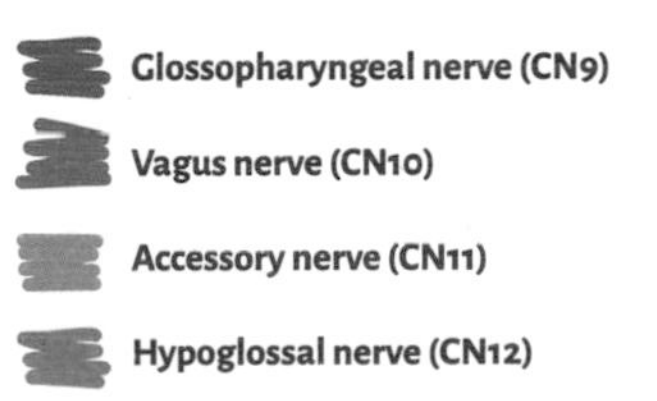
Glossopharyngeal nerve (CN9)
Vagus nerve (CN10)
Accessory nerve (CN11)
Hypoglossal nerve (CN12)

NERVES OF THE SHOULDER AND UPPER LIMB

The skin and muscles of the upper limb are supplied by the brachial plexus (attached to cervical spinal segment C5 to thoracic segment T1). There are five major nerves from the brachial plexus: radial, axillary, musculocutaneous, median, and ulnar.

Brachial plexus

The **brachial plexus** forms from spinal nerves C5 to T1. Roots from C5 and C6 join to form a superior trunk, root C7 forms the middle trunk, and roots C8 and T1 form the inferior trunk. Posterior divisions from each trunk form a posterior cord, anterior divisions from the superior and middle trunk form the lateral cord, while the inferior trunk continues as the medial cord.

The **radial nerve** is a branch from the posterior cord that supplies the extensor of the elbow (triceps brachii), brachioradialis, extensors of the wrist and fingers, and sensation of the dorsum (back) of the hand.

The **ulnar nerve** is a branch from the medial cord that supplies the flexor carpi ulnaris, medial half of flexor digitorum profundus, hypothenar muscles, and sensation to skin over the medial one and a half digits.

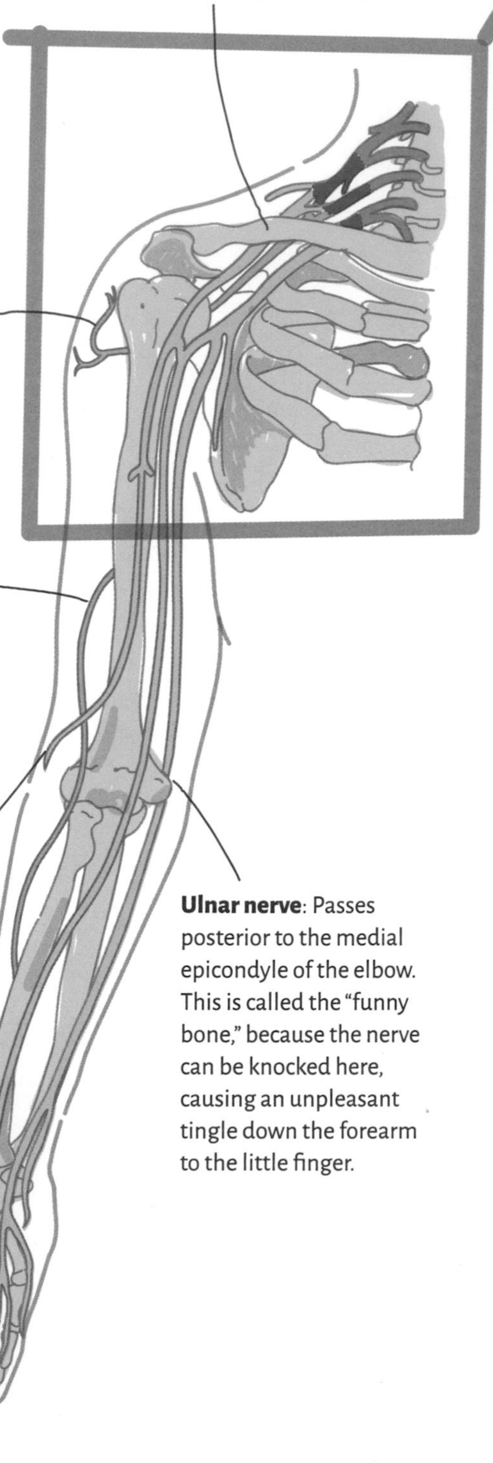

Nerves of the upper limb
The brachial plexus passes between the clavicle and the first rib.
Axillary nerve: A branch of the posterior cord that supplies the deltoid and teres minor muscles, and sensation for skin over the shoulder prominence.
Radial nerve: Here, it spirals around the posterior surface of the humerus, where it can be damaged by a fracture of the midshaft of the humerus.
Musculocutaneous nerve: A branch from the lateral cord that supplies the flexors of the elbow (biceps brachii and brachialis) as well as sensation to skin over the forearm.
Ulnar nerve: Passes posterior to the medial epicondyle of the elbow. This is called the "funny bone," because the nerve can be knocked here, causing an unpleasant tingle down the forearm to the little finger.
Median nerve: Passes here through a tunnel formed by the carpal bones of the wrist (carpal tunnel) and can be compressed here by increased pressure in the tunnel.

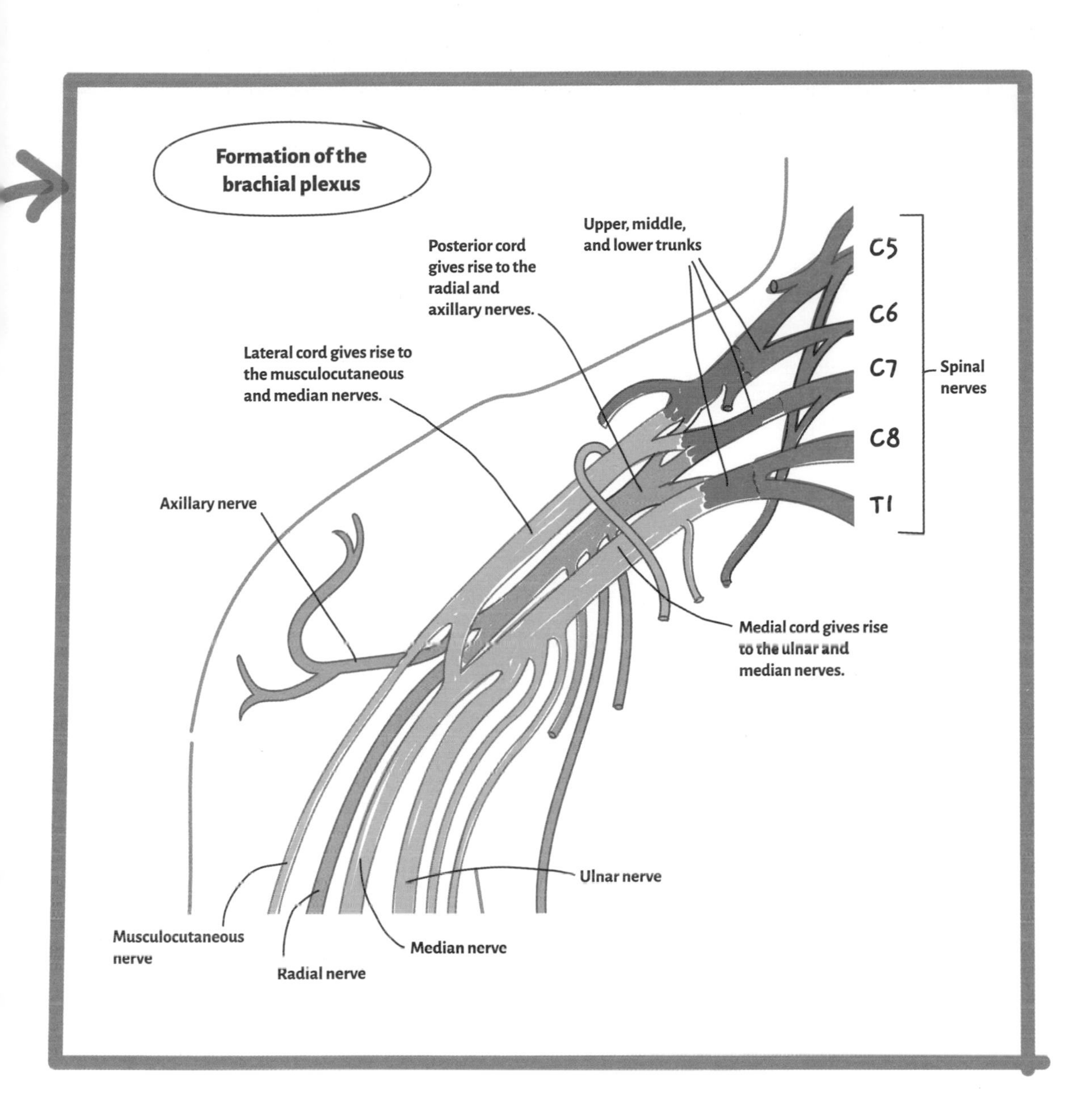

Formation of the brachial plexus
Upper, middle, and lower trunks
Posterior cord gives rise to the radial and axillary nerves.
Lateral cord gives rise to the musculocutaneous and median nerves.
C5
C6
C7
C8
T1
Spinal nerves
Axillary nerve
Medial cord gives rise to the ulnar and median nerves.
Ulnar nerve
Musculocutaneous nerve
Radial nerve
Median nerve

The **median nerve** (see page 92 and above) is formed by the junction of branches from the lateral and medial cords. It supplies most flexors of the fingers and wrist (except flexor carpi ulnaris and the medial half of flexor digitorum profundus), pronators of forearm, thenar muscles at the thumb base, and the sensation of the skin of the lateral palm and the lateral three and a half digits.

Nerves and trauma

Many of the nerves shown above and opposite are vulnerable to fractures, cuts, and compressions.

The radial nerve can often be damaged by a fractured midshaft of the humerus.

The axillary nerve can be damaged by fracture of the proximal end of the humerus.

The median nerve can be damaged by fracture of the distal humerus. It can also be squeezed by pressure in the bony carpal tunnel of the wrist.

The ulnar nerve passes posterior to the medial distal humerus. It can be cut by falling onto or through glass.

NERVES OF THE BUTTOCK AND LOWER LIMB

Nerves from the lumbar and sacral plexuses of the peripheral nervous system supply the skin and muscles of the lower limb. Important nerves include the femoral, sciatic, and obturator. The sciatic nerve divides into tibial and fibular branches.

Lumbar and sacral plexuses

The **lumbar plexus** arises from spinal nerves L1 to L4. Small branches innervate the abdominal wall and psoas major. Its biggest branches are the **femoral** and **obturator** nerves.

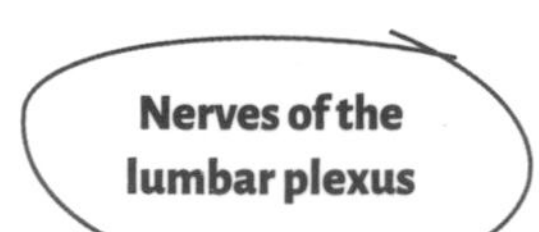
Nerves of the lumbar plexus

Iliohypogastric and ilioinguinal nerves: Supply the lower abdomen and groin.

Lumbosacral trunk: A branch from L4 and L5 that descends to join the sacral plexus.

Femoral nerve: A branch of the lumbar plexus (L2 to L4). It supplies the muscles that flex the thigh (pectineus, iliacus, and sartorius) and extend the knee (quadriceps femoris). It also supplies sensation to the anterior and lower medial side of the thigh and the medial surface of the leg and foot (**saphenous nerve**).

Obturator nerve: A branch of the lumbar plexus (mainly L3 and L4). It supplies thigh muscles that adduct the hip (adductor group) and skin over the upper medial surface of the thigh.

Lateral femoral cutaneous nerve: Supplies the skin of the upper lateral thigh.

Saphenous nerve

The **sacral plexus** is a network of nerves from spinal nerves L4 to S3. Its branches include the **sciatic**, **superior** and **inferior gluteal nerves**, and the **pudendal nerve**.

Nerves of the sacral plexus

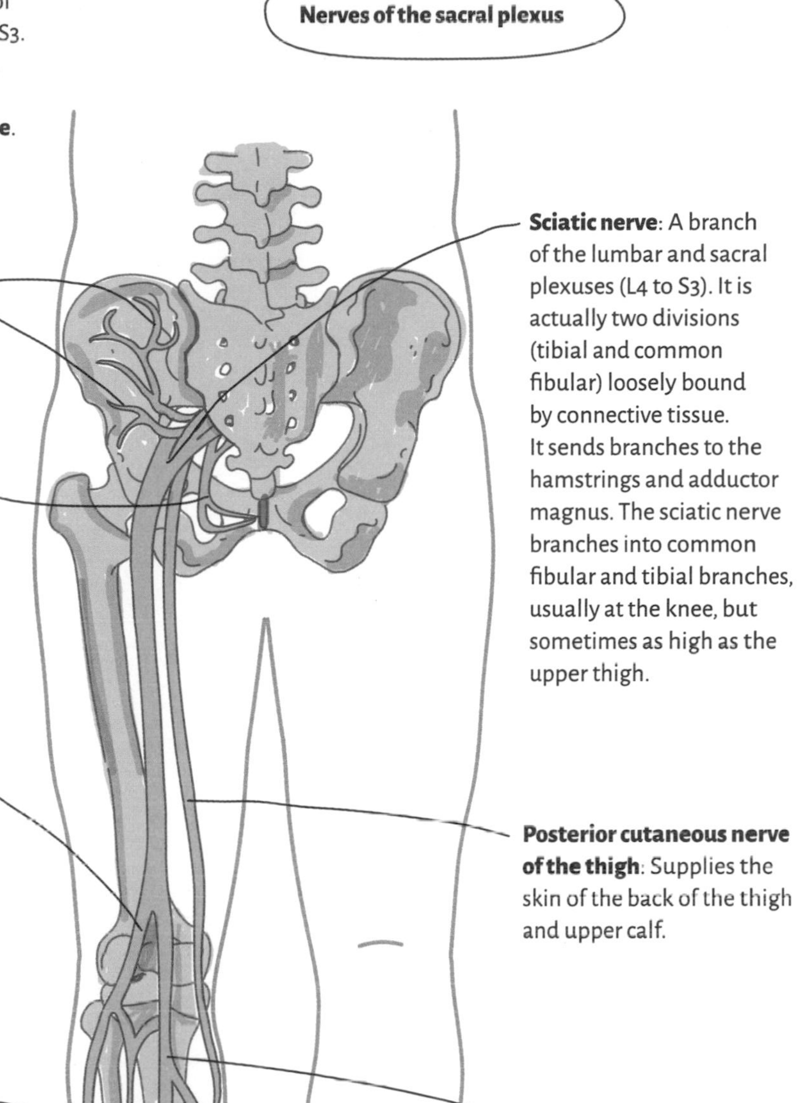

Gluteal nerves: Supply the muscles of the gluteal region (buttock).

Pudendal nerve: Supplies the muscles and skin of the **perineum** (space between the thigh bases) and controls defecation and micturition.

Common fibular nerve: Curves around the proximal end of the fibula and divides into **superficial** and **deep fibular nerves**.

Deep fibular nerve: Supplies the muscles of the anterior compartment of the leg (extensors of toes and dorsiflexors of foot).

Superficial fibular nerve: Supplies the muscles of the lateral compartment of the leg (fibular muscles).

Sciatic nerve: A branch of the lumbar and sacral plexuses (L4 to S3). It is actually two divisions (tibial and common fibular) loosely bound by connective tissue. It sends branches to the hamstrings and adductor magnus. The sciatic nerve branches into common fibular and tibial branches, usually at the knee, but sometimes as high as the upper thigh.

Posterior cutaneous nerve of the thigh: Supplies the skin of the back of the thigh and upper calf.

Tibial nerve: Passes through the **popliteal fossa** at the back of the knee. It supplies the muscles of the posterior compartment of the leg and the sole of the foot.

EYE AND VISION

Vision requires the formation of an image on the neural retina, the transmission of that visual information to the brain by the optic nerve, and central processing of visual data in the cerebral cortex to extract behaviorally important information.

Structure of the eye

The globe of the eye is divided into two basic regions: anterior and posterior. The **anterior segment** is filled with liquid aqueous humor. The **posterior segment** is filled with jelly-like vitreous humor.

The eye has three layers. The outer fibrous layer consists of the **sclera** (white of eye) in the posterior segment and the **cornea** in the anterior segment.

The middle vascular layer consists of a vascular **choroid** in the posterior segment and a **ciliary body** in the anterior segment.

The inner sensory layer consists of the **retina**, which is confined to the posterior three-quarters of the eyeball.

The optical surfaces of the eye are the cornea and the **lens**. The cornea has the most refractive power, but the focus of the lens can be adjusted for near or far vision. The shape of the lens can be changed by contracting the ciliary muscle in the ciliary body, thereby releasing the tension on the **zonular ligaments** that hold the lens and allow the lens to return to its natural spherical shape.

The **optic nerve** carries sensory information from the retina to the brain. The nerve is composed of axons of retinal ganglion cells.

The human retina is supplied by a **central retinal artery**. The central artery emerges into the eye at the **optic disk**. Blockage of the central artery can cause death of the retina and blindness.

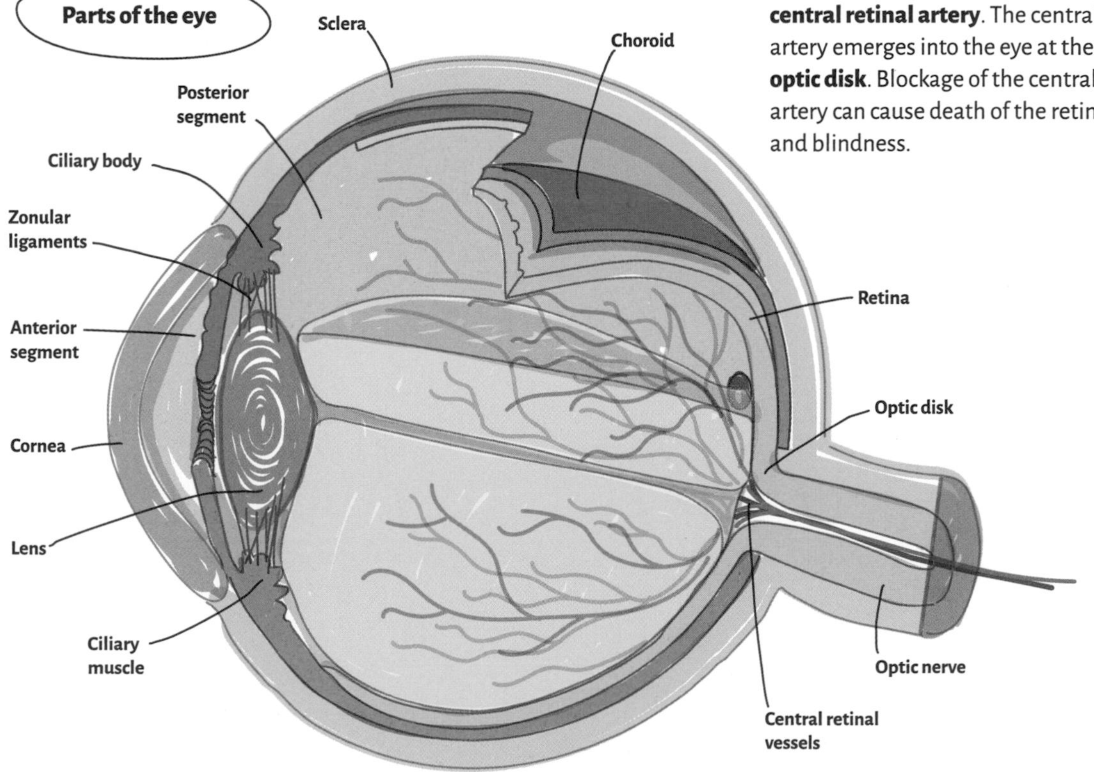

Parts of the eye
Sclera
Choroid
Posterior segment
Ciliary body
Zonular ligaments
Anterior segment
Cornea
Lens
Ciliary muscle
Retina
Optic disk
Optic nerve
Central retinal vessels

Retina and optic nerve

The **retina** has pigmented and neural layers. The **pigmented layer** is a sheet of melanin-containing epithelial cells between the neural retina and the choroid to reduce light scattering.

The **neural retina** has three layers of retinal neurons (photoreceptors, bipolar cells, ganglion cells), each separated by zones of synapses.

Light passes through the **ganglion cell** and **bipolar cell** layers before reaching the photoreceptors.

There are two types of photoreceptors: **rods** for dim light vision in black and white and **cones** for acuity color vision in good light.

Axons from ganglion cells gather at the optic disk (corresponding to the blind spot) before leaving the eye.

The optic nerve passes to the **optic chiasm**, where axons from the nasal retina cross. Axons continue into the **optic tract** (see below).

The nerve fibers in the optic tract mainly terminate in the lateral geniculate nucleus of the thalamus. Some continue to the **pretectum** and **superior colliculus** of the midbrain for visual reflexes.

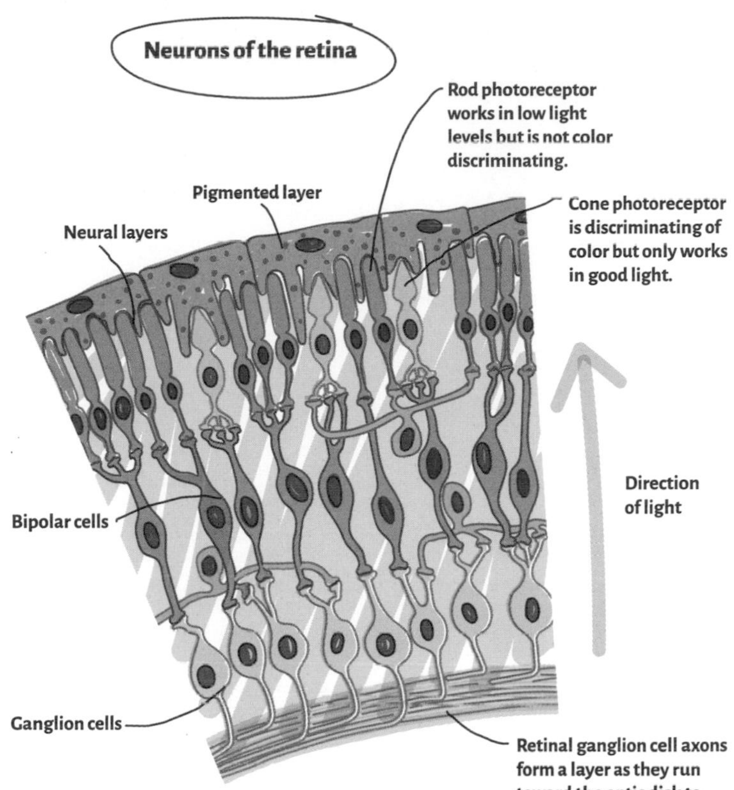

Neurons of the retina
Rod photoreceptor works in low light levels but is not color discriminating.
Pigmented layer
Neural layers
Cone photoreceptor is discriminating of color but only works in good light.
Direction of light
Bipolar cells
Ganglion cells
Retinal ganglion cell axons form a layer as they run toward the optic disk to leave the eye.

Visual form, space, and reflexes

Analysis of visual form occurs in the temporal lobe (from the ventral visual stream), where recognition of the shape, color, and visual texture of everyday familiar objects, such as people's faces and animals, takes place.

The **lateral geniculate nucleus** sends axons to the **primary visual cortex** in the occipital lobe, where visual space is mapped out. Information is then streamed to secondary visual cortical areas concerned with the position and form of objects.

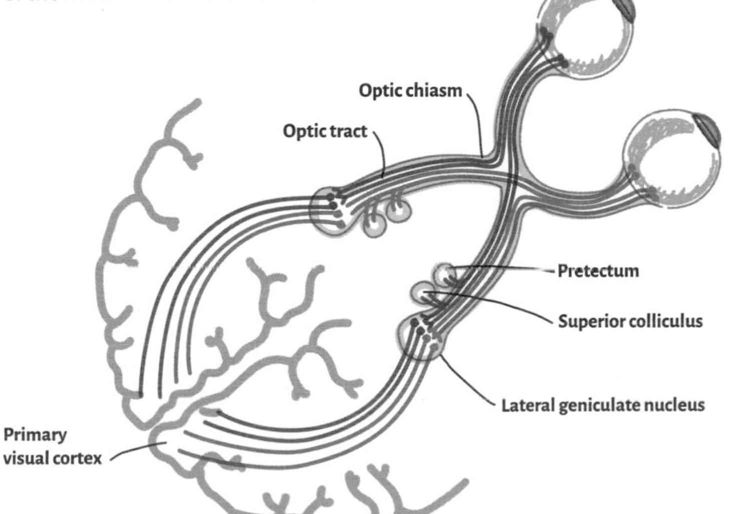

Optic chiasm
Optic tract
Pretectum
Superior colliculus
Lateral geniculate nucleus
Primary visual cortex
Visual pathways

EAR AND HEARING

The ear is divided into external, middle, and inner components. The sensory apparatuses for hearing, balance, and acceleration are located within the inner ear.

External ear

The external ear consists of the cartilaginous **auricle**, a **cartilage tube** for the outer part of the ear canal, a bony inner **external ear canal**, and the **eardrum** (tympanic membrane/tympanum). Glands in the ear canal produce **cerumen**, a waxy substance that inhibits bacterial and fungal growth.

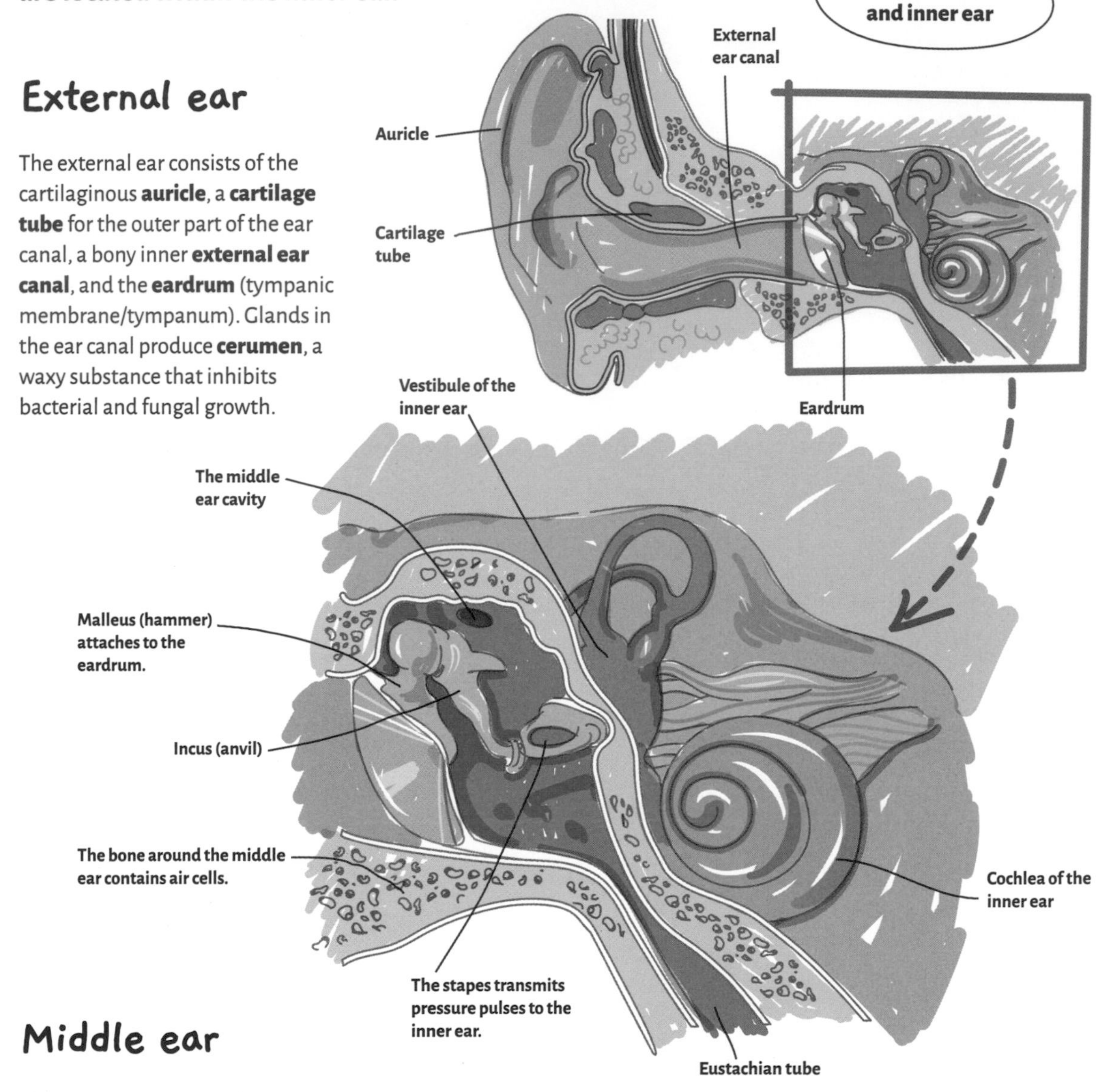

External, middle, and inner ear
External ear canal
Auricle
Cartilage tube
Eardrum
Vestibule of the inner ear
The middle ear cavity
Malleus (hammer) attaches to the eardrum.
Incus (anvil)
The bone around the middle ear contains air cells.
Cochlea of the inner ear
The stapes transmits pressure pulses to the inner ear.
Eustachian tube

Middle ear

The middle ear is an air-filled cavity that connects with the nasopharynx through the auditory (**Eustachian** or **pharyngotympanic**) tube. The connection with the nasopharynx allows equalization of pressure on both sides of the tympanic membrane during changes in altitude.

A chain of three tiny bones (auditory ossicles) crosses the space between the tympanic membrane and oval window. These bones are the **malleus**, **incus**, and **stapes** (hammer, anvil, and stirrup) and amplify by 20 times the vibration of the tympanic membrane and transmit that vibratory information to the inner ear.

The middle ear communicates with the mastoid air cells within the temporal bone.

Inner ear

The inner ear is a series of fluid-filled canals (the labyrinth) embedded within the petrous temporal bone. An outer **bony labyrinth** encloses an inner **membranous labyrinth**.

There are two functional divisions within the inner ear: for hearing and for balance.

The **cochlea** consists of the cochlear duct, scala tympani, and scala vestibuli, which are used to help hearing.

The **vestibular apparatus** consists of the utricle, saccule, and semicircular ducts and is used for balance and linear and angular acceleration.

The **utricle** and **saccule** have sensory zones (**maculae**) that detect force due to gravity and linear acceleration.

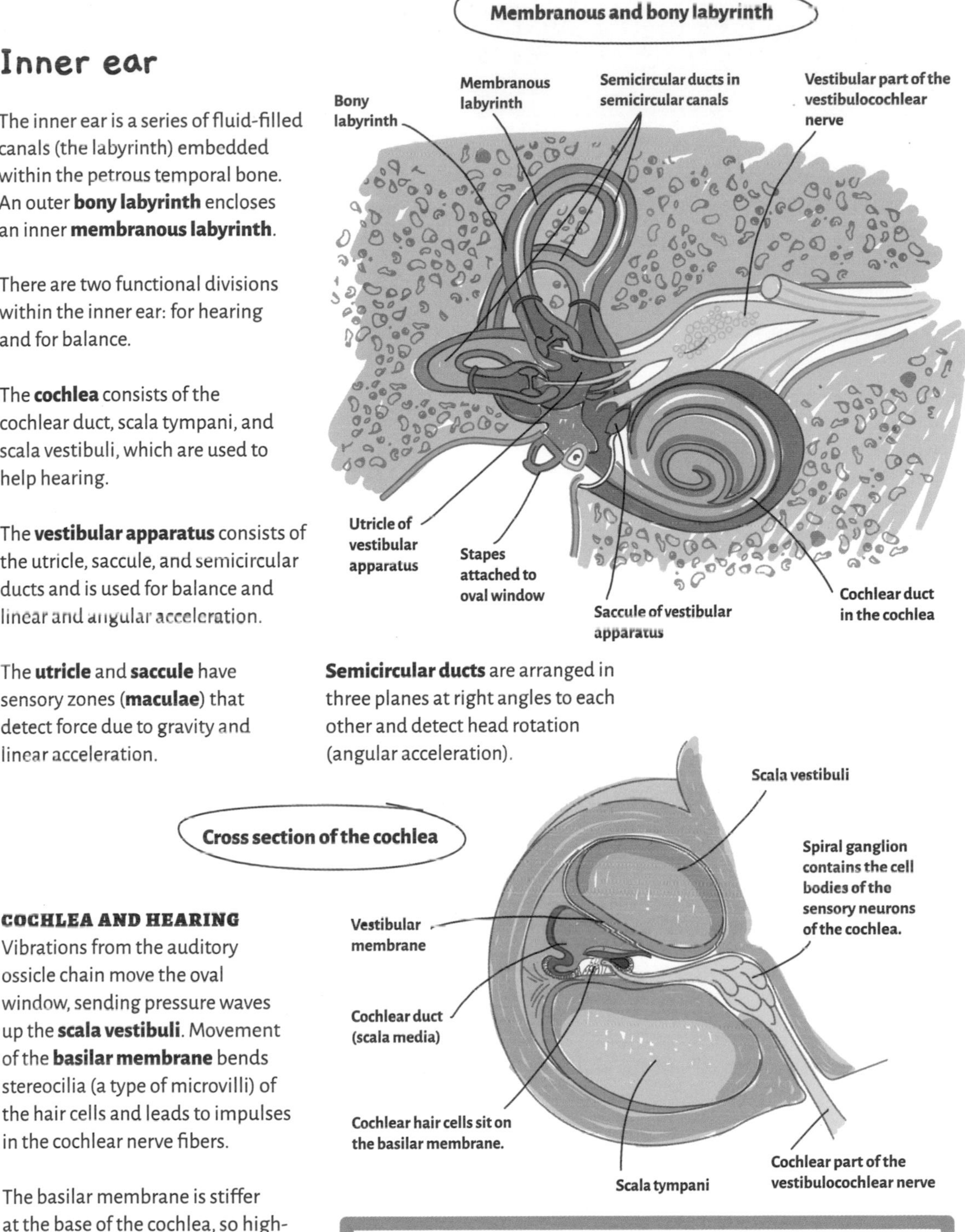

Membranous and bony labyrinth
Bony labyrinth
Membranous labyrinth
Semicircular ducts in semicircular canals
Vestibular part of the vestibulocochlear nerve
Utricle of vestibular apparatus
Stapes attached to oval window
Saccule of vestibular apparatus
Cochlear duct in the cochlea
Cross section of the cochlea
Scala vestibuli
Spiral ganglion contains the cell bodies of the sensory neurons of the cochlea.
Vestibular membrane
Cochlear duct (scala media)
Cochlear hair cells sit on the basilar membrane.
Scala tympani
Cochlear part of the vestibulocochlear nerve

Semicircular ducts are arranged in three planes at right angles to each other and detect head rotation (angular acceleration).

COCHLEA AND HEARING

Vibrations from the auditory ossicle chain move the oval window, sending pressure waves up the **scala vestibuli**. Movement of the **basilar membrane** bends stereocilia (a type of microvilli) of the hair cells and leads to impulses in the cochlear nerve fibers.

The basilar membrane is stiffer at the base of the cochlea, so high-frequency sounds are most effective at stimulating hair cells there.

Auditory information passes along the cochlear division of the **vestibulocochlear nerve** (CN8) to the brain stem, to finally reach the auditory cortex on the temporal lobe of the cerebral cortex.

BALANCE AND ACCELERATION

Vestibular information passes to the vestibular neuron groups in the brain stem. These neurons communicate with the cerebellum to coordinate eye movement and with the spinal cord to control postural muscles. Conscious awareness of balance is felt in the parietal lobe.

SENSE OF TASTE

The perception of taste is a chemical sense, meaning that a chemical (the tastant) must lock into a specific receptor to generate a signal. These receptors are located on taste buds in the oral cavity.

Taste bud structure

Taste buds are oval structures embedded in the oral epithelium, with a central **taste pore** and three cell types.

Supporting cells surround the gustatory receptor cells. They develop into gustatory receptor cells.

Basal cells are stem cells in the edge of the taste bud. They produce supporting cells.

Gustatory receptor cells have tiny microvilli that project through the taste pore to the external surface. The microvilli have receptors for tastant molecules. These cells last only 10 days and are replaced by supporting cells.

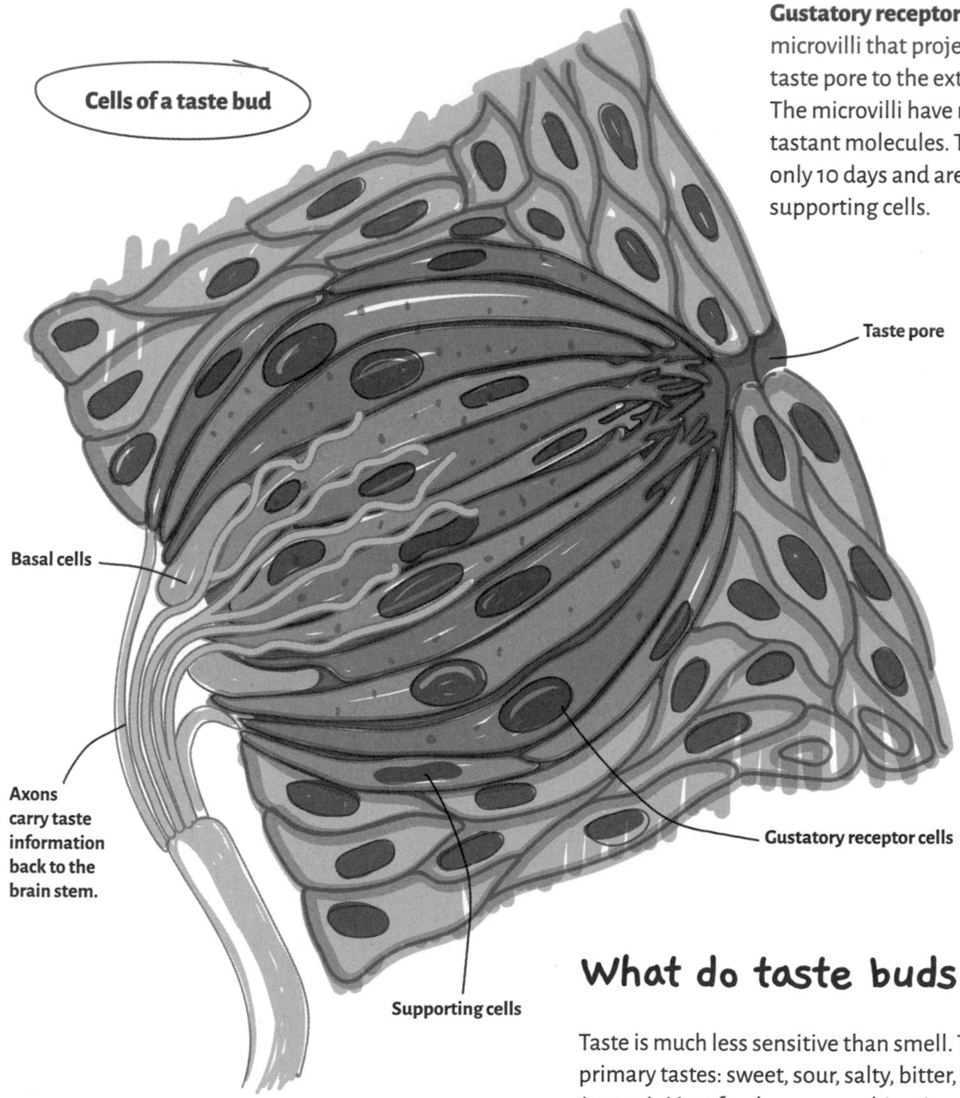

Cells of a taste bud

Taste pore

Basal cells

Axons carry taste information back to the brain stem.

Gustatory receptor cells

Supporting cells

What do taste buds detect?

Taste is much less sensitive than smell. There are five primary tastes: sweet, sour, salty, bitter, and umami (savory). Most foods are a combination of the primary tastes, plus odors detected by the nose.

Where are taste buds?

There are nearly 10,000 taste buds in young adults, but this declines with age. They are mainly on the tongue, soft palate, pharynx, and epiglottis.

Papillae are elevations on the tongue where taste buds are found. They may be on the side of the tongue (**foliate papillae**), on the tongue dorsum (mushroom-shaped **fungiform papillae**), or castle-shaped structures with tiny moats on the back of the tongue (**circumvallate papillae**).

Papillae of the tongue

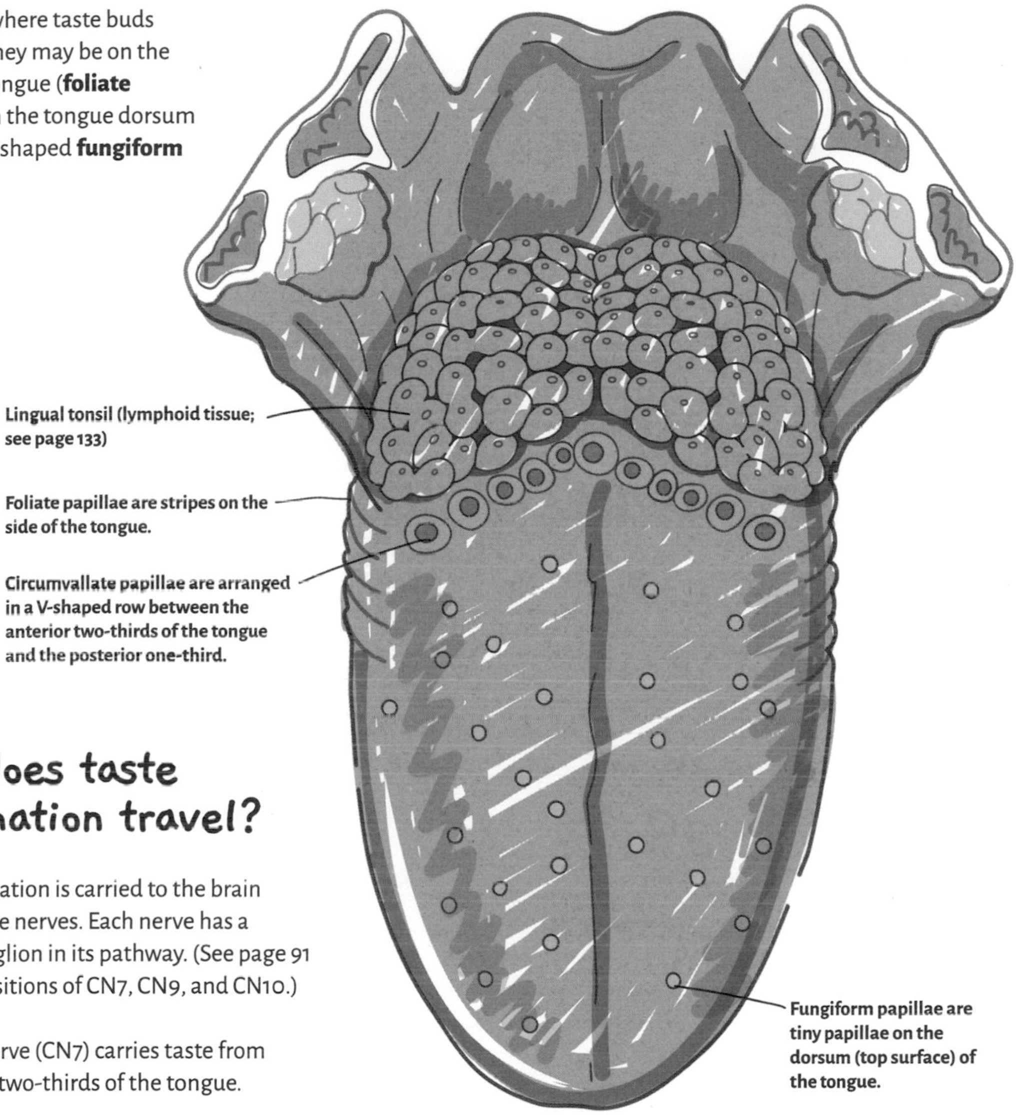

Lingual tonsil (lymphoid tissue; see page 133)

Foliate papillae are stripes on the side of the tongue.

Circumvallate papillae are arranged in a V-shaped row between the anterior two-thirds of the tongue and the posterior one-third.

Fungiform papillae are tiny papillae on the dorsum (top surface) of the tongue.

How does taste information travel?

Taste information is carried to the brain stem by three nerves. Each nerve has a sensory ganglion in its pathway. (See page 91 for nerve positions of CN7, CN9, and CN10.)

The facial nerve (CN7) carries taste from the anterior two-thirds of the tongue.

The glossopharyngeal nerve (CN9) carries taste from the posterior one-third of the tongue.

The vagus nerve (CN10) carries taste from the throat and the epiglottis.

Taste information is carried to the medulla. Signals pass from there to the thalamus and then to the primary gustatory area of the parietal lobe for conscious awareness of taste.

Nonconscious taste affects hypothalamic function and emotions. This acts through memories evoked by tastes and smells and is mediated by the temporal lobe of the brain.

SENSE OF SMELL

Olfaction or smell is a chemical sense like taste, so chemicals (odorants) lock into specifically shaped receptors in the olfactory epithelium to generate a signal. Signals reach the olfactory bulb by fine nerve fibers that constitute the olfactory nerve (CN1).

What can we smell?

Humans can distinguish about 10,000 distinct odors. In fact, our sense of smell is much more sensitive than our taste, which has only five different flavors.

Much of our enjoyment of food is due to aromas coming from the food passing from the mouth up the nasopharynx to reach the olfactory epithelium.

The **olfactory epithelium** is about 5 cm^2 (under one square inch) in area. It lies on the underside of the cribriform plate and extends onto the superior nasal concha of the ethmoid bone.

The palate forms the roof of the mouth and is divided into the **hard** and **soft palates**. They separate the oral and nasal cavities, allowing food to be formed into a bolus before swallowing.

Olfactory area of the nasal cavity
Olfactory bulb
Olfactory nerve fibers pass through the bony cribriform plate to reach the olfactory bulb.
The oral and nasal cavity communicate via the nasopharynx.
The hard and soft palates separate the mouth and nasal cavities.
Inhaled air containing odorants

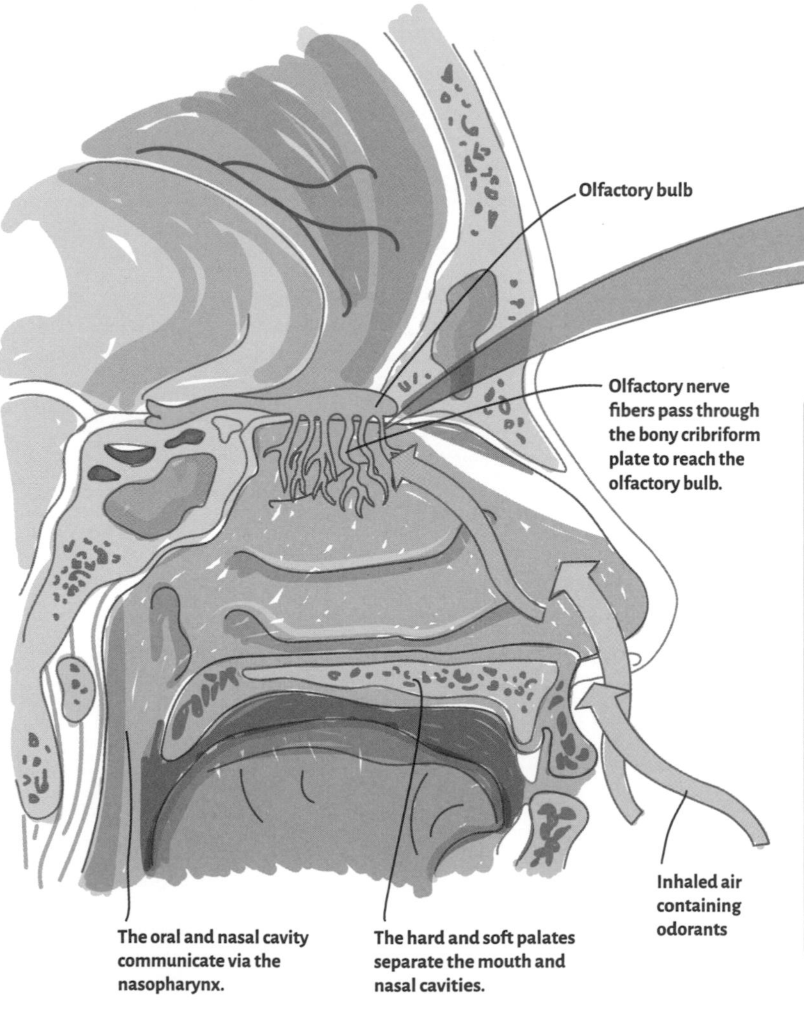

CENTRAL PROCESSING OF OLFACTION

Olfactory tract nerve fibers terminate in the **primary olfactory area**, which is located on the medial surface of the temporal lobe, where the conscious awareness of odors starts.

Olfactory input also affects emotions and reproduction through terminations in the amygdala and hypothalamus.

Olfactory bulb structure and function

The **paired olfactory bulbs** lie inferior to the frontal lobes of the brain and directly superior to the **cribriform plate of the ethmoid**. Olfactory nerve fibers reach the olfactory bulb and terminate on the **dendrites of mitral cells** and other olfactory tract neurons. These give rise to nerve fibers that enter the **olfactory tract** to terminate in the olfactory areas of the forebrain.

Basal cells are stem cells that give rise to olfactory receptor cells. They undergo continuous cell division to replenish the receptor cells that are lost after about 30 days to the toxic environment of the nasal cavity.

Supporting cells provide physical support, nourishment, and electrical insulation for the receptor cells. They also make odorant-binding proteins that convey odorants to the olfactory receptors.

Olfactory (Bowman's) glands produce mucus that dissolves odorants, so they can have easier access to receptors, and protects the receptor cells.

Olfactory epithelium, nerve fibers, and bulb

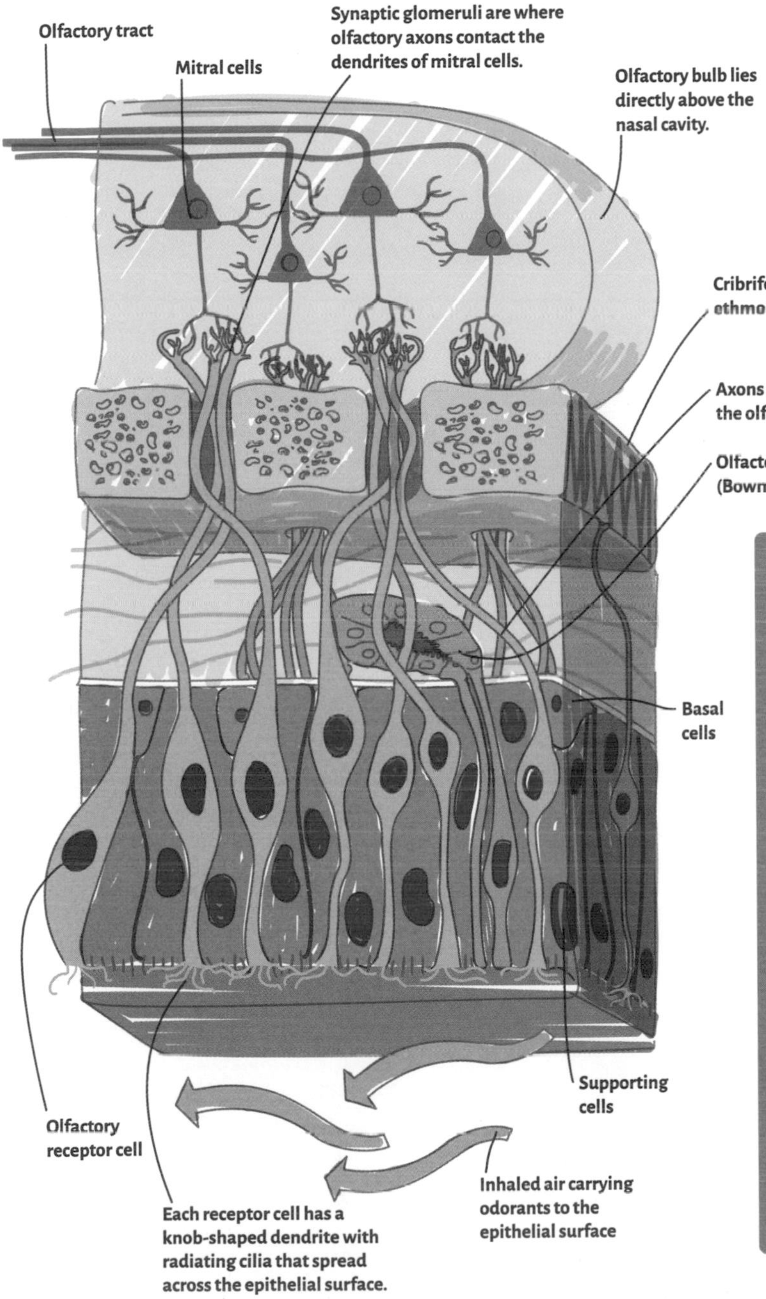

Olfactory tract

Mitral cells

Synaptic glomeruli are where olfactory axons contact the dendrites of mitral cells.

Olfactory bulb lies directly above the nasal cavity.

Cribriform plate of the ethmoid bone

Axons of olfactory receptor cells are the olfactory nerve fibers.

Olfactory (Bowman's) glands

Basal cells

Supporting cells

Olfactory receptor cell

Each receptor cell has a knob-shaped dendrite with radiating cilia that spread across the epithelial surface.

Inhaled air carrying odorants to the epithelial surface

OLFACTORY EPITHELIUM STRUCTURE

Olfactory receptor cells are the sensory cells of the olfactory epithelium. There are between 10 and 100 million olfactory receptor cells in the olfactory epithelium, declining with age. Each receptor cell has a knob-shaped dendrite with radiating cilia that spread across the epithelial surface. Odorants attach to the olfactory receptors on the cilia. This generates an axon potential that travels along the receptor cell axon, through the cribriform plate, to terminate in the olfactory bulb.

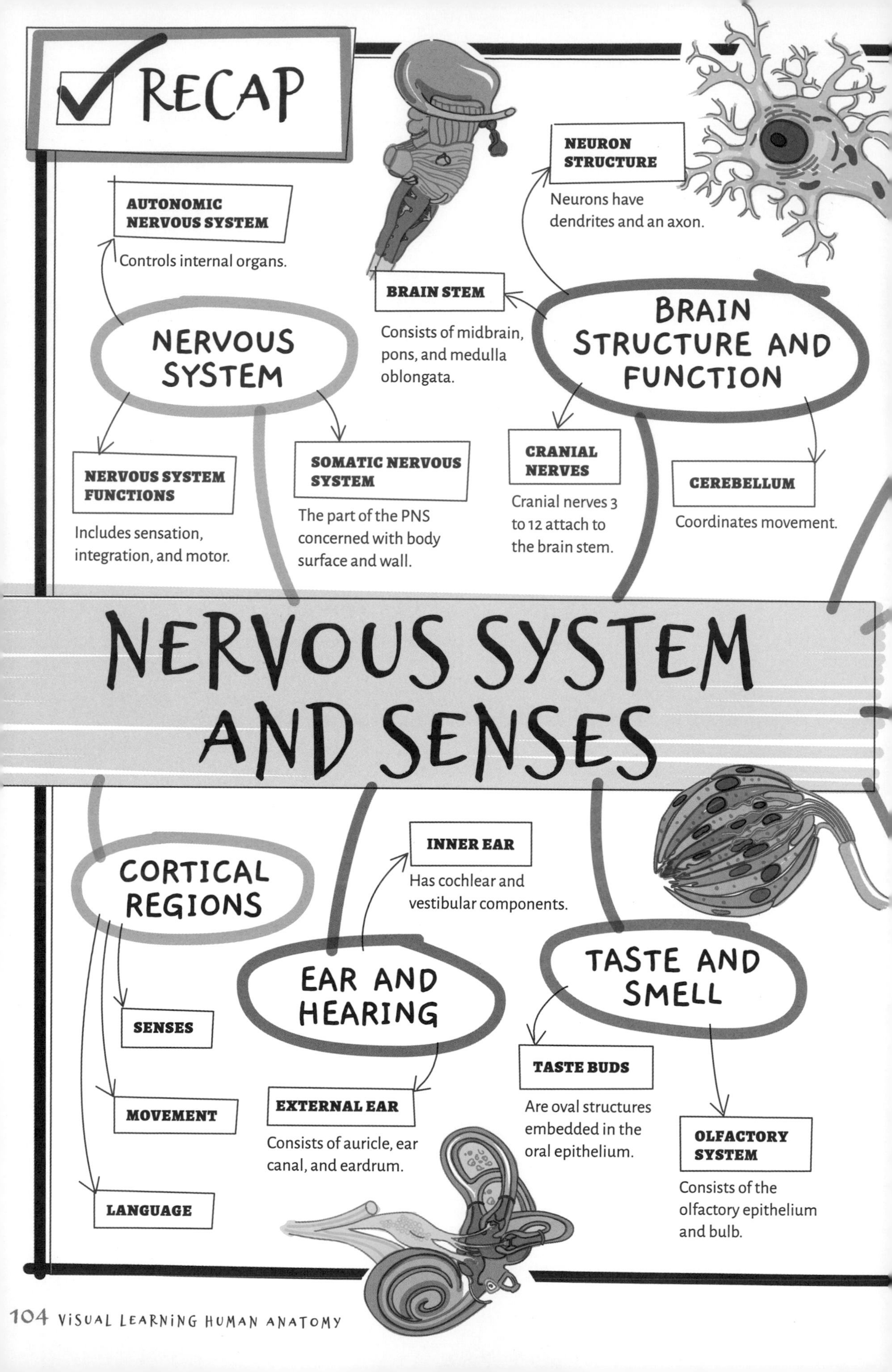
RECAP
NERVOUS SYSTEM AND SENSES
NERVOUS SYSTEM
AUTONOMIC NERVOUS SYSTEM
Controls internal organs.
NERVOUS SYSTEM FUNCTIONS
Includes sensation, integration, and motor.
SOMATIC NERVOUS SYSTEM
The part of the PNS concerned with body surface and wall.
BRAIN STRUCTURE AND FUNCTION
BRAIN STEM
Consists of midbrain, pons, and medulla oblongata.
NEURON STRUCTURE
Neurons have dendrites and an axon.
CRANIAL NERVES
Cranial nerves 3 to 12 attach to the brain stem.
CEREBELLUM
Coordinates movement.
CORTICAL REGIONS
SENSES
MOVEMENT
LANGUAGE
EAR AND HEARING
INNER EAR
Has cochlear and vestibular components.
EXTERNAL EAR
Consists of auricle, ear canal, and eardrum.
TASTE AND SMELL
TASTE BUDS
Are oval structures embedded in the oral epithelium.
OLFACTORY SYSTEM
Consists of the olfactory epithelium and bulb.

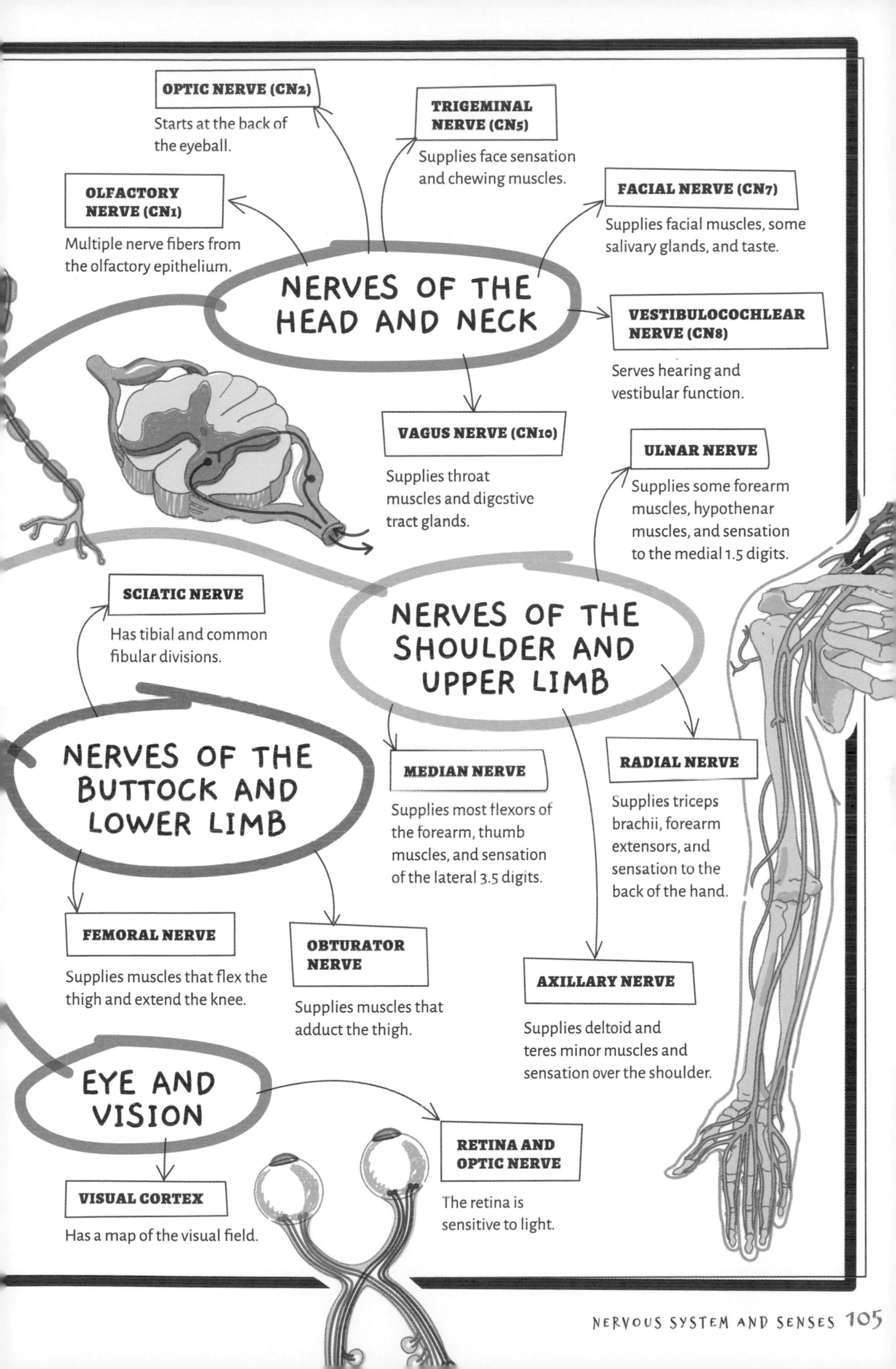
OPTIC NERVE (CN2)
Starts at the back of the eyeball.
TRIGEMINAL NERVE (CN5)
Supplies face sensation and chewing muscles.
OLFACTORY NERVE (CN1)
Multiple nerve fibers from the olfactory epithelium.
FACIAL NERVE (CN7)
Supplies facial muscles, some salivary glands, and taste.
NERVES OF THE HEAD AND NECK
VESTIBULOCOCHLEAR NERVE (CN8)
Serves hearing and vestibular function.
VAGUS NERVE (CN10)
Supplies throat muscles and digestive tract glands.
ULNAR NERVE
Supplies some forearm muscles, hypothenar muscles, and sensation to the medial 1.5 digits.
SCIATIC NERVE
Has tibial and common fibular divisions.
NERVES OF THE SHOULDER AND UPPER LIMB
NERVES OF THE BUTTOCK AND LOWER LIMB
MEDIAN NERVE
Supplies most flexors of the forearm, thumb muscles, and sensation of the lateral 3.5 digits.
RADIAL NERVE
Supplies triceps brachii, forearm extensors, and sensation to the back of the hand.
FEMORAL NERVE
Supplies muscles that flex the thigh and extend the knee.
OBTURATOR NERVE
Supplies muscles that adduct the thigh.
AXILLARY NERVE
Supplies deltoid and teres minor muscles and sensation over the shoulder.
EYE AND VISION
RETINA AND OPTIC NERVE
The retina is sensitive to light.
VISUAL CORTEX
Has a map of the visual field.

CHAPTER 6

CARDIOVASCULAR SYSTEM

The cardiovascular, or circulatory, system allows the efficient transportation of gases, nutrients, waste products, immune cells, and important proteins and minerals around the body. Approximately 5 liters (1.1 gallons) of blood per minute is pumped around the vessels of the body by a heart beating 60 to 70 times per minute. Blood vessel structure is adapted to containing high-pressure fluid (arteries), efficient exchange with tissues (capillaries), or storage and return of blood volume to the heart (veins).

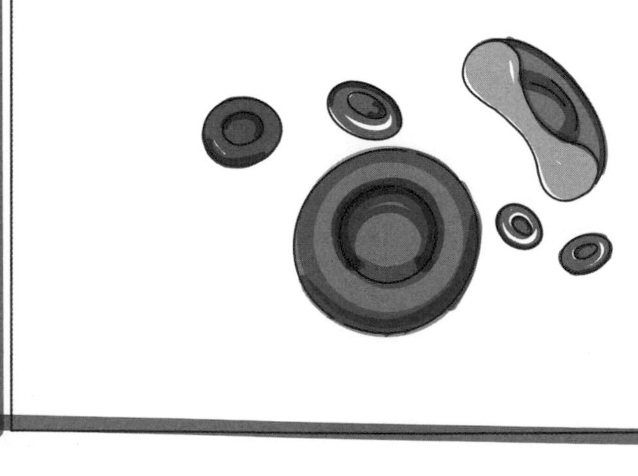

CIRCULATORY SYSTEM

The circulatory system consists of a muscular pump (the heart) and two sequential circulations: systemic and pulmonary.

Systemic and pulmonary circulations

The **systemic circulation** serves gas and nutrient transport to all organs except the lungs. The **pulmonary circulation** serves the process of gas exchange of the blood in the lungs. Blood leaves the ventricles of the heart via arteries, it passes through arterioles, capillaries, venules, and veins to return to the atria of the heart to be pumped out again.

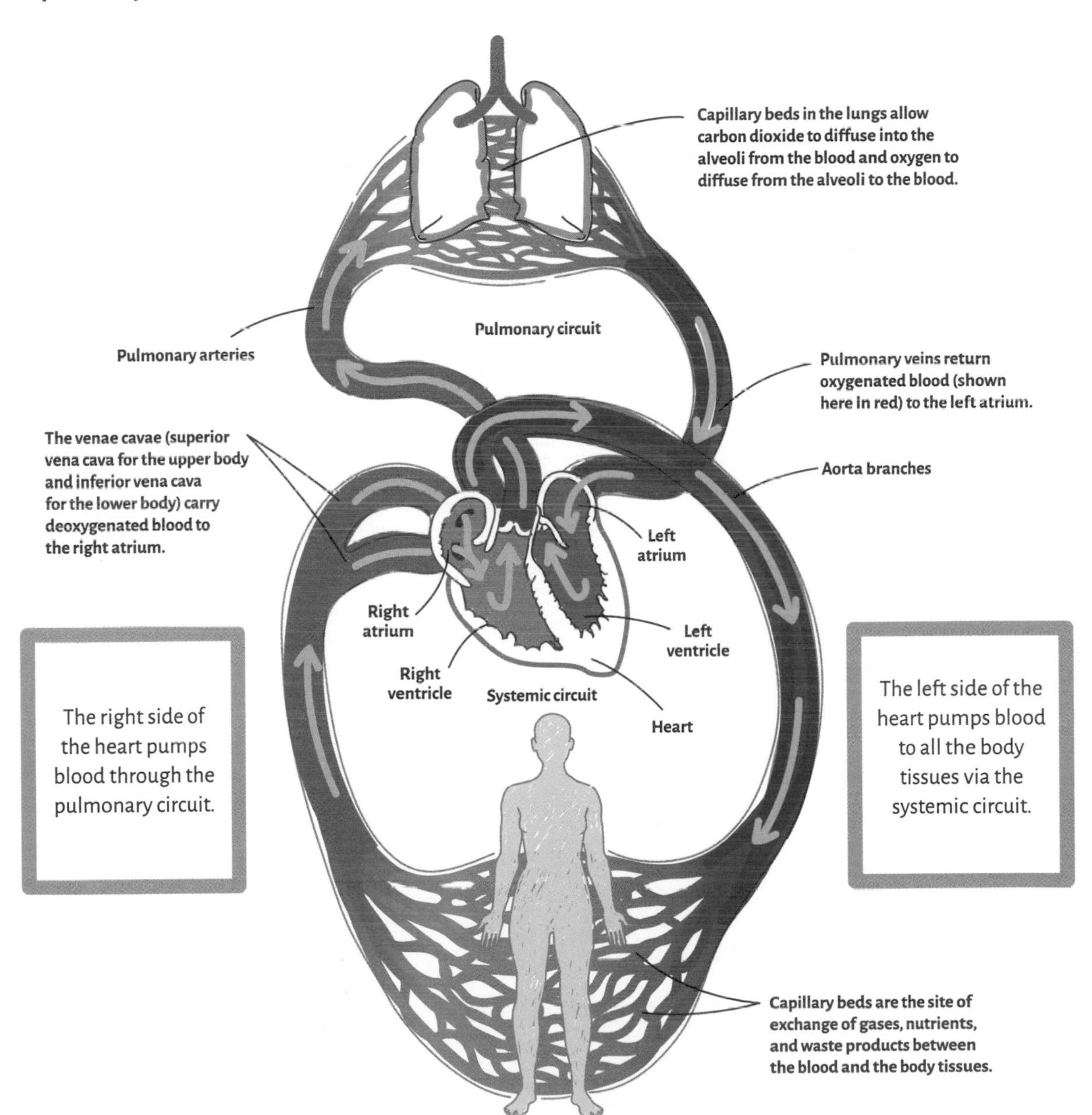
Capillary beds in the lungs allow carbon dioxide to diffuse into the alveoli from the blood and oxygen to diffuse from the alveoli to the blood.
Pulmonary circuit
Pulmonary arteries
Pulmonary veins return oxygenated blood (shown here in red) to the left atrium.
The venae cavae (superior vena cava for the upper body and inferior vena cava for the lower body) carry deoxygenated blood to the right atrium.
Aorta branches
Left atrium
Right atrium
Left ventricle
Right ventricle
Systemic circuit
Heart
The right side of the heart pumps blood through the pulmonary circuit.
The left side of the heart pumps blood to all the body tissues via the systemic circuit.
Capillary beds are the site of exchange of gases, nutrients, and waste products between the blood and the body tissues.

VESSELS IN THE CIRCULATORY SYSTEM

The circulatory system has different vessel types for different functions. Each vessel type is adapted to its role. Arteries have thick walls with smooth muscle and elastic fibers in the middle layer of their walls (tunica media) to hold high pressure. Very small arterioles control blood pressure and regulate flow by the contraction of smooth muscle in their walls. All vessels have three layers and are lined by a layer of squamous cells called endothelium.

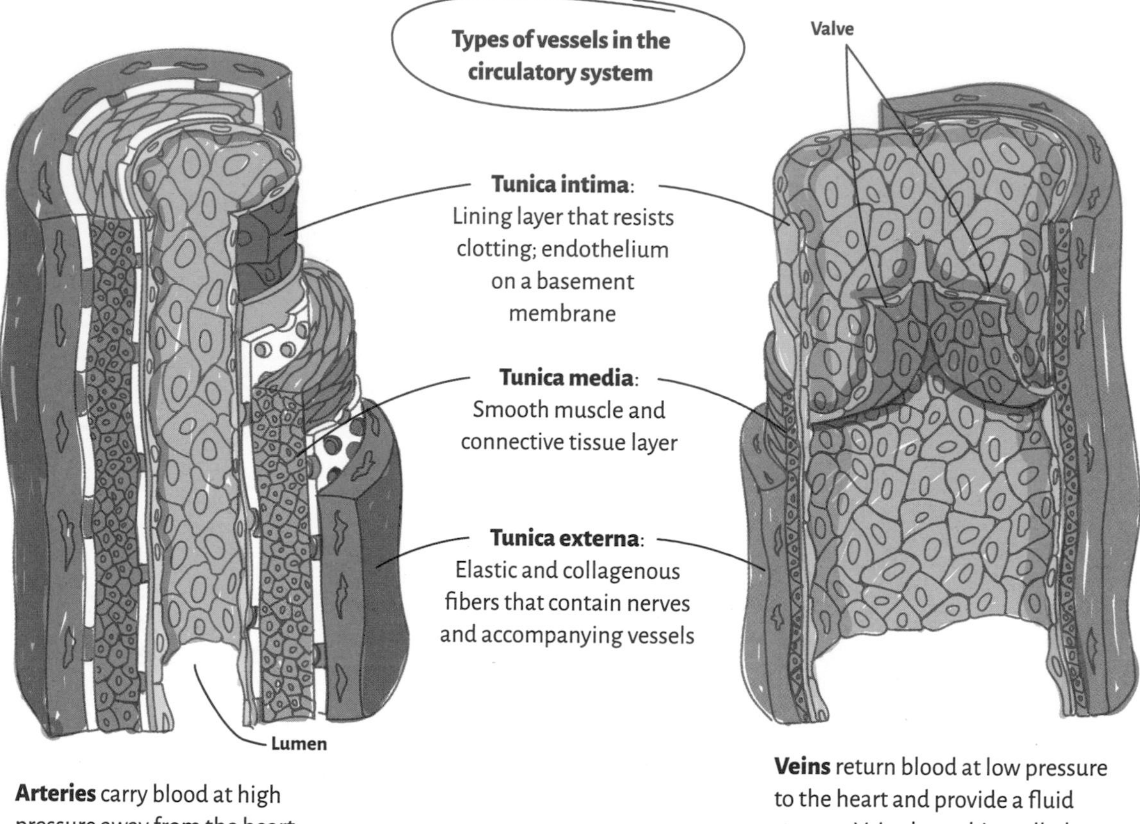
Types of vessels in the circulatory system
Valve
Tunica intima: Lining layer that resists clotting; endothelium on a basement membrane
Tunica media: Smooth muscle and connective tissue layer
Tunica externa: Elastic and collagenous fibers that contain nerves and accompanying vessels
Lumen

Arteries carry blood at high pressure away from the heart.

Veins return blood at low pressure to the heart and provide a fluid reserve. Veins have thin walls, but a large internal diameter to store blood. Veins also contain valves to prevent backflow of blood.

Lumen

Basement membrane

Endothelium

Capillaries provide a bed of large surface area for exchange and have very thin walls so that gases, nutrients, and waste products can easily cross the vessel wall.

Cardiac cycle

The **cardiac cycle** is the sequence of events that occurs with each heartbeat. The cycle starts with the return of systemic venous blood to the right atrium and pulmonary venous blood to the left atrium. The atria then contract (**atrial systole**), pushing blood through the open atrioventricular valves into the respective ventricles.

Ventricular contraction (ventricular systole) begins after atrial contraction has ended. As pressure builds in the ventricles, the atrioventricular valves close (first heart sound), the semilunar valves open (aortic valve from the left ventricle, pulmonary valve from the right ventricle) allowing blood flow from the ventricles to the aorta and pulmonary trunk.

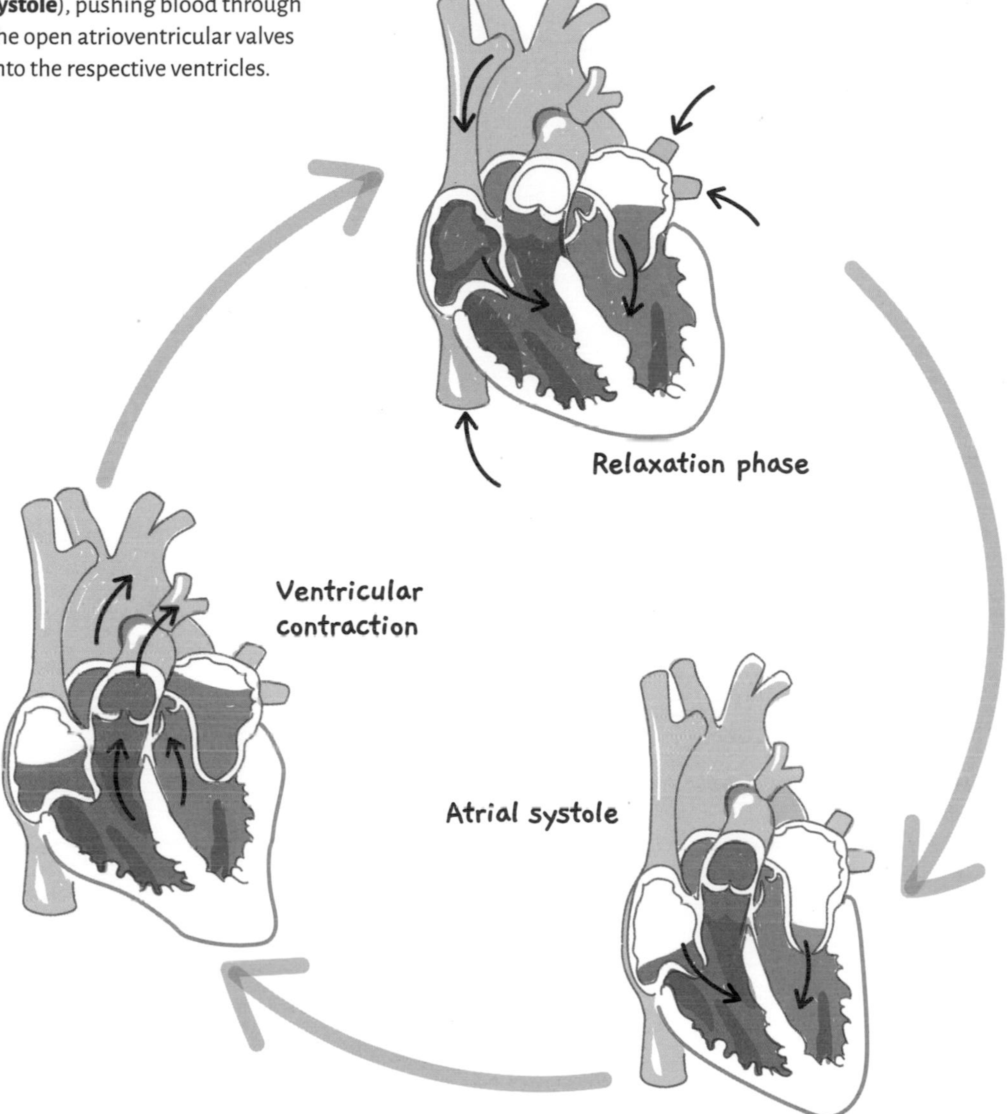

Relaxation phase
Ventricular contraction
Atrial systole

When ventricular contraction ceases, pressure in the ventricles drops below that in the outflow arteries, and the aortic and pulmonary valves close (second heart sound).

During the later part of the ventricular **relaxation phase**, the pressure in the ventricles drops further to less than the atria, which results in the opening of the atrioventricular valves. This allows venous blood to flow from the pulmonary and systemic veins into the atria and on into the ventricles (see arrows).

At the end of the ventricular relaxation phase, the atria will contract (atrial systole), forcing the last 30% of venous blood to return into the ventricles.

STRUCTURE OF THE HEART AND CARDIAC MUSCLE

The heart beats continuously throughout life, so mechanisms for synchronizing cardiac muscle activity and the adjustment of the heart rate and muscle strength for changing circumstances are needed.

The heart

The heart is a four-chambered pump (two atria, two ventricles).

Atria receive venous blood and pass it to the ventricles.

Ventricles receive blood from the atria and pump it to the arteries.

The activity of the heart is under control of the autonomic nervous system and circulating hormones called **catecholamines**.

Cardiac muscle is intensely active and needs a steady supply of oxygen. **Coronary arteries** (left and right) are the initial branches of the aorta, immediately after it arises from the left ventricle. Cardiac veins mainly drain into the right atrium.

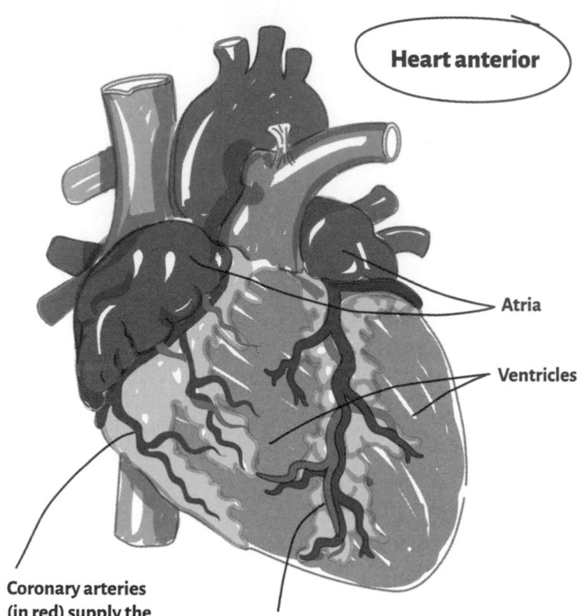
Heart anterior
Atria
Ventricles
Coronary arteries (in red) supply the heart muscle with oxygenated blood.
Cardiac veins (in blue) drain deoxygenated blood from the cardiac muscle to the right side of the heart.

CARDIAC VALVES

Cardiac valves are key to the control of blood flow. There are four valves in the heart. **Atrioventricular valves** prevent regurgitation of blood from the ventricles to the atria when the ventricles contract. **Semilunar valves** prevent regurgitation of blood from the aorta and pulmonary trunk when the ventricles relax.

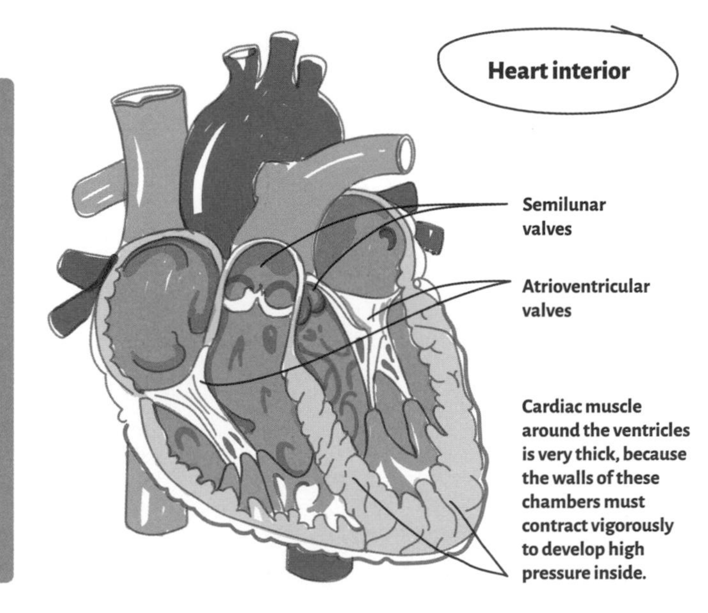
Heart interior
Semilunar valves
Atrioventricular valves
Cardiac muscle around the ventricles is very thick, because the walls of these chambers must contract vigorously to develop high pressure inside.

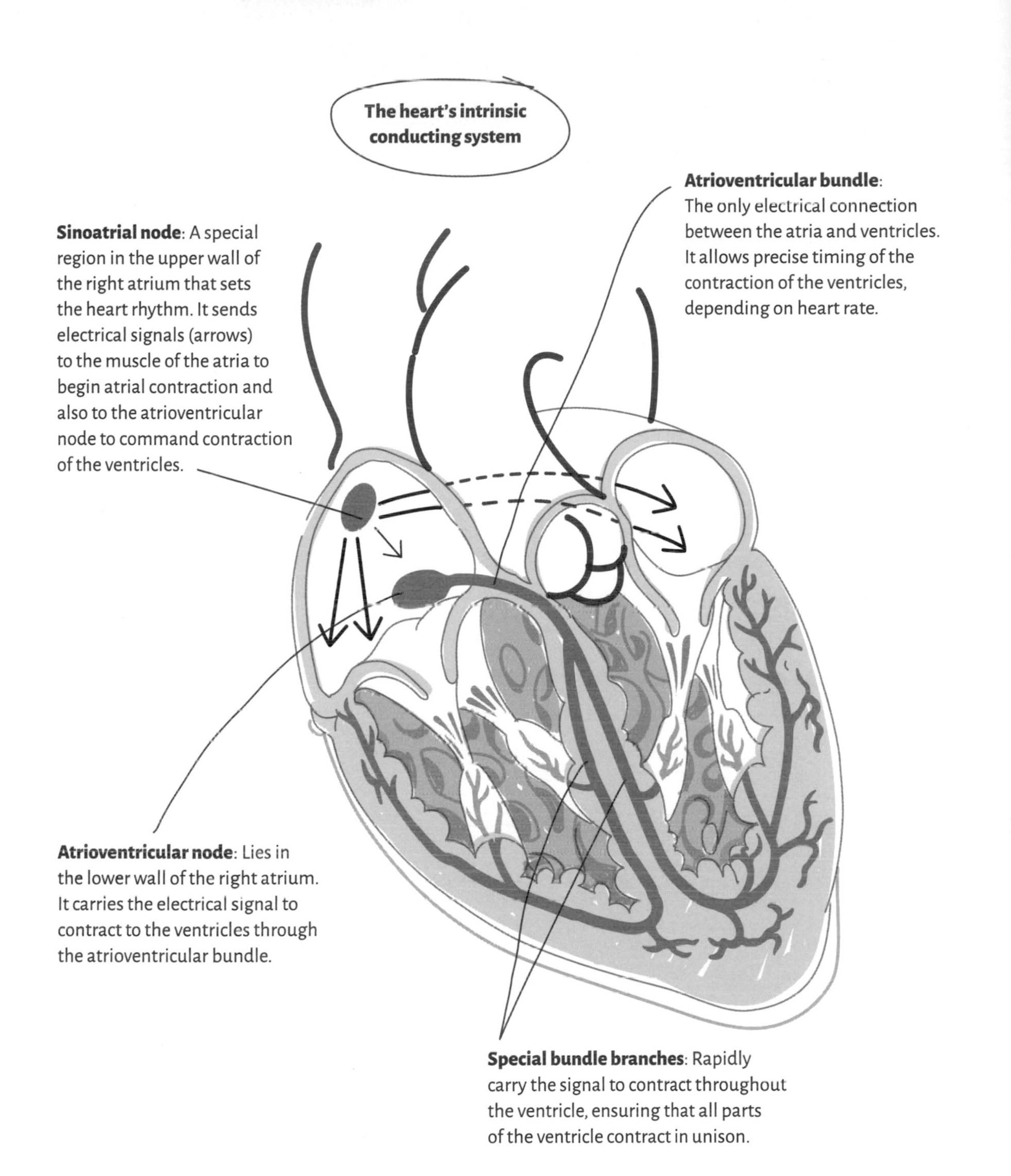
The heart's intrinsic conducting system
Atrioventricular bundle: The only electrical connection between the atria and ventricles. It allows precise timing of the contraction of the ventricles, depending on heart rate.
Sinoatrial node: A special region in the upper wall of the right atrium that sets the heart rhythm. It sends electrical signals (arrows) to the muscle of the atria to begin atrial contraction and also to the atrioventricular node to command contraction of the ventricles.
Atrioventricular node: Lies in the lower wall of the right atrium. It carries the electrical signal to contract to the ventricles through the atrioventricular bundle.
Special bundle branches: Rapidly carry the signal to contract throughout the ventricle, ensuring that all parts of the ventricle contract in unison.

Cardiac muscle

Cardiac muscle is electrically connected. It is striated like skeletal muscle, but cardiac muscle is involuntary. Cardiac muscle cells are also unusual in that intercalated disks electrically connect them. This connection ensures that all the cardiac muscles of the ventricle are activated and contract simultaneously for a smooth pumping action.

ARTERIES AND VEINS

All arteries carry blood away from the heart and have thick muscular walls. Veins carry blood toward the heart or between capillary beds, i.e., the portal systems of the liver and pituitary.

Pulmonary trunk

The **pulmonary trunk** is the outflow from the right ventricle to the alveoli of the lungs. It branches into left and right pulmonary arteries to each lung (see page 113).

Pulmonary veins carry oxygenated blood from the lungs to the left atrium. There are two on each side (superior and inferior).

Superior and **inferior venae cavae** drain venous blood from almost the entire body (except the lungs) into the right atrium. The superior vena cava is formed by the junction of the left and right brachiocephalic veins.

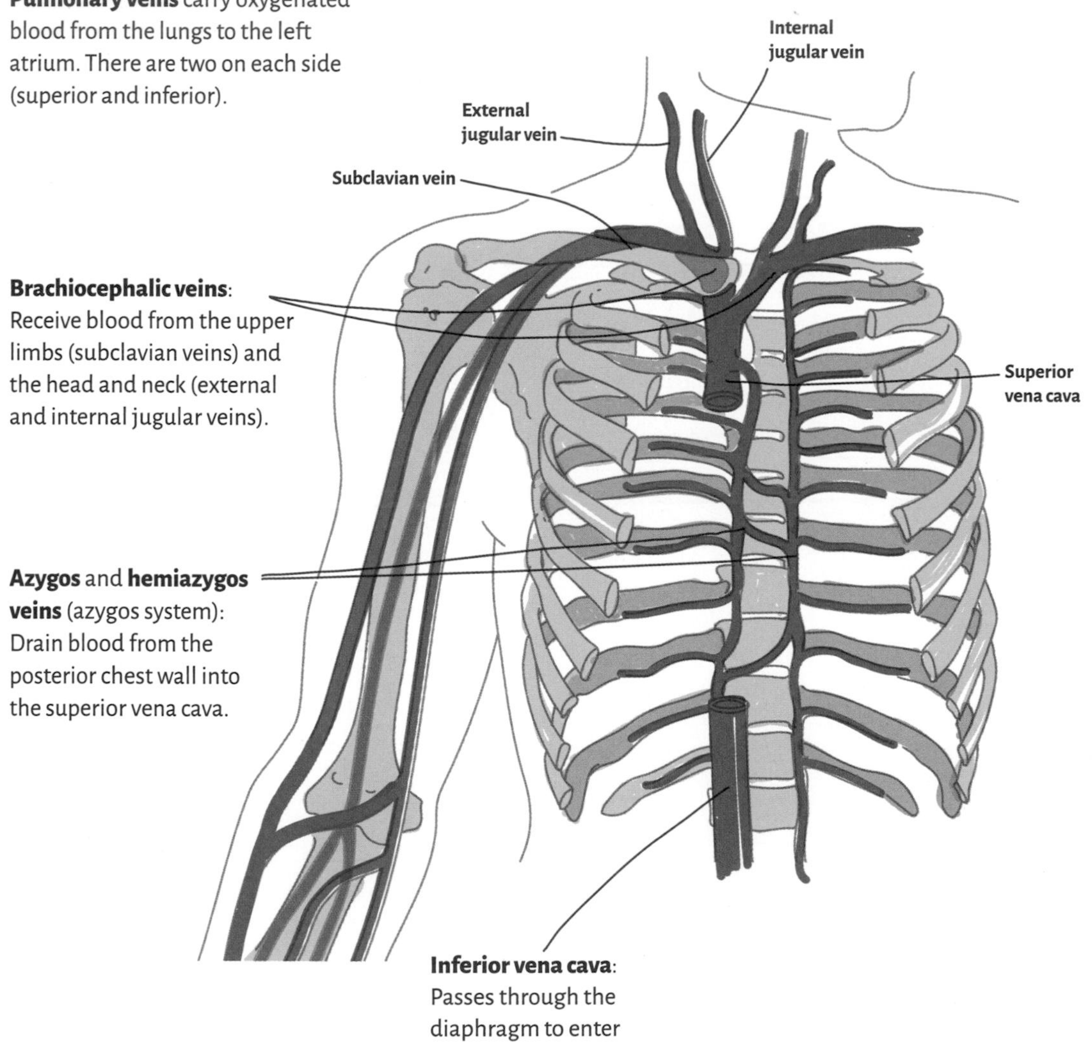
Internal jugular vein
External jugular vein
Subclavian vein
Superior vena cava

Brachiocephalic veins: Receive blood from the upper limbs (subclavian veins) and the head and neck (external and internal jugular veins).

Azygos and **hemiazygos veins** (azygos system): Drain blood from the posterior chest wall into the superior vena cava.

Inferior vena cava: Passes through the diaphragm to enter the right atrium.

Arteries and veins of the chest

The largest vessels of the body are located in the mediastinum of the chest, a midline region between the two lungs and their pleural sacs.

The aorta is the largest artery and carries all oxygenated blood from the left ventricle to the body. It has an ascending portion (with **coronary artery** branches), an arch (with **brachiocephalic trunk**, **left common carotid** and **left subclavian artery** branches), and a descending portion (with branches to the chest wall, large airway walls, esophagus, and spinal cord). The aorta passes through the aortic opening (aortic hiatus) in the diaphragm.

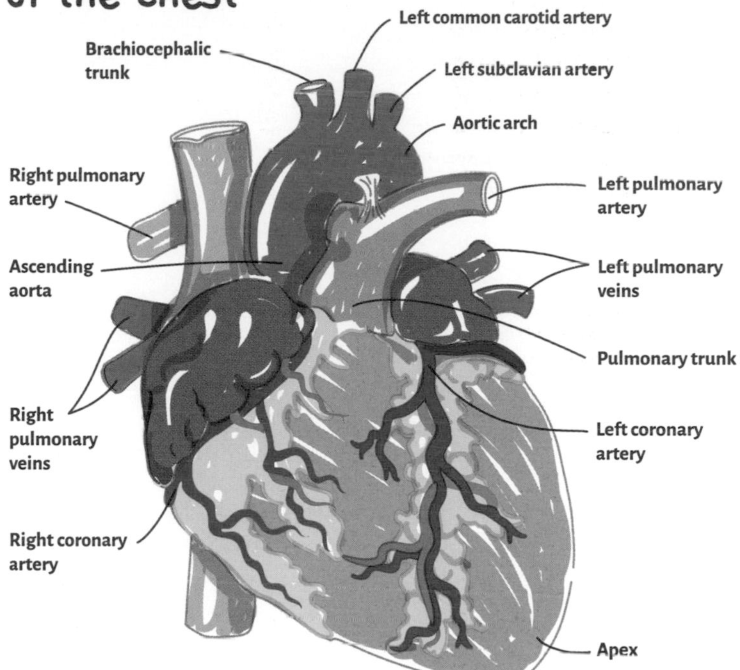
Left common carotid artery
Brachiocephalic trunk
Left subclavian artery
Aortic arch
Right pulmonary artery
Left pulmonary artery
Ascending aorta
Left pulmonary veins
Pulmonary trunk
Right pulmonary veins
Left coronary artery
Right coronary artery
Apex

Arteries and veins of the abdomen

The largest artery of the abdomen is the abdominal aorta, which lies on the lumbar vertebrae. It is accompanied by the inferior vena cava. The abdominal aorta supplies viscera before dividing into the **common iliac arteries**. There are three aortic branches to the alimentary canal.

The aorta also supplies the other viscera and the posterior abdominal wall. The **suprarenal arteries** on each side supply the suprarenal glands. The **renal artery** supplies the kidneys and the lower parts of the suprarenal gland. The **lumbar arteries** supply the posterior abdominal wall. The **gonadal arteries** supply the gonads (ovaries or testes). The **celiac trunk** supplies the stomach, upper duodenum, liver, gallbladder, and the upper part of the pancreas. The **superior mesenteric artery** supplies the lower duodenum, small intestine, lower part of the pancreas, and large bowel to the left colic flexure. The **inferior mesenteric artery** supplies the descending colon, sigmoid colon, and the rectum.

Branches of the abdominal aorta

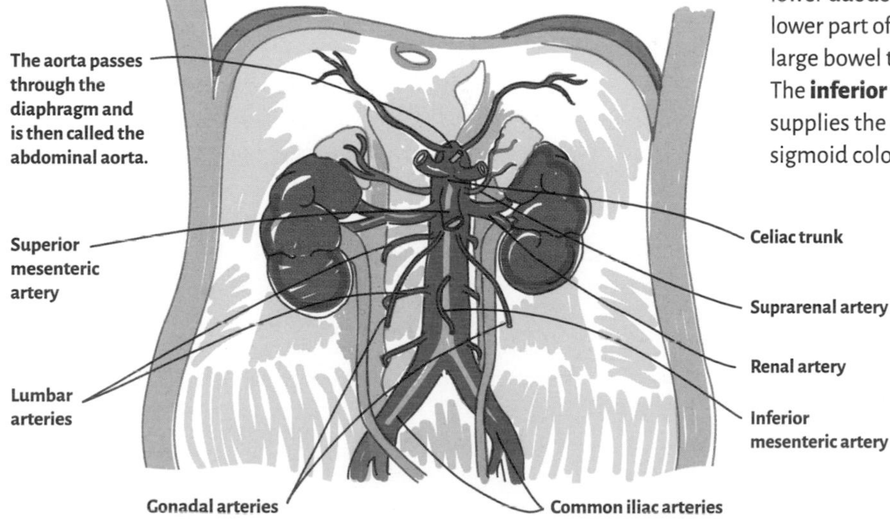
The aorta passes through the diaphragm and is then called the abdominal aorta.
Superior mesenteric artery
Celiac trunk
Suprarenal artery
Renal artery
Lumbar arteries
Inferior mesenteric artery
Gonadal arteries
Common iliac arteries

Veins from the body wall and lower limb drain into the inferior vena cava (IVC). The IVC is formed from the junction of the two **common iliac veins**. The IVC also receives **hepatic**, **renal**, **lumbar**, and **right suprarenal veins**.

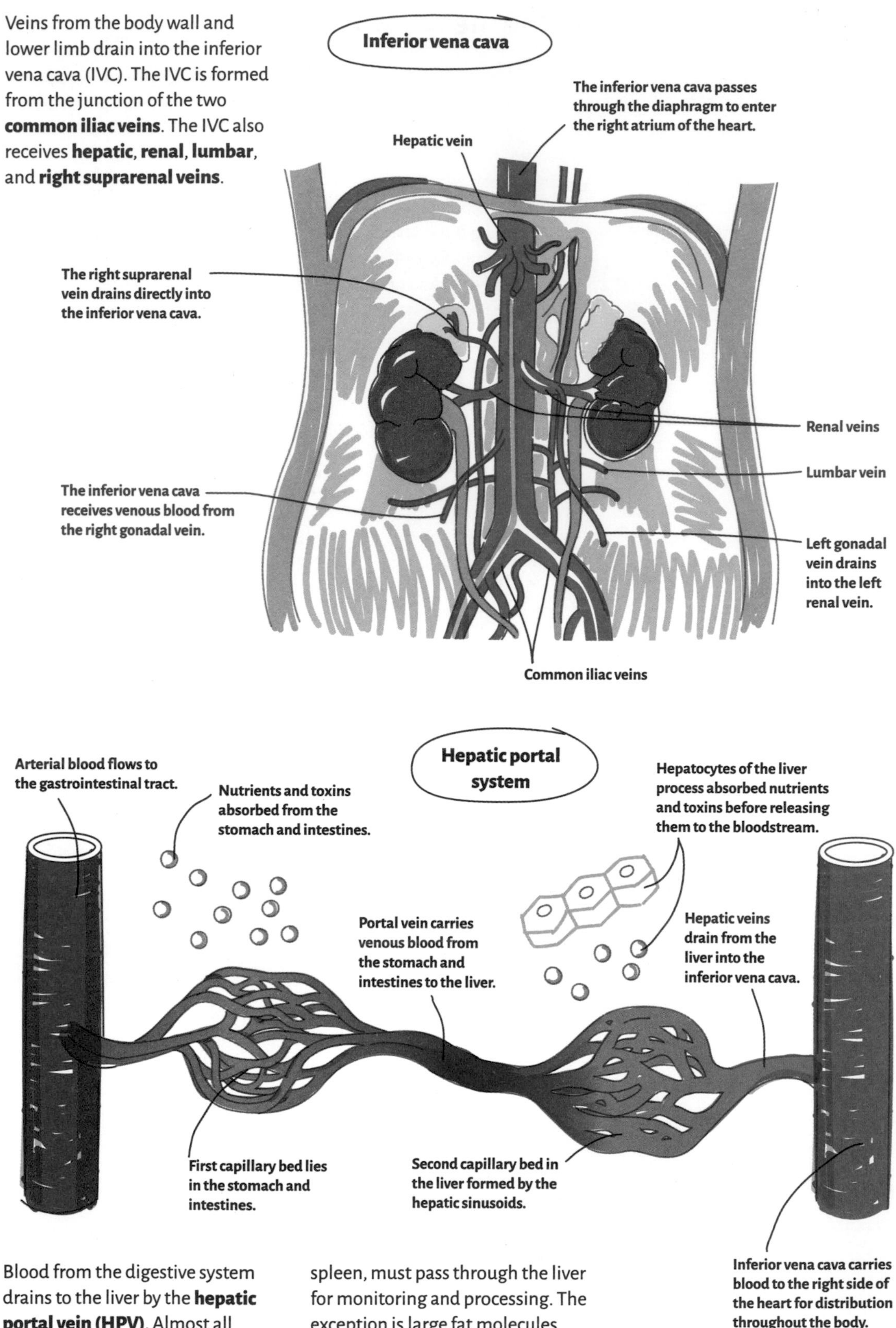

Inferior vena cava
The inferior vena cava passes through the diaphragm to enter the right atrium of the heart.
Hepatic vein
The right suprarenal vein drains directly into the inferior vena cava.
Renal veins
Lumbar vein
The inferior vena cava receives venous blood from the right gonadal vein.
Left gonadal vein drains into the left renal vein.
Common iliac veins
Hepatic portal system
Arterial blood flows to the gastrointestinal tract.
Nutrients and toxins absorbed from the stomach and intestines.
Hepatocytes of the liver process absorbed nutrients and toxins before releasing them to the bloodstream.
Portal vein carries venous blood from the stomach and intestines to the liver.
Hepatic veins drain from the liver into the inferior vena cava.
First capillary bed lies in the stomach and intestines.
Second capillary bed in the liver formed by the hepatic sinusoids.
Inferior vena cava carries blood to the right side of the heart for distribution throughout the body.

Blood from the digestive system drains to the liver by the **hepatic portal vein (HPV)**. Almost all nutrients and all the toxins from the entire gastrointestinal tract, as well as from the pancreas and spleen, must pass through the liver for monitoring and processing. The exception is large fat molecules that leave the digestive system by lymphatic channels.

Arteries and veins of the head and neck

The brain has a very high rate of oxidative metabolism, even at rest, so arteries of the head and neck carry 0.75 liters (0.16 gallon) of blood to the brain each minute (about 15% of the cardiac output). Other arteries supply the face, oral cavity, tongue, and pharynx.

The aorta is the large artery emerging from the left ventricle.

The aortic arch gives off the **brachiocephalic trunk**, the **left common carotid**, and the **left subclavian arteries**. The brachiocephalic trunk divides into the **right common carotid** and the **right subclavian arteries**.

The brachiocephalic trunk supplies the right side of the face and brain and the right upper limb.

The common carotid arteries on each side branch into internal and external carotid arteries. **Internal carotid arteries** enter the skull to supply the brain, pituitary gland, and eye. Branches include the ophthalmic artery, and the anterior and middle cerebral arteries. The **external carotid artery** supplies the tongue and face by lingual and facial arteries.

Vertebral arteries branch from the subclavian arteries on each side. They ascend through the cervical vertebral column to enter the skull via the foramen magnum, where they branch to supply the brain stem and the occipital lobe.

Branches of the aortic arch

Internal carotid artery
External carotid artery
Vertebral artery
Right common carotid artery
Left common carotid artery
Right subclavian artery supplies the right upper limb.
Left subclavian artery supplies the left upper limb.
Brachiocephalic trunk
Aortic arch

The **external jugular vein** receives blood from the scalp and the base of the neck, the muscles of the face, the oral cavity, and the pharynx. The external jugular drains into the **subclavian veins**.

The **internal jugular veins** drain blood from the brain, pituitary, and eye. They join the subclavian veins to form the **brachiocephalic veins** in the chest.

Jugular veins

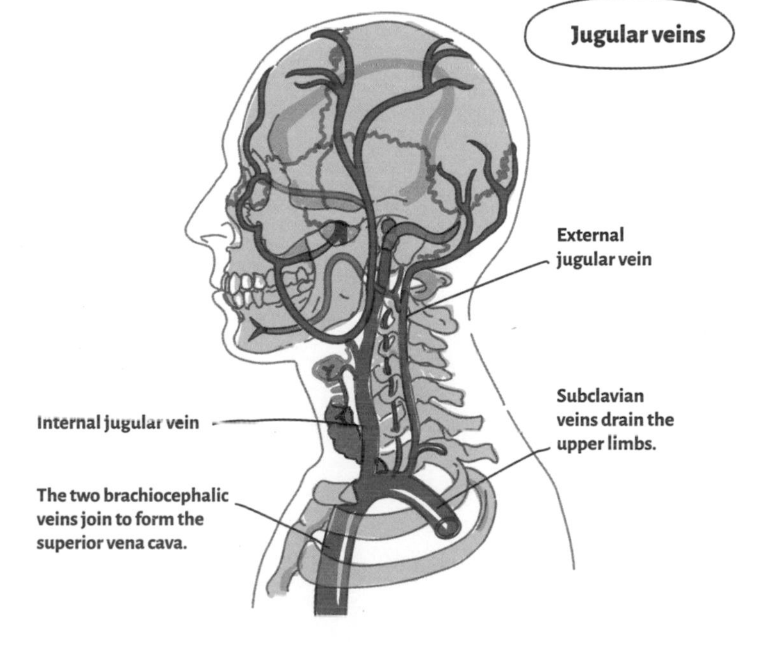
External jugular vein
Subclavian veins drain the upper limbs.
Internal jugular vein
The two brachiocephalic veins join to form the superior vena cava.

Arteries and veins of the upper limb

Arterial supply to the upper limb comes from branches of the aortic arch. The subclavian arteries pass over the first ribs, where they can be compressed in an emergency. Venous drainage of the upper limb reaches the superior vena cava.

The subclavian artery becomes the **axillary artery** when it crosses the outer border of the first rib. The axillary artery becomes the **brachial artery** when it passes the lower margin of the teres major. The brachial artery can be felt when it is pressed against the distal humerus, just medial to the biceps brachii tendon at the front of the elbow.

The brachial artery divides at the elbow into **radial** and **ulnar** arteries.

The radial artery can be felt at the thumb side of the wrist, just proximal to the first metacarpal base.

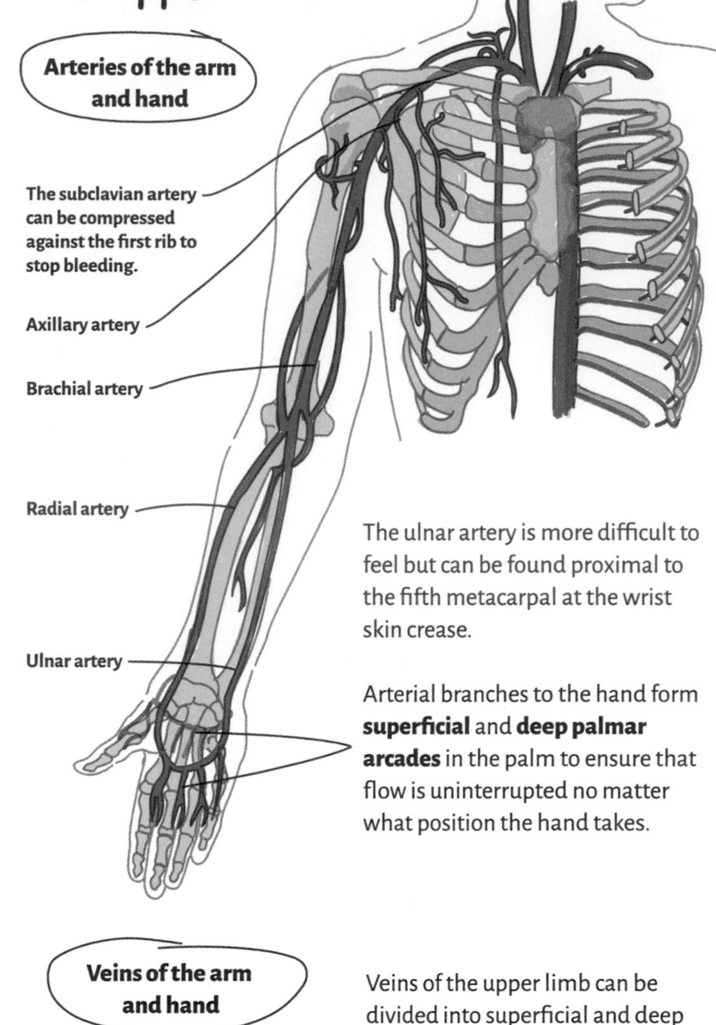
Arteries of the arm and hand
The subclavian artery can be compressed against the first rib to stop bleeding.
Axillary artery
Brachial artery
Radial artery
Ulnar artery

The ulnar artery is more difficult to feel but can be found proximal to the fifth metacarpal at the wrist skin crease.

Arterial branches to the hand form **superficial** and **deep palmar arcades** in the palm to ensure that flow is uninterrupted no matter what position the hand takes.

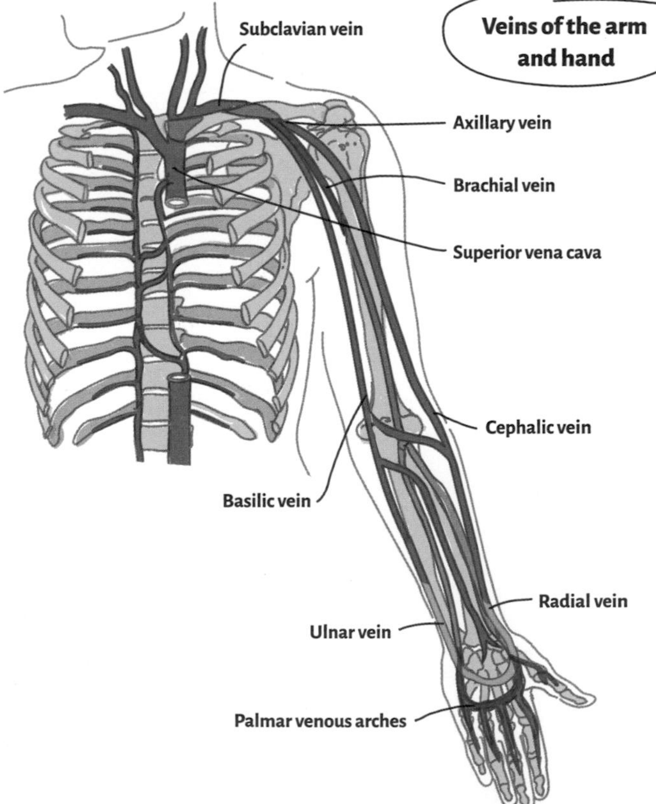
Veins of the arm and hand
Subclavian vein
Axillary vein
Brachial vein
Superior vena cava
Cephalic vein
Basilic vein
Radial vein
Ulnar vein
Palmar venous arches

Veins of the upper limb can be divided into superficial and deep groups, where superficial drains toward deep veins.

Deep veins start at the **palmar venous arches**, which drain into the **radial** and **ulnar veins** that in turn join to form the **brachial vein**.

The dorsal venous arch of the hand drains into the superficial **cephalic** and **basilic veins**.

The brachial vein becomes the **axillary vein**, which will drain into the subclavian vein and ultimately into the superior vena cava.

The cephalic and basilic veins drain into the deep axillary and brachial veins, respectively.

Arteries and veins of the lower limb

Arterial supply for the lower limb is mainly from the branches of the external iliac artery. Venous drainage of the lower limb is by superficial and deep veins, with robust valves to direct flow against gravity.

Arteries of the lower limb: anterior view (left) and posterior view (right)

The aorta divides inside the abdomen to form two **common iliac arteries**.

Each common iliac artery then divides to give an **external** and **internal iliac artery**.

The external iliac artery supplies most blood to the lower limb. It becomes the femoral artery when the **inguinal ligament** at the groin crease is crossed.

The **femoral artery** enters the thigh and runs downward in a region called the **femoral triangle**. It gives off a deep branch to the back of the thigh called the **deep artery of the thigh**. Both the femoral and deep thigh arteries supply the nearby muscles.

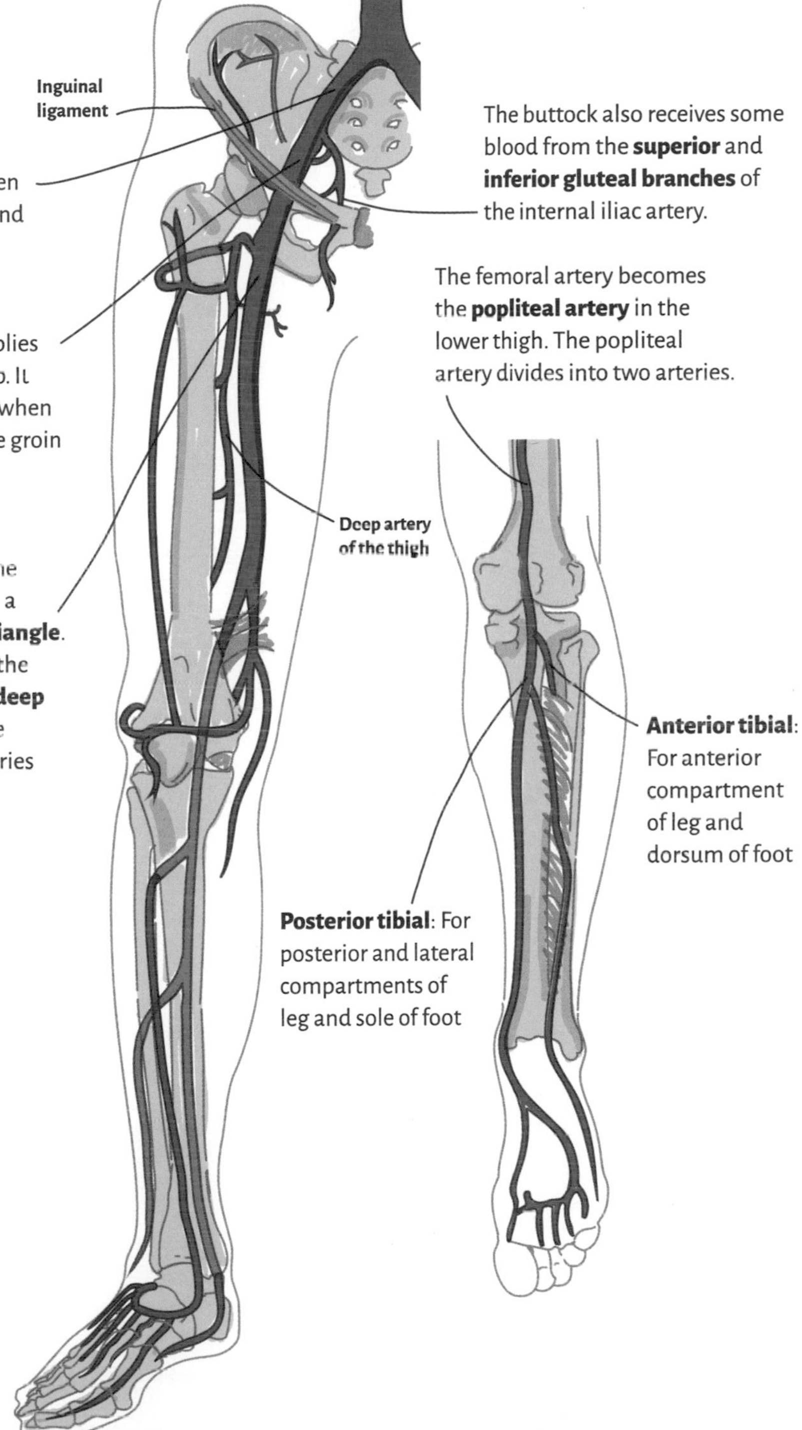

Inguinal ligament
The buttock also receives some blood from the **superior** and **inferior gluteal branches** of the internal iliac artery.
The femoral artery becomes the **popliteal artery** in the lower thigh. The popliteal artery divides into two arteries.
Deep artery of the thigh
Anterior tibial: For anterior compartment of leg and dorsum of foot
Posterior tibial: For posterior and lateral compartments of leg and sole of foot

VENOUS DRAINAGE FROM THE FEET AND LEG

We spend most of our day sitting, standing, or walking, so blood from the feet must be lifted as much as 1.5 meters (5 feet) to reach the heart. A system of muscle pumps and **valves** is necessary to return that blood.

There are superficial and deep veins in the lower limb. Calf muscles compress deep veins allowing blood to be pumped uphill past successive valves. Superficial veins (**great** and **small saphenous vein**) drain to the deep veins by communicating veins, and valves prevent regurgitation.

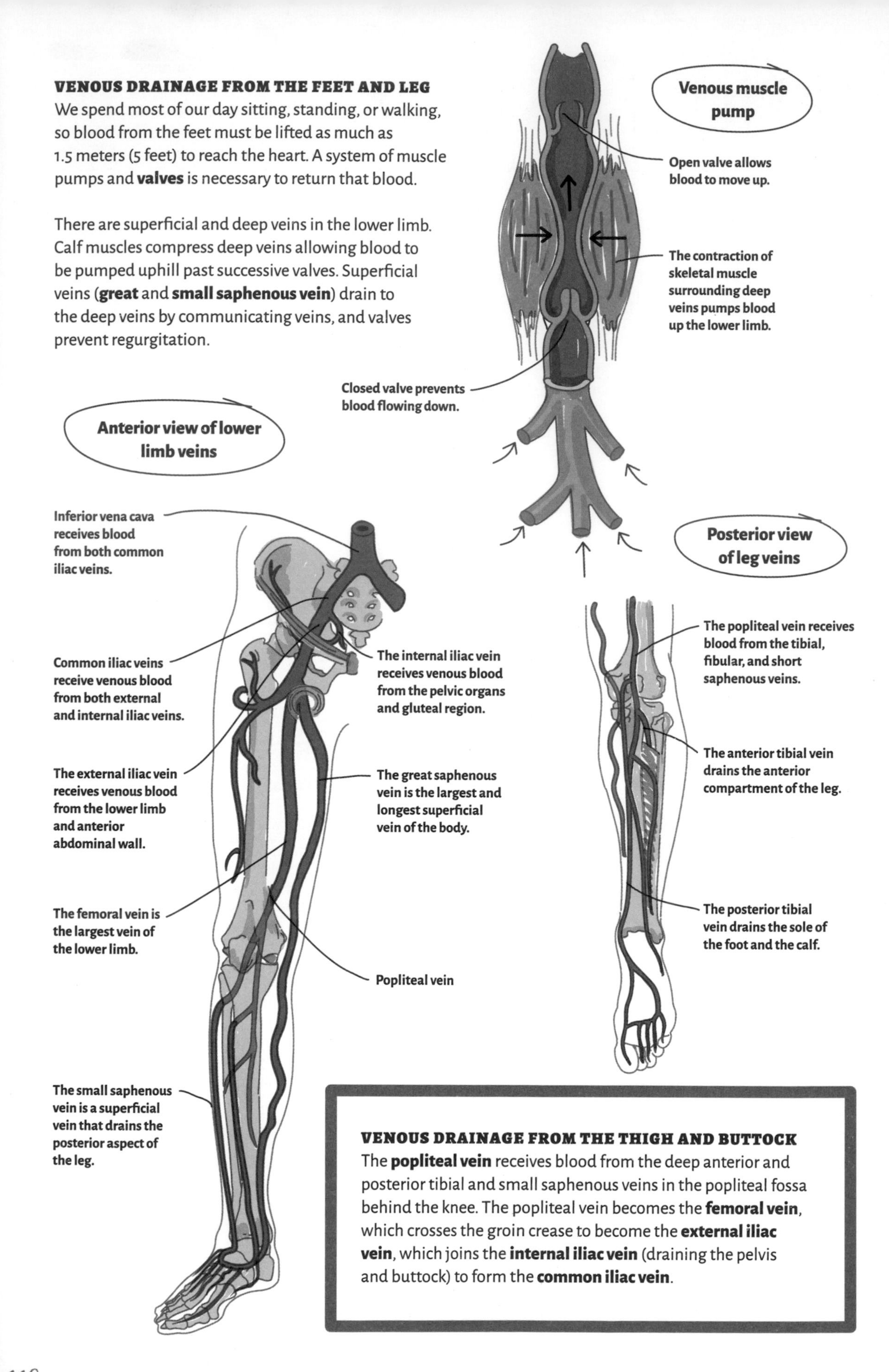

Venous muscle pump
Open valve allows blood to move up.
The contraction of skeletal muscle surrounding deep veins pumps blood up the lower limb.
Closed valve prevents blood flowing down.
Anterior view of lower limb veins
Inferior vena cava receives blood from both common iliac veins.
Common iliac veins receive venous blood from both external and internal iliac veins.
The internal iliac vein receives venous blood from the pelvic organs and gluteal region.
The external iliac vein receives venous blood from the lower limb and anterior abdominal wall.
The great saphenous vein is the largest and longest superficial vein of the body.
The femoral vein is the largest vein of the lower limb.
Popliteal vein
The small saphenous vein is a superficial vein that drains the posterior aspect of the leg.
Posterior view of leg veins
The popliteal vein receives blood from the tibial, fibular, and short saphenous veins.
The anterior tibial vein drains the anterior compartment of the leg.
The posterior tibial vein drains the sole of the foot and the calf.

VENOUS DRAINAGE FROM THE THIGH AND BUTTOCK

The **popliteal vein** receives blood from the deep anterior and posterior tibial and small saphenous veins in the popliteal fossa behind the knee. The popliteal vein becomes the **femoral vein**, which crosses the groin crease to become the **external iliac vein**, which joins the **internal iliac vein** (draining the pelvis and buttock) to form the **common iliac vein**.

CAPILLARIES

Tiny vessels called **capillaries** allow the exchange of gases, nutrients, cells, proteins, and waste products between blood and tissues. The common feature of all capillaries is a thin wall composed of endothelium on a basement membrane.

Where do capillaries sit in the circulation?

There are about 20 billion capillaries in the body, and they lie between the arterioles and venules in the circulation.

Flow into the capillaries is regulated by contraction or relaxation of **precapillary sphincters**, which are found at the junction of a metarteriole and a capillary.

They are formed into **capillary beds**, which are groups of 10 to 100 capillaries that receive blood from a single **metarteriole**.

Structure of a typical capillary bed

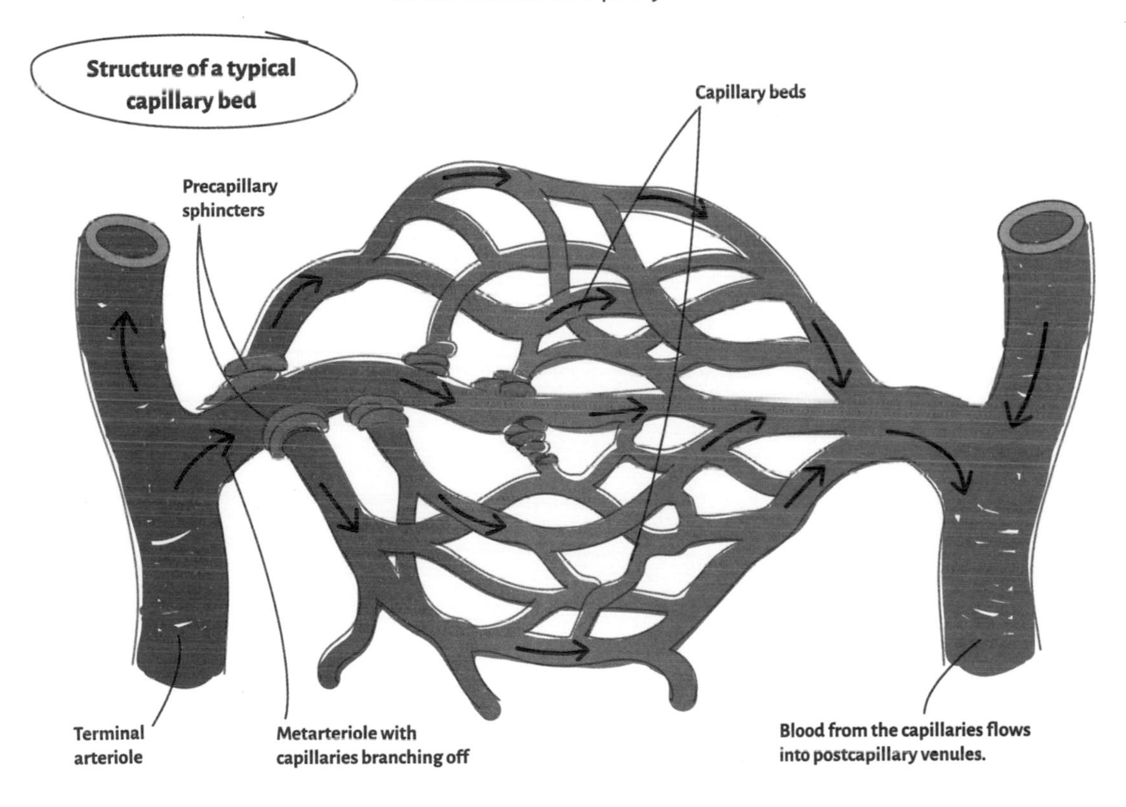
Capillary beds
Precapillary sphincters
Terminal arteriole
Metarteriole with capillaries branching off
Blood from the capillaries flows into postcapillary venules.

Functions of capillaries

The primary function of capillaries is as exchange vessels. They are suited for this because capillaries have walls only one endothelial cell thick, with a thin underlying basement membrane. They also branch extensively, so capillaries present a very large surface area for exchange with the surrounding tissues.

In most tissues at rest (e.g., inactive muscle), only a small proportion of capillaries have active flow, but additional capillaries can open as needed (e.g., muscles during exercise).

Three types of capillaries

Capillaries can be continuous, fenestrated, or sinusoidal.

Continuous capillaries are endothelial cells that form a continuous tube. They are found in the brain and spinal cord tissue, lungs, skin, and skeletal muscle. They do not allow proteins or cells to cross the wall.

Types of capillaries

Red blood cell in the lumen of the capillary

Endothelial cell

Tight intercellular cleft does not allow proteins or cells to cross.

Red blood cell in the lumen of the capillary

Nucleus of endothelial cell

Fenestrations

Fenestrated capillaries are endothelial cells that have tiny pores in their plasma membranes. They are found in the kidney, the villi of the small intestine, brain ventricles, and in many endocrine glands. Fenestrations (literally tiny windows) allow larger molecules and sometimes cells to cross from the blood to surrounding tissue.

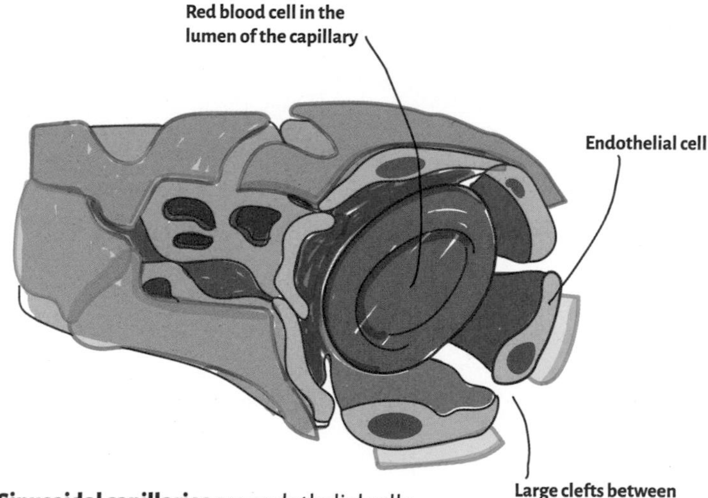
Red blood cell in the lumen of the capillary
Endothelial cell
Large clefts between endothelial cells

Sinusoidal capillaries are endothelial cells that have very wide openings (clefts) that allow large proteins and cells to cross easily. They are found in red bone marrow, the spleen, and the pituitary, parathyroid, and adrenal glands. In bone marrow, the gaps in the sinusoid wall allow newly formed blood cells to cross into the bloodstream.

Portal systems

Usually blood passes from arterioles to capillaries to venules. In some body systems, blood can pass from one capillary bed to another through a portal venous channel.

Hepatic portal system: Drains blood from the capillary bed of the digestive system. It allows the liver to monitor and process nutrients and toxins from the digestive system (see page 114).

Hypophyseal portal system: Drains blood from the hypothalamus to the anterior pituitary gland. It allows passage of releasing hormones from the brain to the pituitary gland.

BLOOD FUNCTION AND COMPOSITION

Blood is a connective tissue and consists of blood cells in a fluid matrix. The blood cells make up as much as half the volume of blood, but the fluid contains important proteins.

Functions of blood

Blood has three functions.

* **Transportation**: Blood transports oxygen and carbon dioxide, nutrients from the digestive system, hormones from the endocrine glands, toxins, heat, and waste products.
* **Regulation**: Blood regulates the pH of body fluids, adjusts body temperature, and influences the osmotic pressure of cells.
* **Protection**: Blood clots to prevent fluid loss. It carries white blood cells, immune proteins such as antibodies, an assortment of blood solutes, complement, and interferons to protect against disease.

Components of blood

Whole blood consists of plasma and formed elements. These components can be separated when the blood is prevented from clotting and is centrifuged.

Plasma is 55% by volume and has a watery extracellular matrix with dissolved substances. Plasma is 91.5% water and 8.5% dissolved substances (solutes). Most solutes are proteins. Other solutes are electrolytes, nutrients, enzymes, hormones, vitamins, gases, creatine, creatinine, urea, uric acid, ammonia, and bilirubin.

The **buffy coat** of centrifuged blood is made up of platelets and white blood cells.

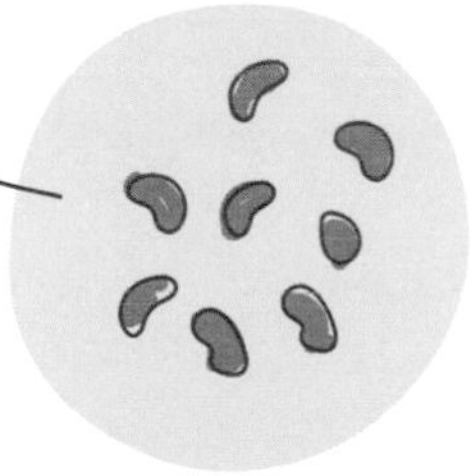

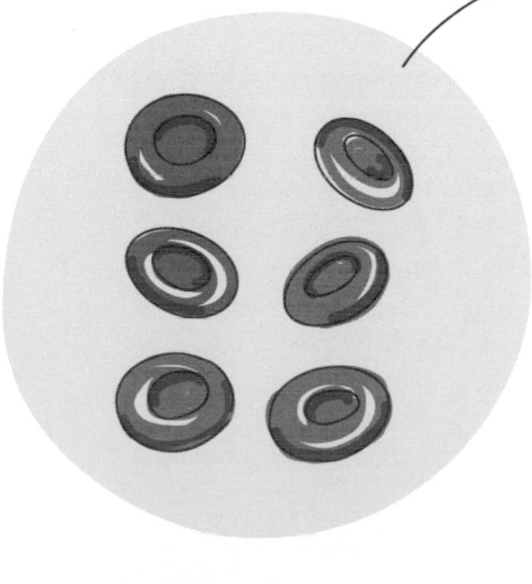

White blood cells (leukocytes) and platelets (thrombocytes) make up less than 1% of blood volume.

Formed elements are 45% by volume with cells and cell fragments. Most formed elements (cells) are red blood cells (erythrocytes), about 42% of blood volume in adult females and 47% in adult males.

Blood proteins

Three blood proteins are found in blood plasma.

- ★**Albumins**: Made by the liver and make up 54% of plasma proteins. They provide colloid osmotic pressure for the blood to ensure that water returns from tissue spaces at the end of capillaries. They also act as a buffer for pH, and transport steroid hormones and fatty acids.
- ★**Globulins**: Include immunoglobulins made by plasma cells and make up 38% of plasma proteins. They attack viruses and bacteria. Alpha and beta globulins transport iron and fats.
- ★**Fibrinogens**: Made by the liver and comprise 7% of plasma proteins. They are essential for blood clotting.

Blood clotting (hemostasis)

Hemostasis is the natural process of stopping blood loss from vessels. There are four components.

- ★**Vascular spasm**: Constricts arterioles supplying the area where vessels have been damaged.
- ★**Platelet activation**: Sticky platelet plug forms at the site of vessel damage.
- ★**Coagulation**: Blood coagulates in a clotting factor cascade, converting fibrinogen to fibrin in order to bind the blood cells into a solid gel.
- ★**Clot retraction**: Actin and myosin (contractile proteins) in platelets pull the wound edges together and squeeze the liquid of the coagulated blood out as serum.

Red blood cell

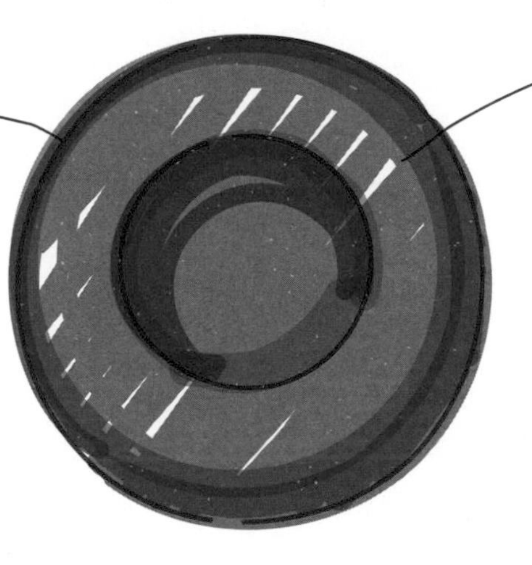

Antigens: Red blood cells have specific molecules (antigens) on their surface that are recognized by the immune system.

Red blood cells: Have no nuclei, but their cytoplasm is packed with hemoglobin to carry oxygen and (to a lesser extent) carbon dioxide.

Blood cells

Most blood cells are **red blood cells** (RBC; erythrocytes), with white blood cells and platelets making up less than 1% of blood volume. Blood cells are made in the red bone marrow of the axial skeleton and long bones.

Mature red blood cells are biconcave disks and do not have a nucleus or other organelles. They are 7 to 8 µm (1/4,000 inch) in diameter and packed with hemoglobin (33% by weight).

Hemoglobin is a globin protein with four peptide chains. Each peptide encloses a ringlike heme molecule and one iron ion. Oxygen reversibly binds with the iron ion. Carbon dioxide can also bind reversibly with hemoglobin.

BLOOD GROUPS

The **ABO system** is the most important of the blood groups. There are four possible types—type A, type B, type AB, and type O—based on the presence or absence of two antigens (A antigen, B antigen) on the RBC surface.

Individuals may have the **Rhesus factor** on their red blood cells (Rh+) or not (Rh-).

White blood cells

White blood cells (**leukocytes**) make up less than 1% of blood volume, but they play vital roles. There are about 4,800 to 11,000 white blood cells in every cubic millimeter of blood.

White blood cells include **neutrophil** granulocytes (66%), **lymphocytes** (23%), **monocytes** (7%), eosinophil granulocytes (3%), and basophil granulocytes (1%).

Eosinophils destroy parasitic worms and regulate allergic reactions. **Basophils** release heparin, histamine, and serotonin in allergic reactions.

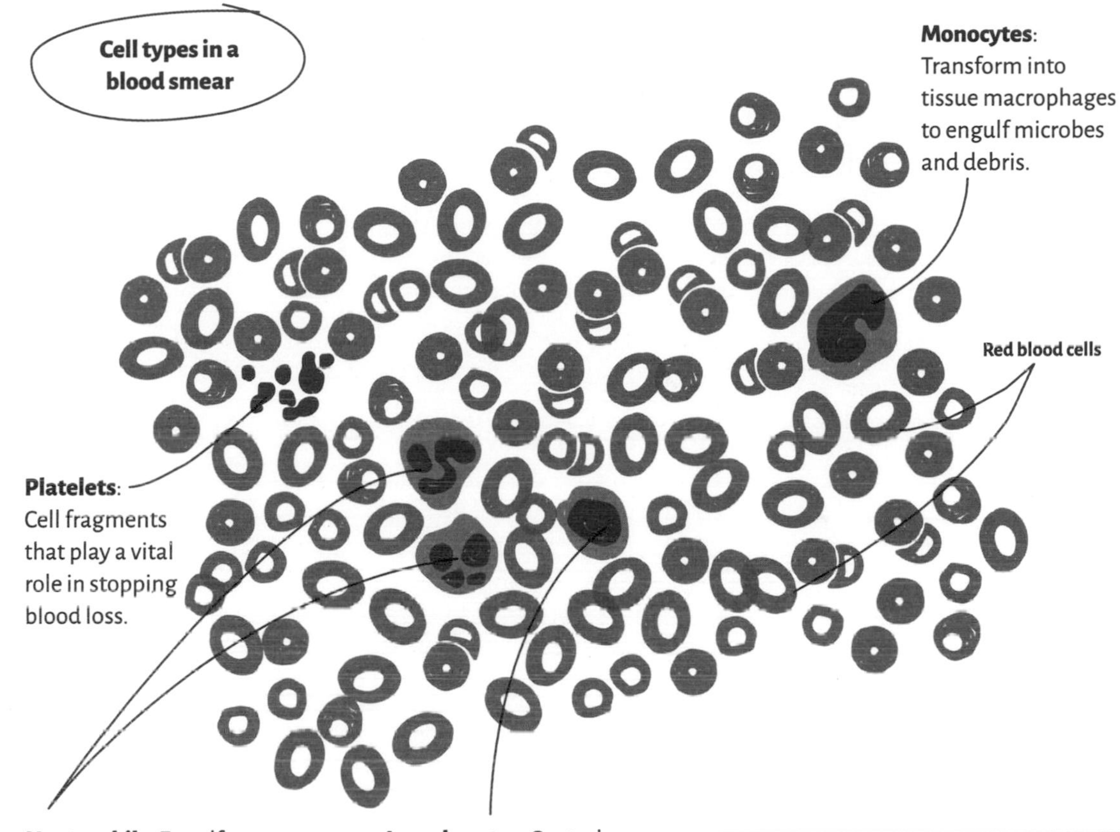

Cell types in a blood smear
Monocytes: Transform into tissue macrophages to engulf microbes and debris.
Red blood cells
Platelets: Cell fragments that play a vital role in stopping blood loss.

Neutrophils: Engulf (phagocytose) bacteria and destroy them with lysozyme (an antimicrobial enzyme).

Lymphocytes: Control immune responses, B cells develop into plasma cells to make antibodies, T cells attack viruses and cancer cells.

PLATELETS

Platelets (also called thrombocytes) are 2 to 4 μm (1/12,000 to 1/6,000 inch) diameter cell fragments, without nuclei, which last for only five to nine days. They form a platelet plug in the vessel wall during hemostasis. This is particularly effective for small breaks in the wall.

Platelets are made from megakaryocytes in the red bone marrow.

How are red blood cells made?

Red blood cells last only 120 days. **Erythropoiesis** is the process of making new red blood cells. Like all other blood cells, red blood cells are derived from a pluripotent (the capacity to become many different cells) stem cell. The nucleated stem cell becomes a myeloid stem cell, then proerythroblast, reticulocyte, and RBC. The reticulocyte stage has lost its nucleus but may retain some organelles (ribosomes, mitochondria).

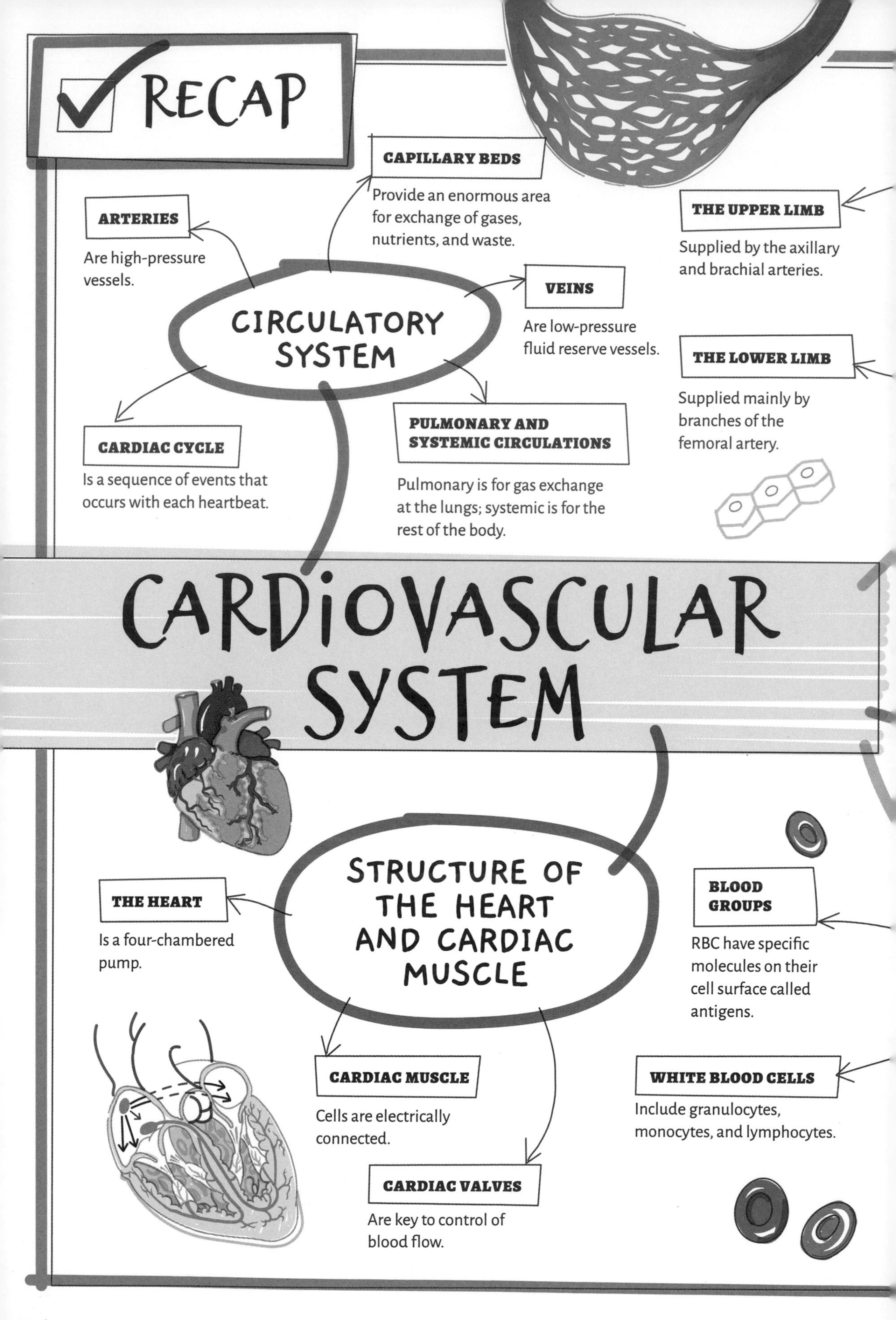
RECAP
CARDIOVASCULAR SYSTEM
CIRCULATORY SYSTEM
ARTERIES
Are high-pressure vessels.
CAPILLARY BEDS
Provide an enormous area for exchange of gases, nutrients, and waste.
VEINS
Are low-pressure fluid reserve vessels.
THE UPPER LIMB
Supplied by the axillary and brachial arteries.
THE LOWER LIMB
Supplied mainly by branches of the femoral artery.
CARDIAC CYCLE
Is a sequence of events that occurs with each heartbeat.
PULMONARY AND SYSTEMIC CIRCULATIONS
Pulmonary is for gas exchange at the lungs; systemic is for the rest of the body.
STRUCTURE OF THE HEART AND CARDIAC MUSCLE
THE HEART
Is a four-chambered pump.
CARDIAC MUSCLE
Cells are electrically connected.
CARDIAC VALVES
Are key to control of blood flow.
BLOOD GROUPS
RBC have specific molecules on their cell surface called antigens.
WHITE BLOOD CELLS
Include granulocytes, monocytes, and lymphocytes.

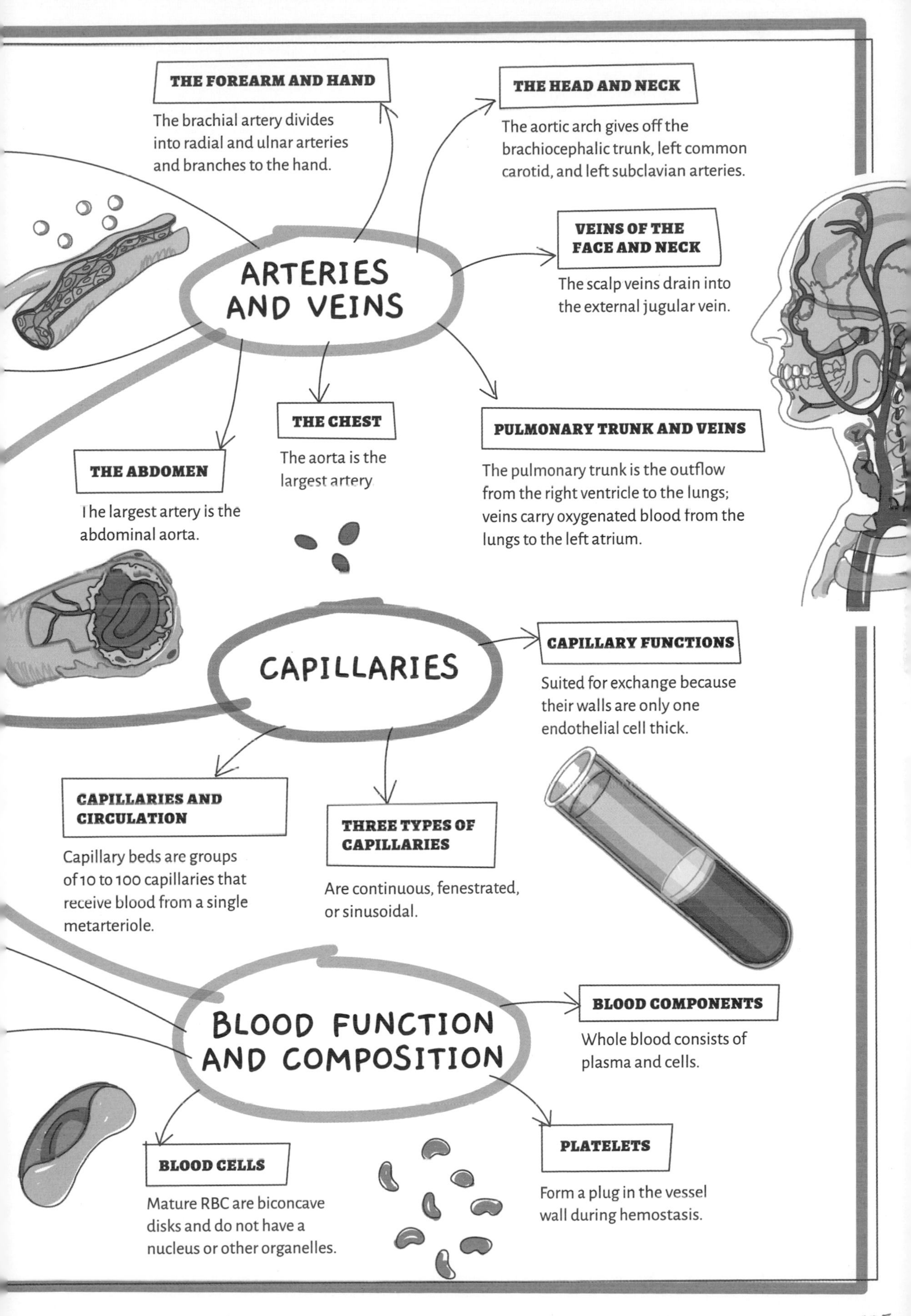
THE FOREARM AND HAND
The brachial artery divides into radial and ulnar arteries and branches to the hand.
THE HEAD AND NECK
The aortic arch gives off the brachiocephalic trunk, left common carotid, and left subclavian arteries.
VEINS OF THE FACE AND NECK
The scalp veins drain into the external jugular vein.
ARTERIES AND VEINS
THE CHEST
The aorta is the largest artery.
THE ABDOMEN
The largest artery is the abdominal aorta.
PULMONARY TRUNK AND VEINS
The pulmonary trunk is the outflow from the right ventricle to the lungs; veins carry oxygenated blood from the lungs to the left atrium.
CAPILLARIES
CAPILLARY FUNCTIONS
Suited for exchange because their walls are only one endothelial cell thick.
CAPILLARIES AND CIRCULATION
Capillary beds are groups of 10 to 100 capillaries that receive blood from a single metarteriole.
THREE TYPES OF CAPILLARIES
Are continuous, fenestrated, or sinusoidal.
BLOOD FUNCTION AND COMPOSITION
BLOOD COMPONENTS
Whole blood consists of plasma and cells.
BLOOD CELLS
Mature RBC are biconcave disks and do not have a nucleus or other organelles.
PLATELETS
Form a plug in the vessel wall during hemostasis.

CHAPTER 7

IMMUNE/LYMPHATIC SYSTEM

Lymph is the excess fluid that accumulates in the spaces between cells when arterial flow exceeds venous return. The lymphatic system drains this fluid back toward the major veins of the chest and monitors it for the presence of foreign invaders.

The macrophages of the lymphatic system remove cell debris and invaders (e.g., bacteria, viruses, and parasites), preventing their entry into the bloodstream. Lymph nodes situated along lymph channels also produce lymphocytes and control immune responses.

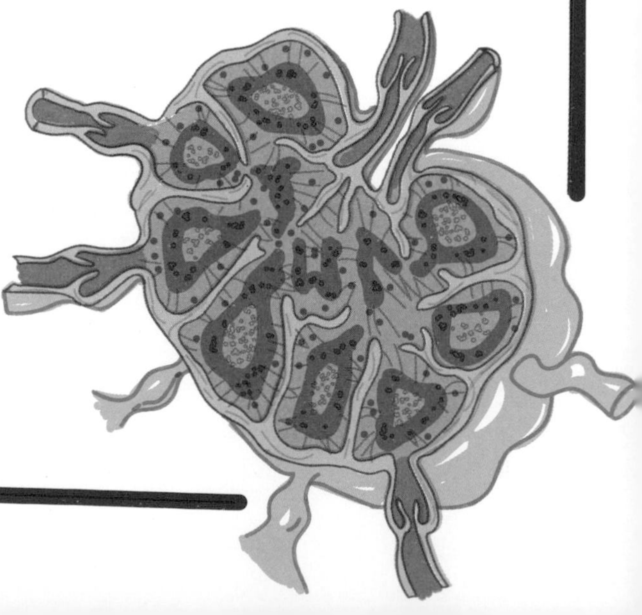

OVERVIEW OF THE LYMPHATIC SYSTEM

Tissue fluid between the cells drains into the lymphatic channels and on to lymph nodes, which provide monitoring and defense against foreign proteins and microorganisms.

Most lymph nodes are located in the chest cavity along the airways and in the abdominal cavity between the mesenteric membranes supporting the digestive system. Other lymph nodes are located at major joints: the front of the elbow, the armpit, behind the knee, in the groin, and in the neck.

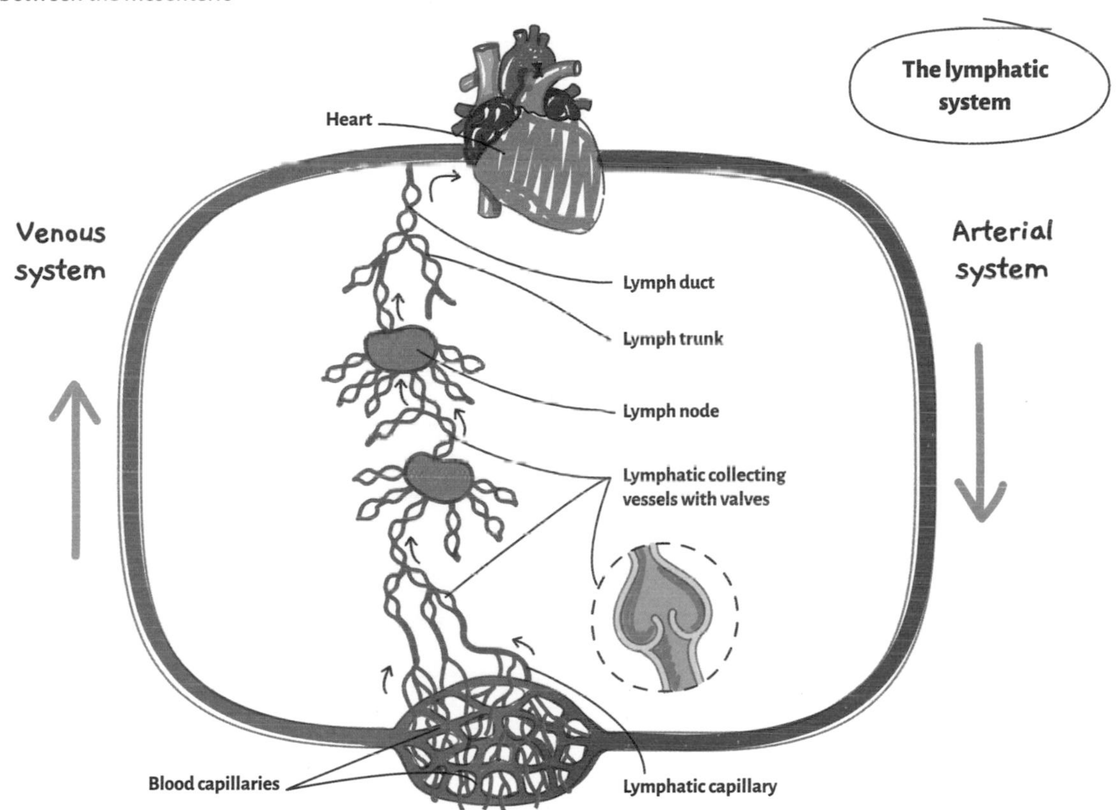

The lymphatic system
Heart
Venous system
Arterial system
Lymph duct
Lymph trunk
Lymph node
Lymphatic collecting vessels with valves
Blood capillaries
Lymphatic capillary

Immune system function depends on secreted immune proteins (antibodies and complement proteins) and cellular immune defense (neutrophil granulocytes, macrophages, and lymphocytes). The two arms work together to control foreign invaders and cancer.

Lymph vessels drain excess tissue fluid from many parts of the body. Twenty liters (about 42 pints) of blood passes out of the capillaries by capillary filtration every day. Seventeen liters (about 36 pints) per day is reabsorbed at the venous end of the capillaries, so 3 liters (6.3 pints) per day is left in the tissue spaces and must be returned to the systemic circulation. It will eventually reach the venous side of the systemic circulation.

The lymph vessels drain through lymph nodes where immune cells are concentrated.

LYMPH NODES AND CHANNELS

The lymph nodes are pea-sized structures that receive lymph by afferent lymph channels and give off lymph by efferent lymph channels. Each lymph node also has a small artery and vein.

Lymph node structure

Lymph nodes may be between 1 and 25 mm (0.039 to 0.98 inches) in length. They are covered by a capsule of dense connective tissue, which sends extensions (called **trabeculae**) into the node. The functional tissue of the lymph node is divided into a superficial cortex and a deep medulla.

The outer cortex contains ovoid collections of B lymphocytes called **lymphatic nodules** (follicles). **Lymphoid follicles** are aggregated in the cortex of the lymph node. Some nodules have a central region called a **germinal center**. These are sites of immune system memory and production of antibodies by plasma cells. After initial exposure to an antigen (e.g., bacterial wall protein), memory B cells persist in lymph nodes and can ramp up a response when the antigen is encountered again.

The inner cortex contains many T lymphocytes and dendritic cells, which present antigen to T cells, stimulating them to divide and to leave the lymph nodes to begin their fight against foreign invaders.

The medulla contains B lymphocytes, antibody-making plasma cells and phagocytic macrophages.

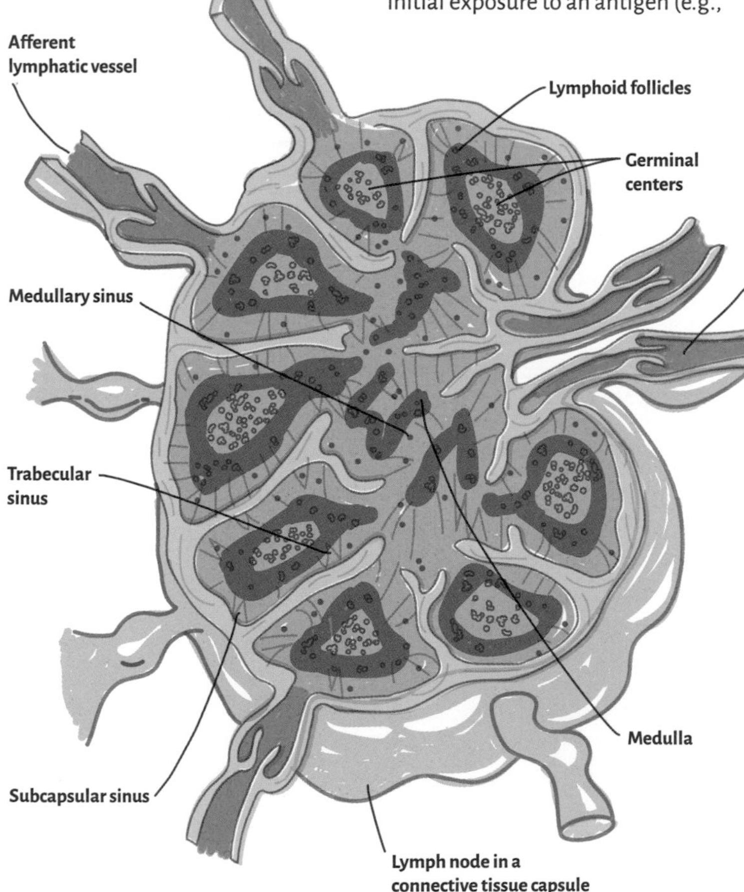

Afferent lymphatic vessel
Lymphoid follicles
Germinal centers
Medullary sinus
Efferent lymphatic vessel
Trabecular sinus
Medulla
Subcapsular sinus
Lymph node in a connective tissue capsule

LYMPH FLOW

Lymph flows through the lymph node. It enters from an afferent lymphatic vessel, passes through the **subcapsular sinus**, **trabecular sinus**, **medullary sinus** and exits through the efferent lymphatic vessel. Afferent channels enter through the capsule. **Efferent lymphatic vessels** carry lymph fluid away from the hilum opening of the lymph node.

Lymph channels

Lymph channels in the body drain from the periphery toward the center. So, lymph from the finger drains toward the axilla (armpit) and then into the chest cavity. Ultimately, all lymph drains into systemic veins in the upper chest.

The **thoracic duct** is the largest lymphatic channel in the body and drains lymph from all the lower half of the body and the left half of the upper trunk, the left upper limb, and the left half of the head.

The thoracic duct drains into the junction of the left subclavian vein and the left internal jugular vein. The right lymphatic duct drains into the junction of the right subclavian vein and the right internal jugular vein.

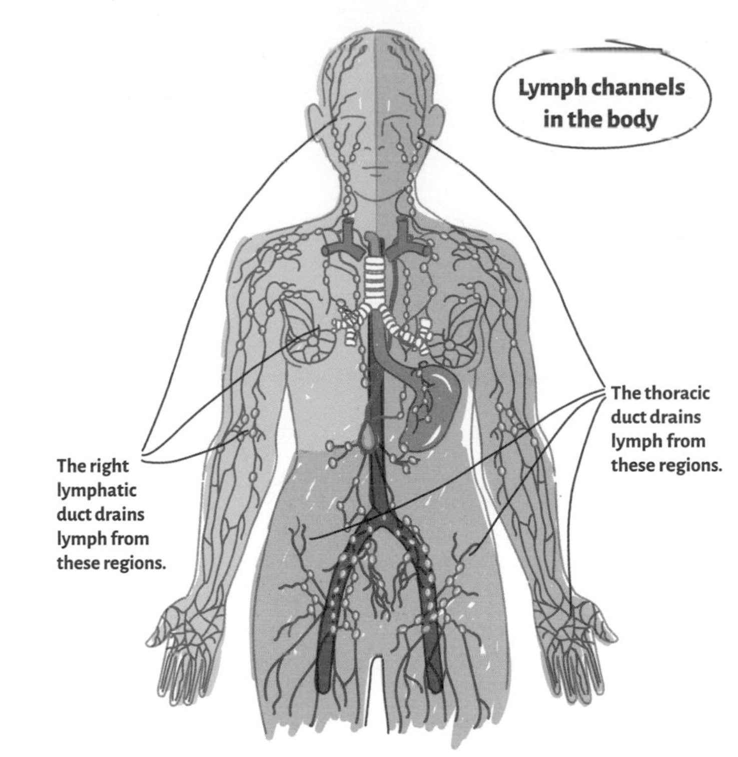

Lymph channels in the body
The right lymphatic duct drains lymph from these regions.
The thoracic duct drains lymph from these regions.

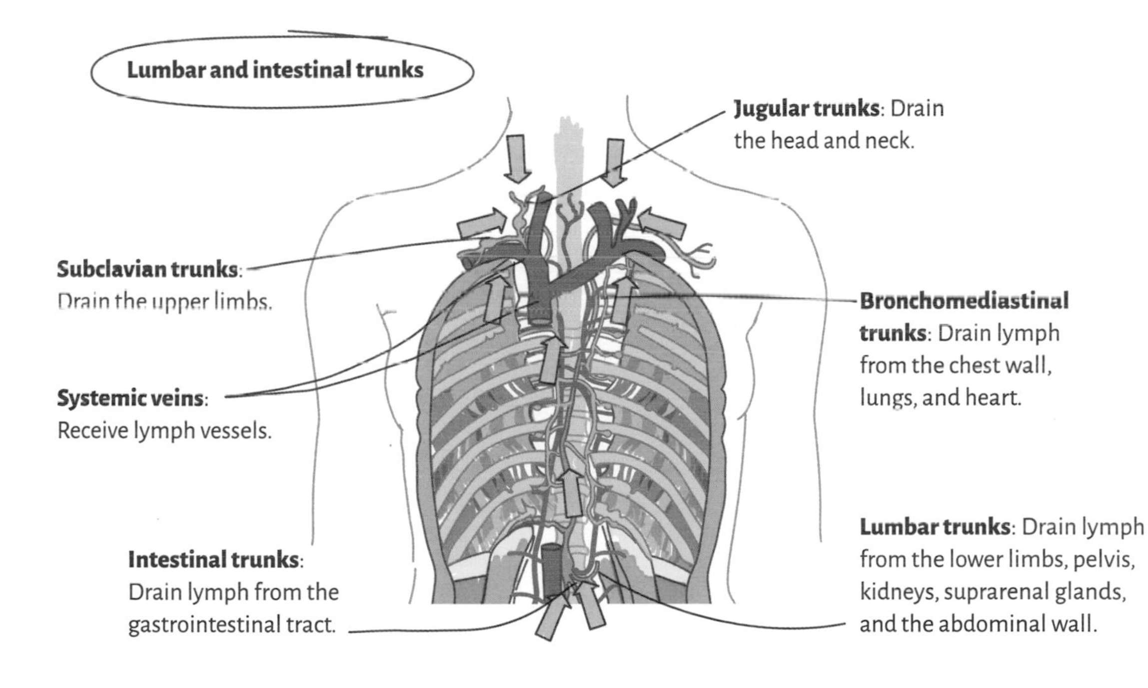

Lumbar and intestinal trunks
Jugular trunks: Drain the head and neck.
Subclavian trunks: Drain the upper limbs.
Bronchomediastinal trunks: Drain lymph from the chest wall, lungs, and heart.
Systemic veins: Receive lymph vessels.
Lumbar trunks: Drain lymph from the lower limbs, pelvis, kidneys, suprarenal glands, and the abdominal wall.
Intestinal trunks: Drain lymph from the gastrointestinal tract.

Lymphatic vessels begin as closed lymphatic capillaries that lie in the spaces between cells (intercellular fluid). Lymphatic capillaries combine to form lymphatic vessels, which have very thin walls and valves to direct lymph centrally.

Lymphatic vessels unite to form **lymph trunks**, which continue the flow of lymph toward the junction with systemic veins.

Lymphatic capillaries from the intestines, called **lacteals**, carry large fat molecules.

Lymph from most tissues is clear and colorless, but from the small intestine it is milky due to the absorption of fat globules (chylomicrons) and is called **chyle**.

INNATE AND ADAPTIVE IMMUNITY

The response of the body to foreign substances and pathogens (disease-causing microorganisms) is called immunity. Immunity can be innate (natural) or adaptive (acquired).

Innate immunity

Innate immunity does not require previous exposure to the pathogen and is rapidly responsive. Innate or natural immunity has four components:

★Physical barrier of the epithelia
★Phagocytic cells (macrophages and neutrophils)
★Natural killer cells
★Blood proteins, such as the cytokines and complement system

BLOOD CELLS
Blood cells are the mobile arm of the immune system. White blood cells (leukocytes) play important roles in both innate and adaptive immunity. Rapid transport of cells and immune proteins by blood is essential to ensure that the immune response is coordinated throughout the body.

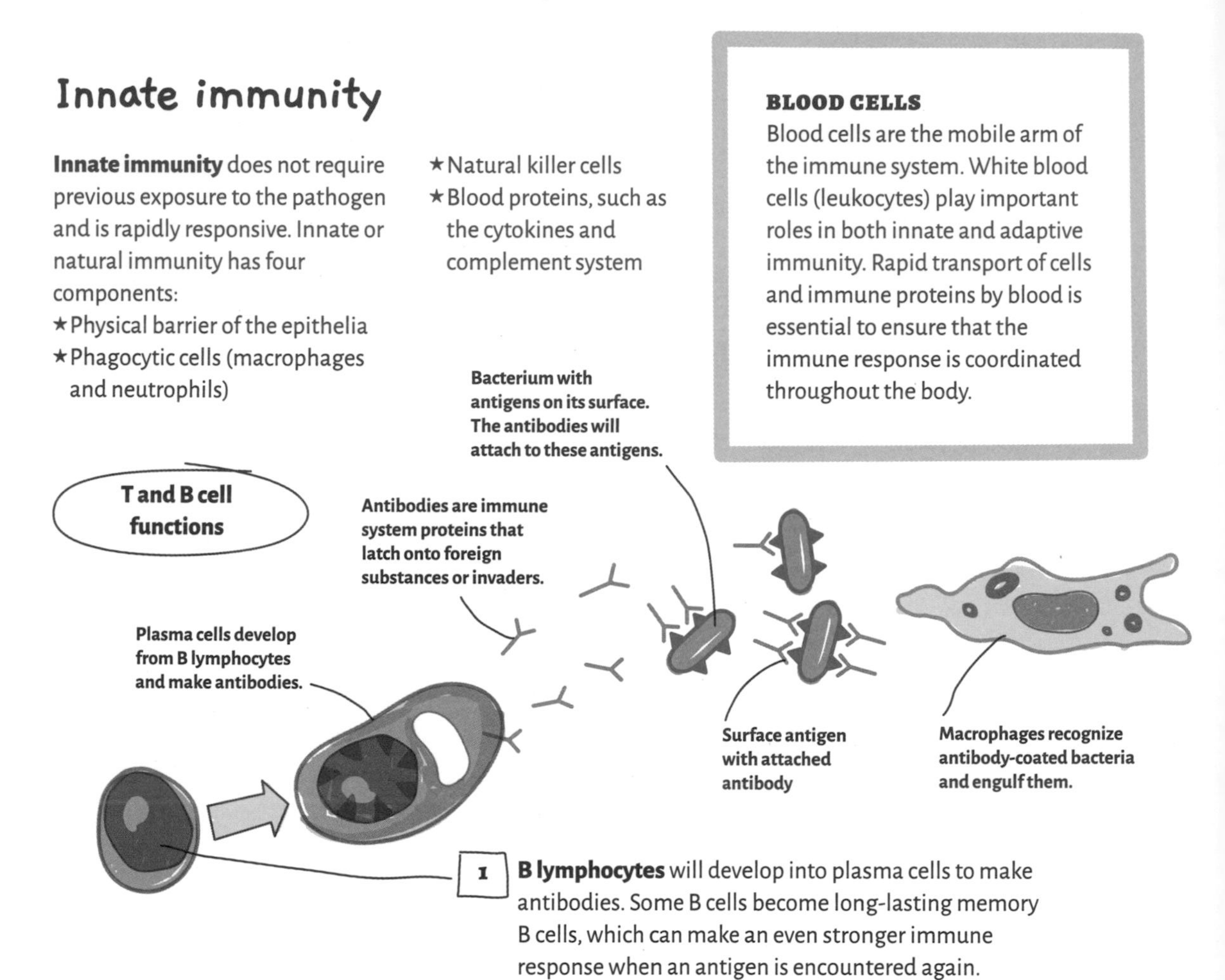

T and B cell functions
Bacterium with antigens on its surface. The antibodies will attach to these antigens.
Antibodies are immune system proteins that latch onto foreign substances or invaders.
Plasma cells develop from B lymphocytes and make antibodies.
Surface antigen with attached antibody
Macrophages recognize antibody-coated bacteria and engulf them.
1

B lymphocytes will develop into plasma cells to make antibodies. Some B cells become long-lasting memory B cells, which can make an even stronger immune response when an antigen is encountered again.

Adaptive immunity

Adaptive immunity develops after exposure to a pathogen and involves cellular learning to recognize antigens and respond to them. Humoral and cellular immunity are types of adaptive immunity.

The first adaptive immune response is **humoral immunity**, which involves antibodies made by plasma cells binding to antigens and toxins produced by pathogens. This binding recruits macrophages to engulf the foreign material.

The second response follows the uptake of a pathogen by a phagocytic cell, or the invasion of body cells by pathogens, such as viruses and rickettsia. A cellular immune response (cell-mediated immunity) destroys pathogens inside the body cells and can also protect against abnormal body cells, such as cancer.

LYMPHOCYTES AND MONOCYTES

Lymphocytes mediate immune responses, including reactions between antigens and antibodies. Lymphocytes are of three types: B lymphocytes, T lymphocytes, and natural killer cells (see below).

The **monocytes** found in blood can leave the circulation by crossing the capillary wall and become tissue macrophages. They engulf foreign matter and may present the antigens of pathogens on their cell surface to trigger other components of the immune system.

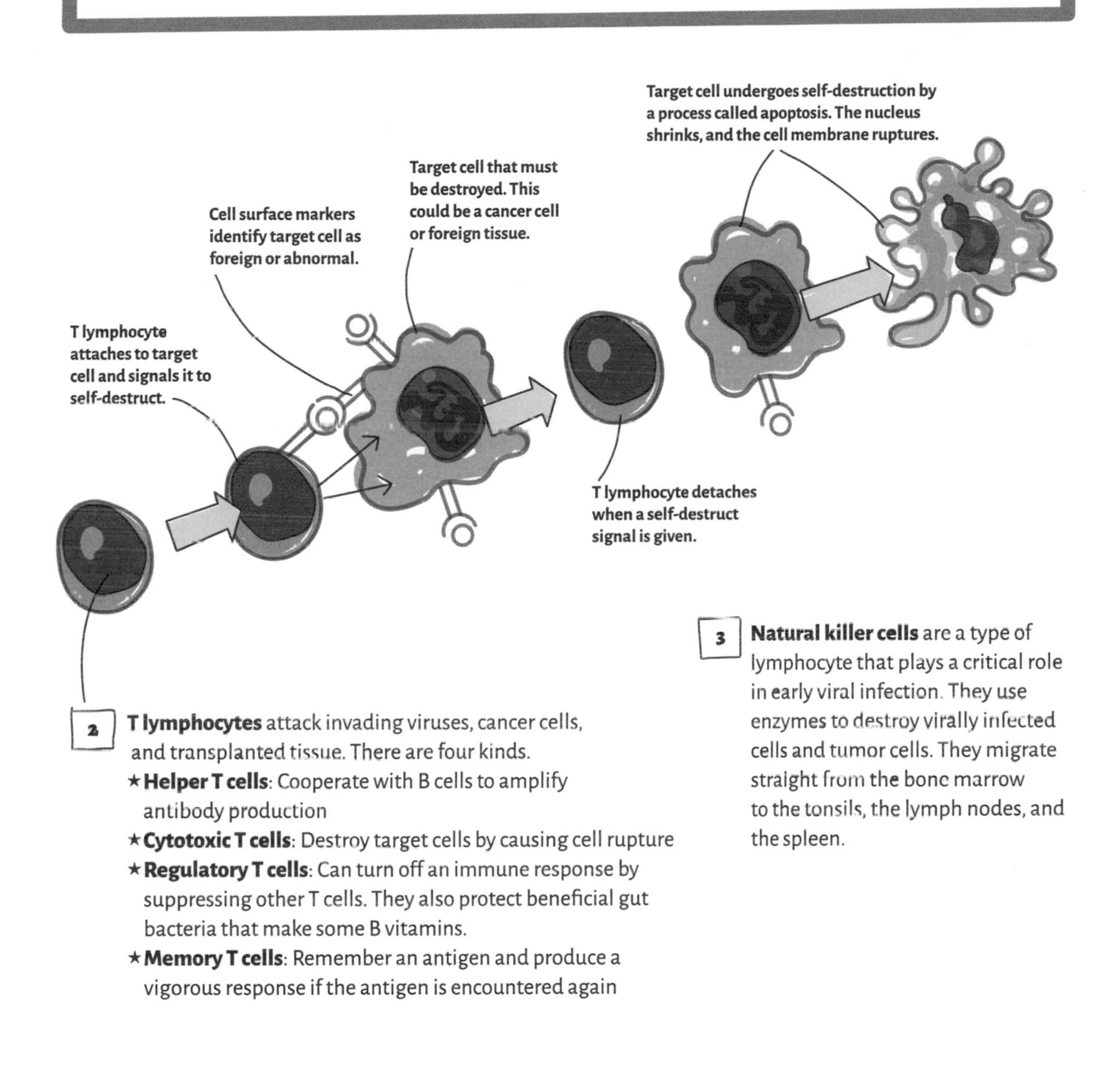

Target cell undergoes self-destruction by a process called apoptosis. The nucleus shrinks, and the cell membrane ruptures.
Target cell that must be destroyed. This could be a cancer cell or foreign tissue.
Cell surface markers identify target cell as foreign or abnormal.
T lymphocyte attaches to target cell and signals it to self-destruct.
T lymphocyte detaches when a self-destruct signal is given.

2 **T lymphocytes** attack invading viruses, cancer cells, and transplanted tissue. There are four kinds.

- ★ **Helper T cells**: Cooperate with B cells to amplify antibody production
- ★ **Cytotoxic T cells**: Destroy target cells by causing cell rupture
- ★ **Regulatory T cells**: Can turn off an immune response by suppressing other T cells. They also protect beneficial gut bacteria that make some B vitamins.
- ★ **Memory T cells**: Remember an antigen and produce a vigorous response if the antigen is encountered again

3 **Natural killer cells** are a type of lymphocyte that plays a critical role in early viral infection. They use enzymes to destroy virally infected cells and tumor cells. They migrate straight from the bone marrow to the tonsils, the lymph nodes, and the spleen.

Immune system cells

Like red blood cells, white blood cells are derived from a pluripotent stem cell in the red bone marrow. The pluripotent stem cell can become either a myeloid stem cell or a lymphoid stem cell. Myeloid stem cells that will become granulocytes or macrophages follow the colony-forming unit-granulocyte macrophage lineage (CFU-GM). CFU-GM become eosinophils, basophils, neutrophils, or monocytes.

Lymphoid stem cells can become T lymphocytes, B lymphocytes, or NK (natural killer) lymphocytes.

THYMUS, TONSILS, AND SPLEEN

The thymus, tonsils, and spleen are large aggregations of lymphoid tissue with diverse functions. The thymus makes T lymphocytes, the tonsils protect against invaders, and the spleen cleanses the blood.

Thymus

The **thymus** is a double-lobed organ sitting in the upper chest. It is responsible for maturing **T lymphocytes** that have migrated there from the bone marrow. The thymus has a connective tissue **capsule** that sends trabeculae into its interior, dividing the tissue into lobules. Each lobule has an outer cortex and an inner medulla.

The **cortex** consists of many T lymphocytes, dendritic cells, epithelial cells, and macrophages. Dendritic cells assist T cell maturation, while epithelial cells provide a framework supporting up to 50 lymphocytes. Most T cells die in the cortex, so the macrophages clear the cellular debris. Surviving T cells enter the medulla. The **medulla** consists of scattered, more mature T cells, dendritic cells, and macrophages.

During childhood, T cells leave the thymus to lodge in lymph nodes, the spleen, and the tonsils. The thymus becomes a fatty remnant after puberty.

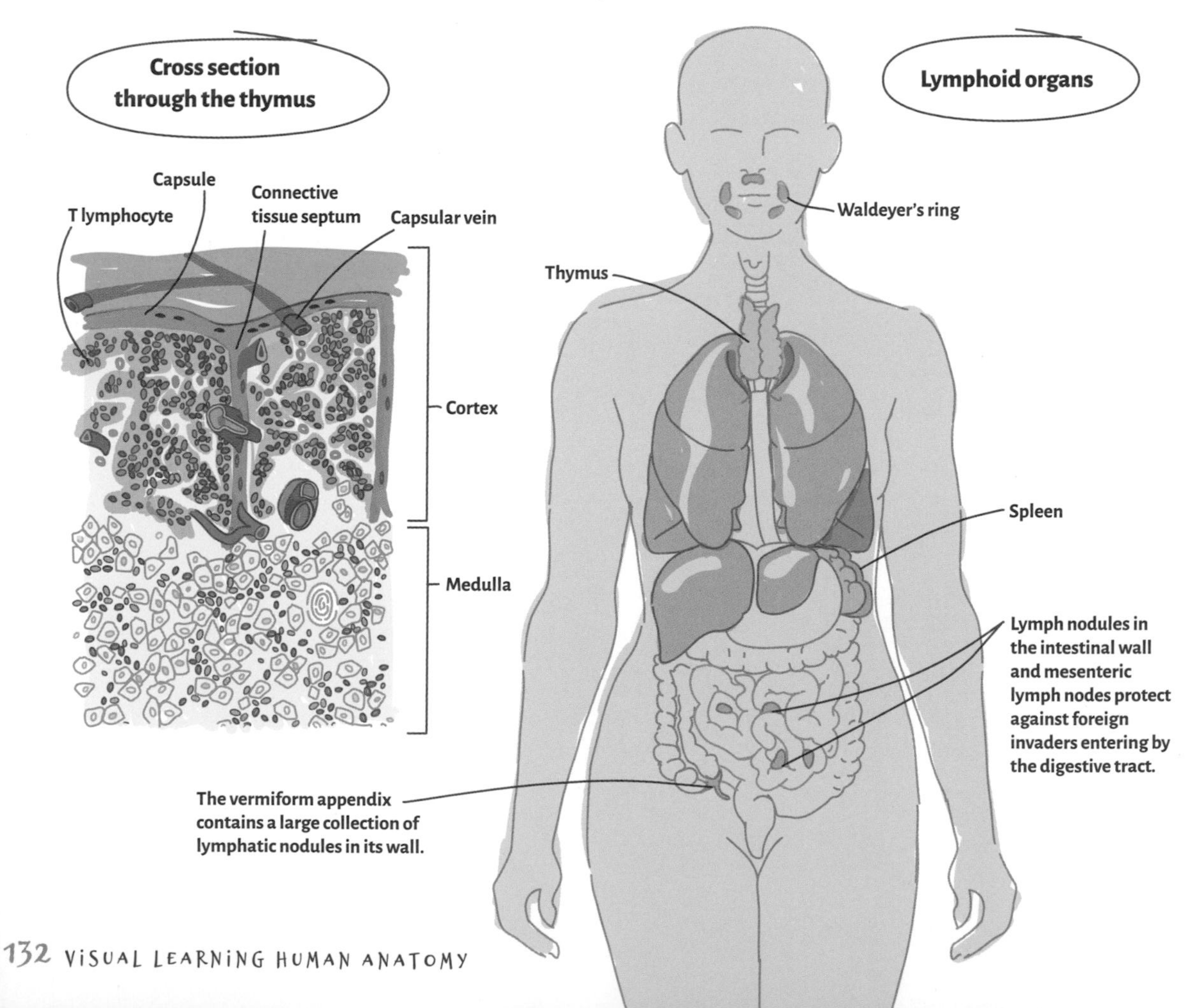
Cross section through the thymus
Capsule
T lymphocyte
Connective tissue septum
Capsular vein
Cortex
Medulla
Lymphoid organs
Waldeyer's ring
Thymus
Spleen
Lymph nodules in the intestinal wall and mesenteric lymph nodes protect against foreign invaders entering by the digestive tract.
The vermiform appendix contains a large collection of lymphatic nodules in its wall.

Tonsils

There are five **tonsils** arranged as a ring, called **Waldeyer's ring**, around the entrance to the respiratory and digestive tracts (see below left). They are masses of lymphoid tissue covered with mucosa and detect disease-causing microbes entering the body.

Pharyngeal tonsil (adenoid): Embedded in the posterior wall of the nasopharynx.

Palatine tonsils: Two structures embedded in the lateral walls of the oropharynx in the palatine fossa.

Lingual tonsils: Paired structures on the surface of the posterior third of the tongue.

Spleen

The **spleen** is the largest lymphoid organ in the body and lies in the upper left side, beneath the diaphragm. It has a capsule of dense connective tissue that sends trabeculae into the interior. Splenic tissue is divided into white pulp and red pulp.

White pulp: Consists of lymphoid tissue, with lymphocytes and macrophages surrounding central arteries. Lymphocytes in the white pulp behave much like lymphoid tissue elsewhere. Splenic macrophages destroy blood-borne microbes.

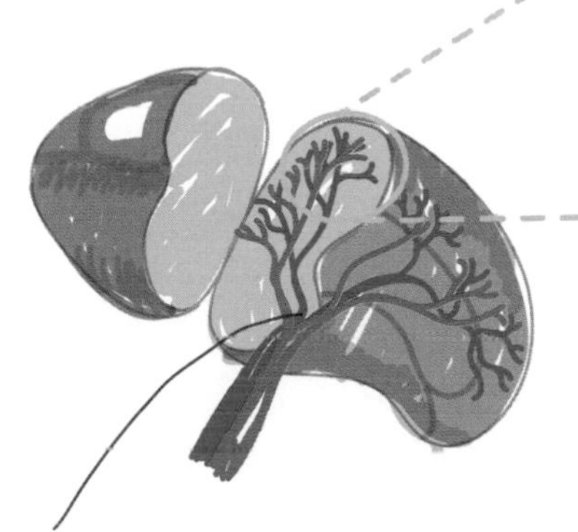

The spleen lies in the upper left abdomen and has a rich vascular supply.

Red pulp: Consists of blood-filled sinuses and cords of splenic tissue (**Billroth's cords**). Splenic cords contain red blood cells, macrophages, lymphocytes, plasma cells, and granulocytes. Red pulp removes old red blood cells and platelets but may also store platelets for release when needed.

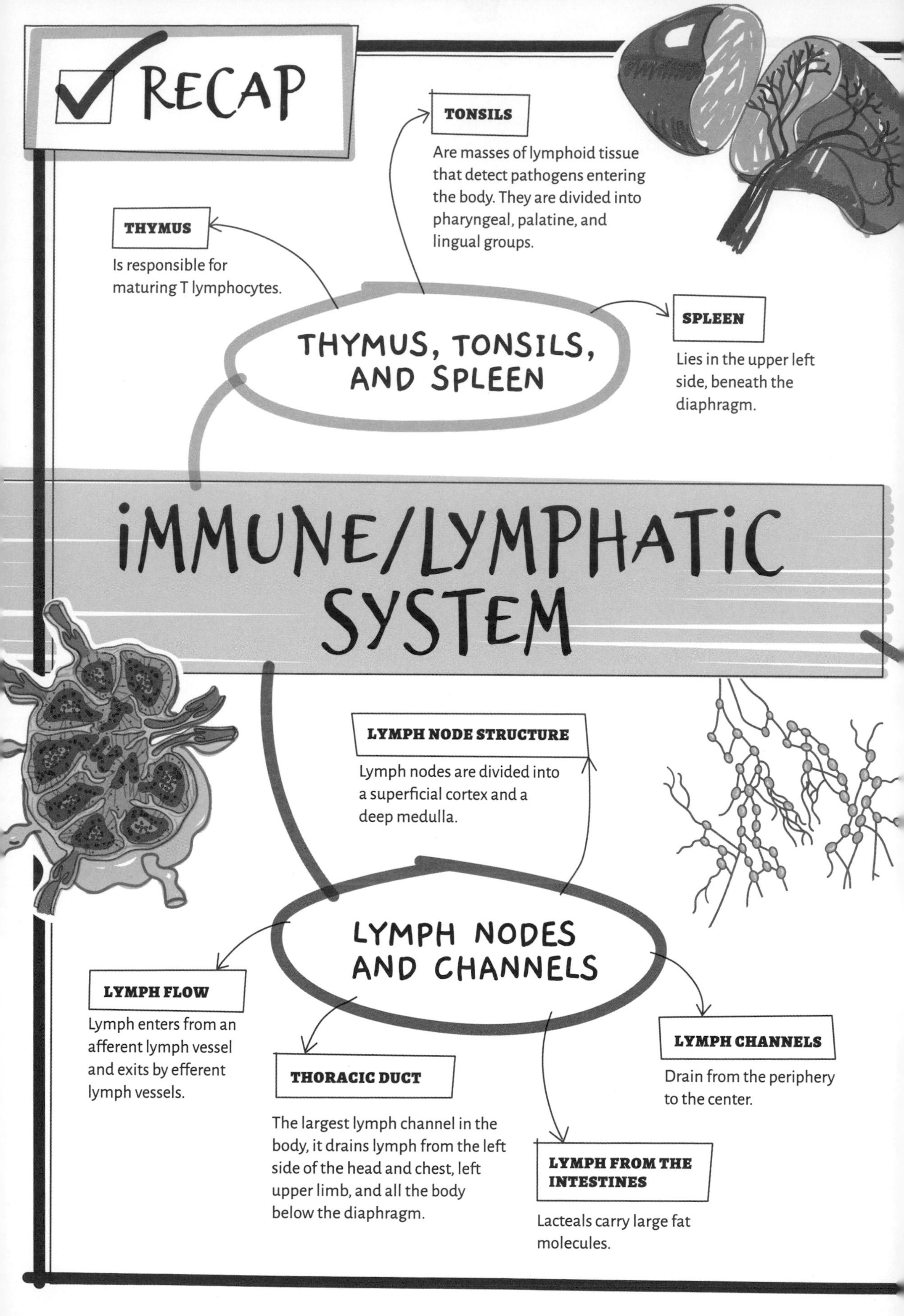
RECAP
TONSILS
Are masses of lymphoid tissue that detect pathogens entering the body. They are divided into pharyngeal, palatine, and lingual groups.
THYMUS
Is responsible for maturing T lymphocytes.
THYMUS, TONSILS, AND SPLEEN
SPLEEN
Lies in the upper left side, beneath the diaphragm.
IMMUNE/LYMPHATIC SYSTEM
LYMPH NODE STRUCTURE
Lymph nodes are divided into a superficial cortex and a deep medulla.
LYMPH NODES AND CHANNELS
LYMPH FLOW
Lymph enters from an afferent lymph vessel and exits by efferent lymph vessels.
THORACIC DUCT
The largest lymph channel in the body, it drains lymph from the left side of the head and chest, left upper limb, and all the body below the diaphragm.
LYMPH FROM THE INTESTINES
Lacteals carry large fat molecules.
LYMPH CHANNELS
Drain from the periphery to the center.

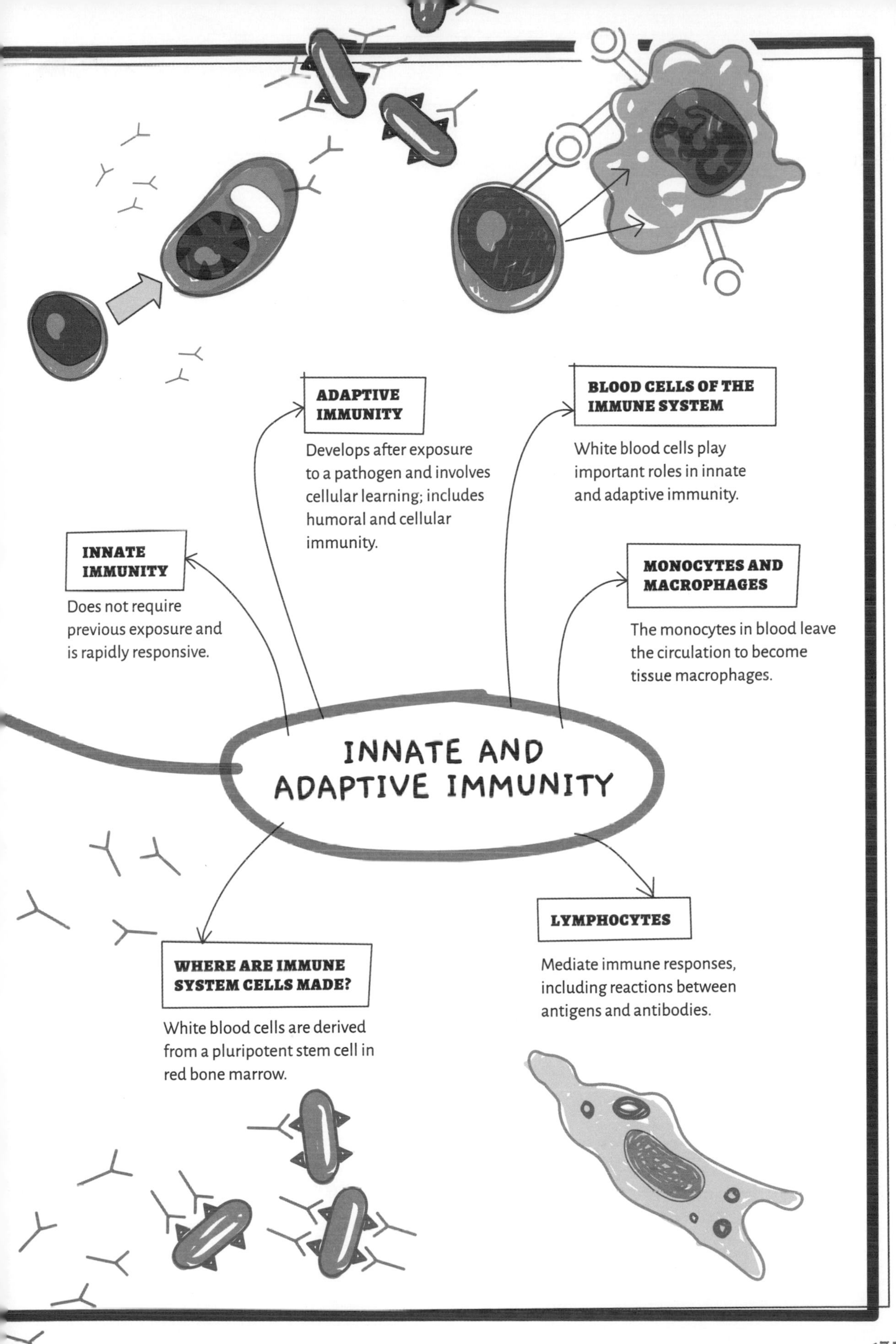
ADAPTIVE IMMUNITY
Develops after exposure to a pathogen and involves cellular learning; includes humoral and cellular immunity.
BLOOD CELLS OF THE IMMUNE SYSTEM
White blood cells play important roles in innate and adaptive immunity.
INNATE IMMUNITY
Does not require previous exposure and is rapidly responsive.
MONOCYTES AND MACROPHAGES
The monocytes in blood leave the circulation to become tissue macrophages.
INNATE AND ADAPTIVE IMMUNITY
LYMPHOCYTES
Mediate immune responses, including reactions between antigens and antibodies.
WHERE ARE IMMUNE SYSTEM CELLS MADE?
White blood cells are derived from a pluripotent stem cell in red bone marrow.

CHAPTER 8

RESPIRATORY SYSTEM

The respiratory system is primarily concerned with gas exchange between the blood and the external environment. For this reason, the lungs have a rich vascular supply via the pulmonary circulation, with a pulmonary capillary bed delivering 5 liters (1.3 gallons) of blood per minute for gas exchange. The rich blood flow passes within 1/2,000 mm (1/50,000 inch) of inhaled air.

Additional roles of the respiratory system include the action of smelling (olfaction), voice production (phonation), the ability to maintain body temperature (thermoregulation), and acid-base balance control.

OVERVIEW OF THE RESPIRATORY SYSTEM

The respiratory system consists of nasal passages, the nasopharynx, larynx, trachea, and bronchi, and progressively smaller airways down to the alveoli.

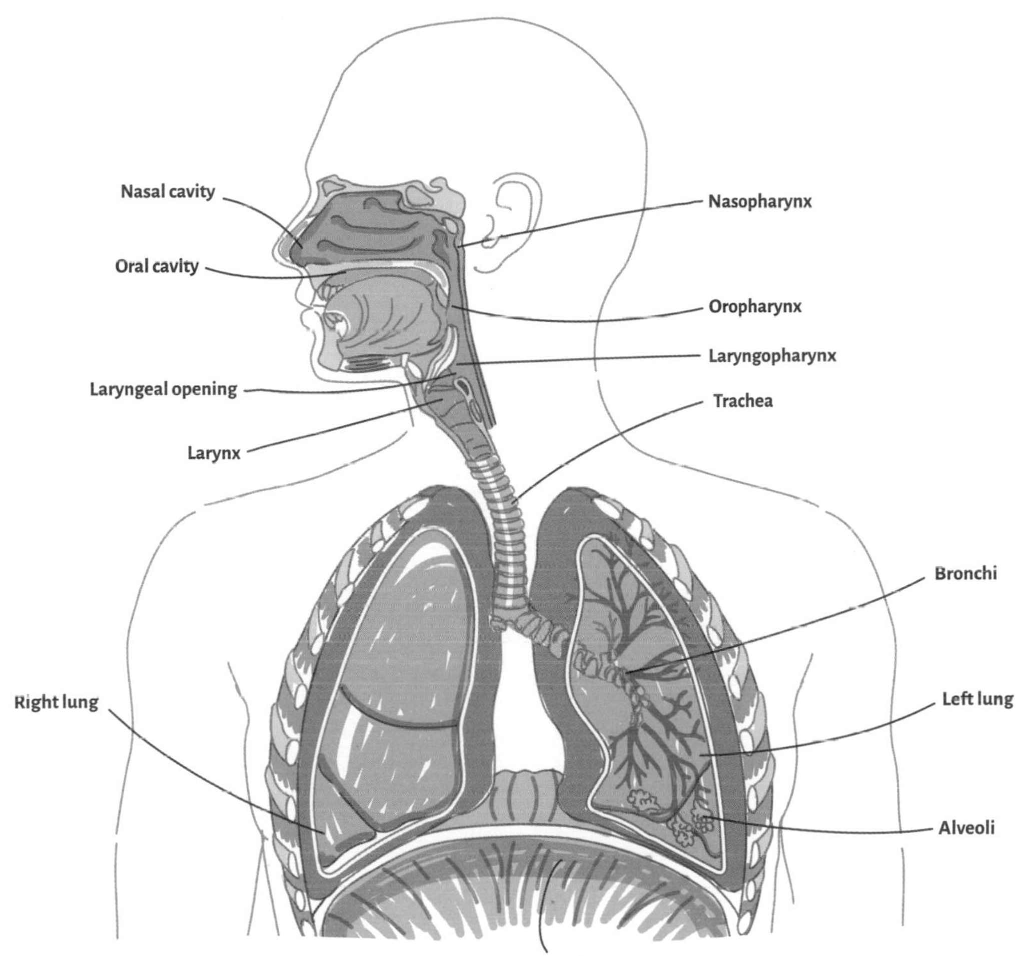

Nasal cavity
Oral cavity
Laryngeal opening
Larynx
Nasopharynx
Oropharynx
Laryngopharynx
Trachea
Bronchi
Right lung
Left lung
Alveoli
Diaphragm

Airflow in the respiratory system

Inhaled air passes across the nasal epithelium for warming and moistening. Air then passes into the **nasopharynx** and **oropharynx**, before entering the **laryngeal opening**. The laryngeal opening can be closed during swallowing and eating because the pathways for food, fluid, and air cross near its entrance.

Air passes down the interior of the larynx to the trachea. The **trachea** (windpipe) divides into the main **bronchi** in the middle of the chest. Airway division continues for as many as 23 branches until the **alveoli** are reached. Gas exchange occurs across the walls of the grapelike alveoli.

NASAL CAVITY AND PARANASAL SINUSES

The nasal cavity has mucous membranes to warm and moisten inhaled air and to filter out dust, pollens, and microorganisms. The top of the nasal cavity includes the olfactory epithelium for the sensation of smell.

Nose structure

The **nose** consists of an external nose and the nasal cavity. The nose has nostrils (**nares**) at the front (anteriorly) and opens at the back (posteriorly) through the choanae into the nasopharynx.

The external nose is composed of a bony base (**nasal** and **maxillary bones**), and pliable hyaline (glassy) cartilages (**lateral**, **septal**, and **alar**) anteriorly, all covered by skin externally and non-keratinized stratified squamous epithelium internally.

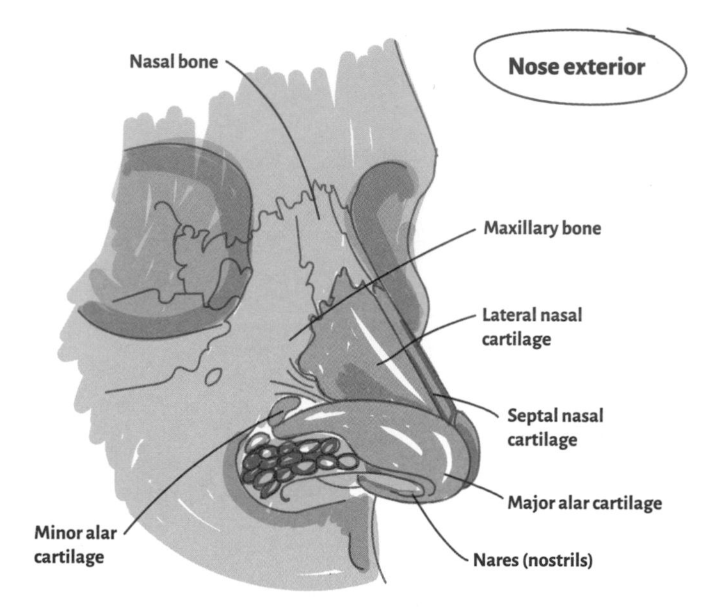

Nose exterior
Nasal bone
Maxillary bone
Lateral nasal cartilage
Septal nasal cartilage
Major alar cartilage
Minor alar cartilage
Nares (nostrils)

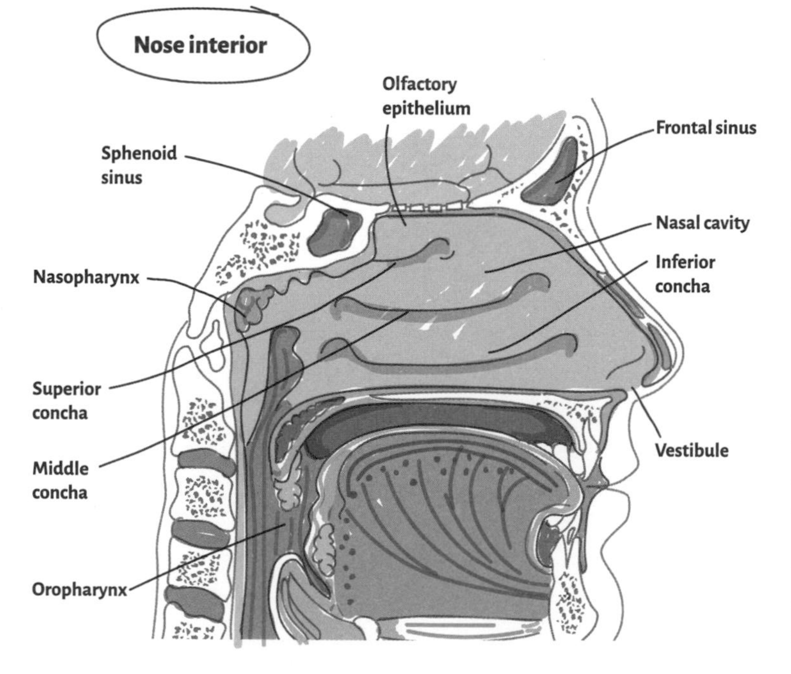

Nose interior
Olfactory epithelium
Sphenoid sinus
Frontal sinus
Nasal cavity
Inferior concha
Nasopharynx
Superior concha
Middle concha
Vestibule
Oropharynx

Nasal cavity

The **nasal cavity** is divided into two equal sides separated by a septum, which consists of the septal cartilage, the ethmoid bone, and the vomer bone. It is mostly a conductive passage for air, but the upper part is sensory.

The **olfactory epithelium** is a special type of sensory tissue at the top of the nasal cavity on the cribriform plate of the ethmoid and the adjacent superior concha.

The initial part of the nasal cavity is called the **vestibule** and may develop hairs with advancing age.

THE LATERAL WALL

The **lateral wall** is composed of bones (nasal, lacrimal, ethmoid, inferior nasal concha, maxilla, palatine, and sphenoid) covered by mucous membranes.

Three elevations on the lateral wall, **superior**, **middle**, and **inferior conchae** (see "Nose interior," page 138), increase surface area for moistening and warming the air. The conchae (also called turbinates) introduce turbulence into the flow of inhaled air to help drop out the dust.

The lateral nasal wall has communications with the frontal, maxillary, and ethmoid paranasal sinuses and receives the nasolacrimal duct, carrying excess tear fluid from the medial angle of the eye (see below).

Paranasal sinuses

The **paranasal sinuses** are air-filled spaces within the skull bones that connect to the nasal cavity. Their function is not certain, but they may lighten the skull or add resonance to the voice.

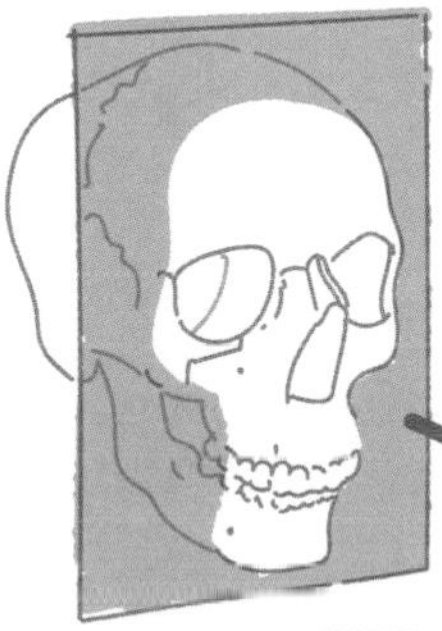

Airway branches

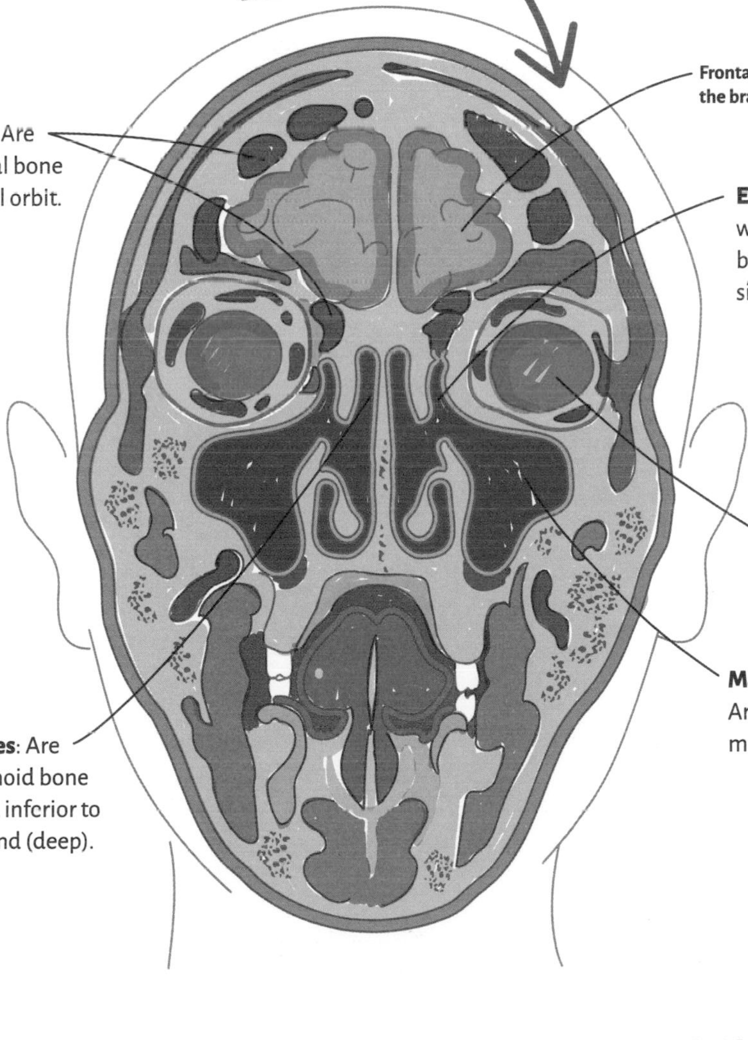

Frontal lobe of the brain

Frontal sinuses: Are within the frontal bone above the medial orbit.

Ethmoid sinuses: Are within the ethmoid bone on the medial side of each orbit.

Eyeball in the bony orbit

Maxillary sinuses: Are within the maxilla of the cheek.

Sphenoid sinuses: Are within the sphenoid bone of the skull base, inferior to the pituitary gland (deep).

LARYNX: INTERIOR AND EXTERIOR

The larynx is commonly known as the voice box. Its function is to protect the airway from the inhalation of water and food by closing during swallowing and to produce the sound of the voice.

Larynx structure

The larynx has a skeleton of cartilage.

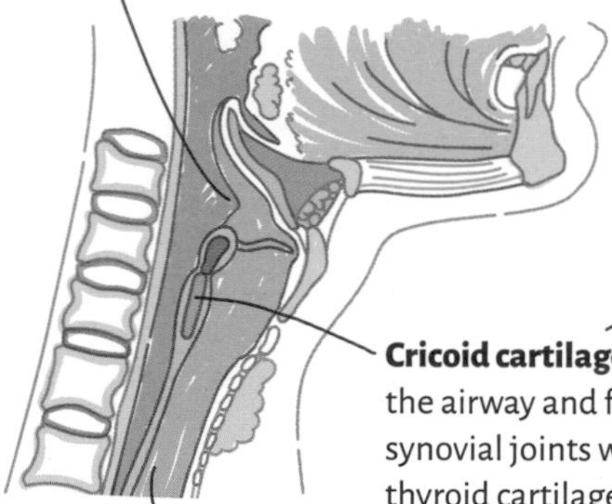

Laryngeal entrance opens into the laryngopharynx.

Arytenoid cartilages: Mobile and pyramidal, each carry one end of the vocal ligament.

Epiglottis cartilage: Leaf-shaped and folds down to close the airway during swallowing.

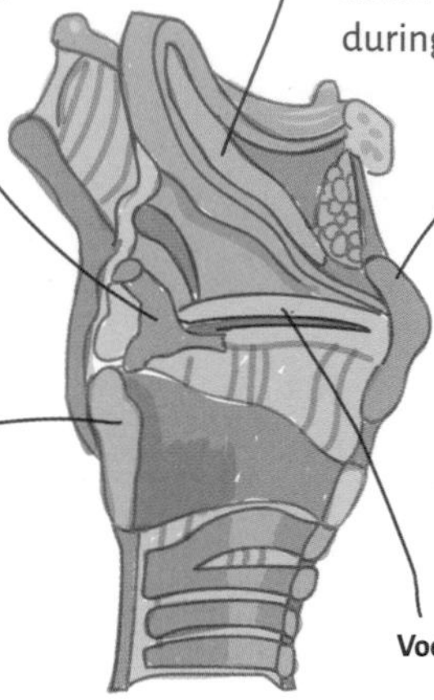

Thyroid cartilage: Largest and consists of two plates joining at an angle anteriorly.

Cricoid cartilage: Surrounds the airway and forms paired synovial joints with the thyroid cartilage.

Vocal ligament

Airway

Swallowing

The larynx can close during swallowing. The entrance to the larynx is bounded by the epiglottis and paired **aryepiglottic folds** that stretch from the epiglottis to the arytenoid cartilages. When swallowing, the arytenoids rise and the epiglottis descends and bends over and closes the laryngeal entrance. Muscles in the aryepiglottic folds (aryepiglotticus) and between the arytenoids (**transverse** and **oblique arytenoid**) close the aperture.

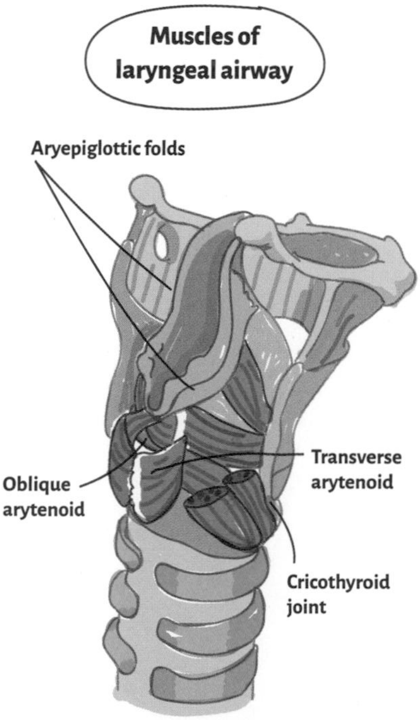

Muscles of laryngeal airway

Aryepiglottic folds

Oblique arytenoid

Transverse arytenoid

Cricothyroid joint

Vocal folds

Vocal folds vibrate to produce the voice. The **vocal ligaments** extend from the arytenoids to the back of the thyroid cartilage. Each vocal ligament is covered by a mucous membrane to form a **vocal fold**. The mobile arytenoids and vocal folds can be brought together and air forced between them to produce vibration and the voice.

The **cricothyroid joint** adjusts tension in the vocal fold. Tilting of the thyroid cartilage forward on the cricoid increases tension in the vocal ligament and fold. This raises the pitch of the sound when the fold vibrates.

TRACHEA, BRONCHI, AND LUNGS

The trachea (windpipe) starts at the lower end of the larynx and enters the thoracic cavity. Both the larynx and bronchi are made of cartilage with smooth muscle and connective tissue to provide elasticity of the airway.

Trachea structure

The **trachea** is composed of 16 to 20 horseshoe-shaped cartilages with a soft posterior wall of smooth muscle (the **trachealis**). The trachea starts at the cricoid cartilage and divides in the chest.

The soft posterior wall of the trachea is in contact with the esophagus, and together they pass through the tight space of the thoracic inlet (bounded by the first rib). When a large bolus is swallowed (a lump of chewed food) and the esophagus expands in the inlet, the soft posterior wall of the trachea can be pushed forward, which prevents the food being stuck at the thoracic inlet.

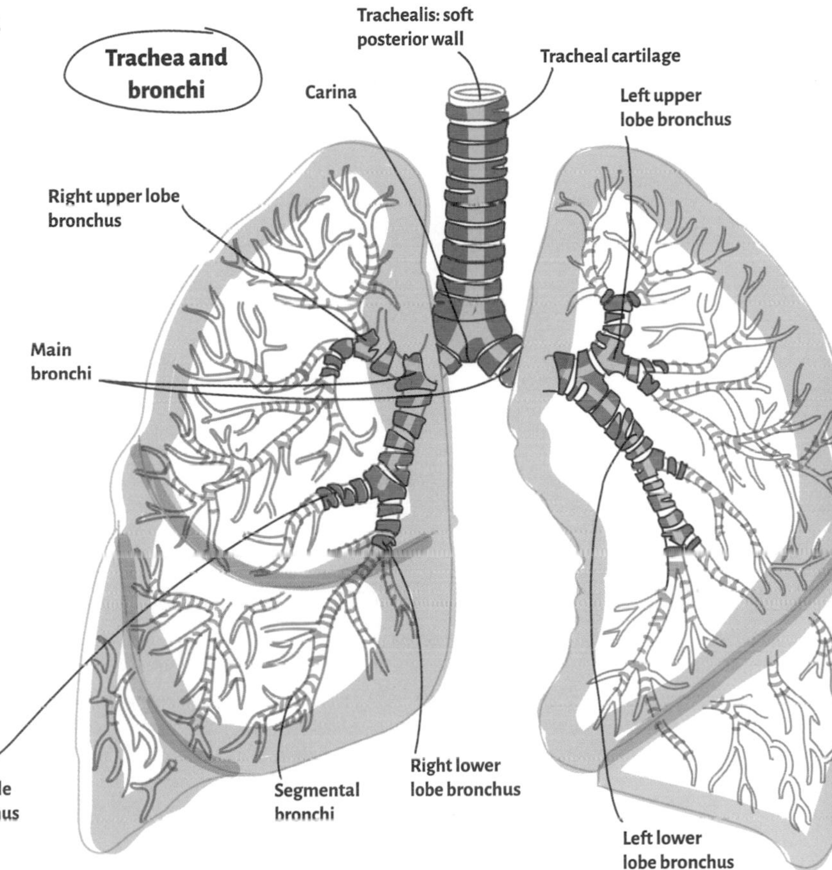

Trachea and bronchi
Trachealis: soft posterior wall
Tracheal cartilage
Carina
Left upper lobe bronchus
Right upper lobe bronchus
Main bronchi
Right middle lobe bronchus
Segmental bronchi
Right lower lobe bronchus
Left lower lobe bronchus

Bronchi structure

The trachea divides into two main **bronchi** in the chest. The inside of the trachea at the division has a sharp ridge called the **carina**, which is sensitive to contact with foreign material. Breathing in dust and foreign substances initiates a cough reflex, which is an explosive exhalation to push air and foreign bodies out of the airway.

The main bronchi divide into **lobar bronchi** for each lobe of the lung—superior and inferior on the left; superior, middle, and inferior on the right. Lobar bronchi then divide into **segmental bronchi**, which supply discrete lung regions separated by connective tissue septa.

Bronchi will divide up to 23 times to reach the tiny air sacs where gas exchange occurs.

The walls of the bronchi are reinforced by cartilage and smooth muscle to prevent collapse when air is drawn rapidly into the lungs.

Lung structure

The final branches of the airways open into the 480 million grapelike alveoli in the lungs, where gas exchange occurs. **Alveoli** are tiny air sacs with a rich blood supply and thin walls. Their total surface area would cover a tennis court.

The lungs are divided into lobes by fissures. The **right lung** has three lobes (upper, middle, and lower), separated by **horizontal** and **oblique fissures**.

The **left lung** has two lobes (upper and lower) that are separated by an oblique fissure.

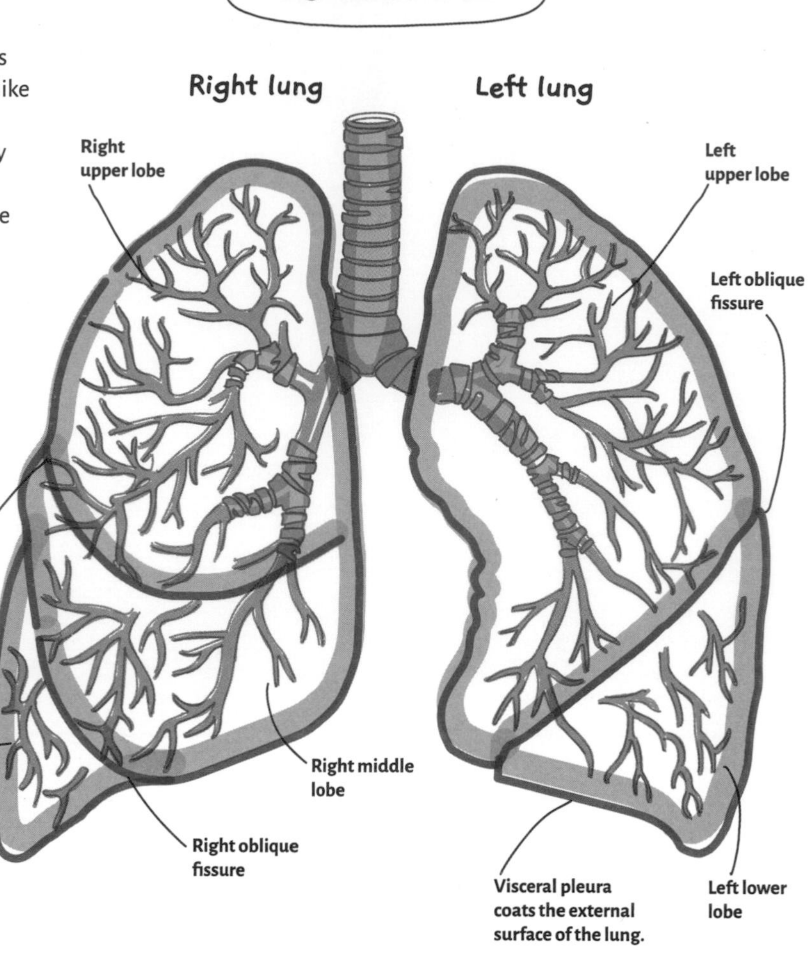
Lung lobes and fissures
Right lung
Left lung
Right upper lobe
Left upper lobe
Left oblique fissure
Transverse or horizontal fissure
Right lower lobe
Right middle lobe
Right oblique fissure
Visceral pleura coats the external surface of the lung.
Left lower lobe

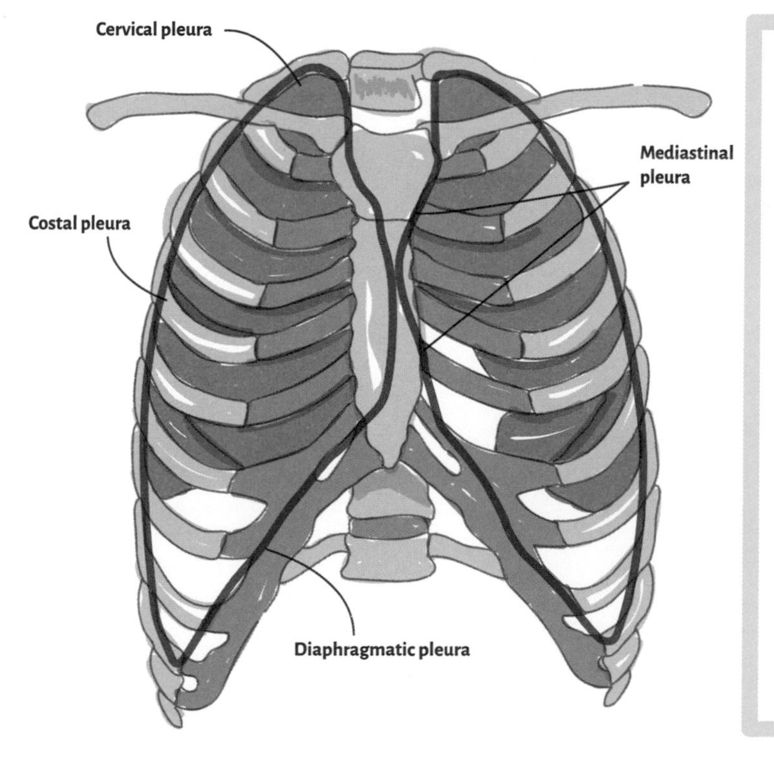
Cervical pleura
Mediastinal pleura
Costal pleura
Diaphragmatic pleura

PLEURAL SACS

The lungs are surrounded by **pleural sacs**, which line the inside of the chest cavity (**parietal pleura**) and cover the lung itself (**visceral** or **pulmonary pleura**). The pleural sac is a low-friction interface that allows the lungs to expand freely inside the chest cavity.

Parietal pleura are divided into **costal**, **mediastinal**, **cervical**, and **diaphragmatic** parts.

Parietal pleura has pain sensation, but visceral pleura does not.

Alveolar structure and cells

Alveoli are hollow grape-shaped structures lined by squamous epithelium. An **alveolar sac** is where two or more alveoli share a common opening.

The walls of alveoli have two cell types.

Type I alveolar cells are simple squamous epithelial cells that form the lining of the alveolus.

Type II alveolar cells are fewer than type I and found between them. They are rounded or cuboidal and secrete alveolar fluid to keep the lining cells moist.

Type II alveolar cells secrete a fatty fluid called **surfactant**, a mixture of phospholipids and lipoproteins. Surfactant reduces the surface tension of the fluid in the alveoli, preventing their collapse when we breathe out.

Alveoli are surrounded by elastic fibers that resist over-distension, and pores connect adjacent alveoli to equalize pressure.

Alveolar macrophages are phagocytic cells that live in the alveolar space and engulf dust and debris. Every hour, 2 million debris-laden and dead alveolar macrophages are wafted by cilia to the larynx where they are swallowed.

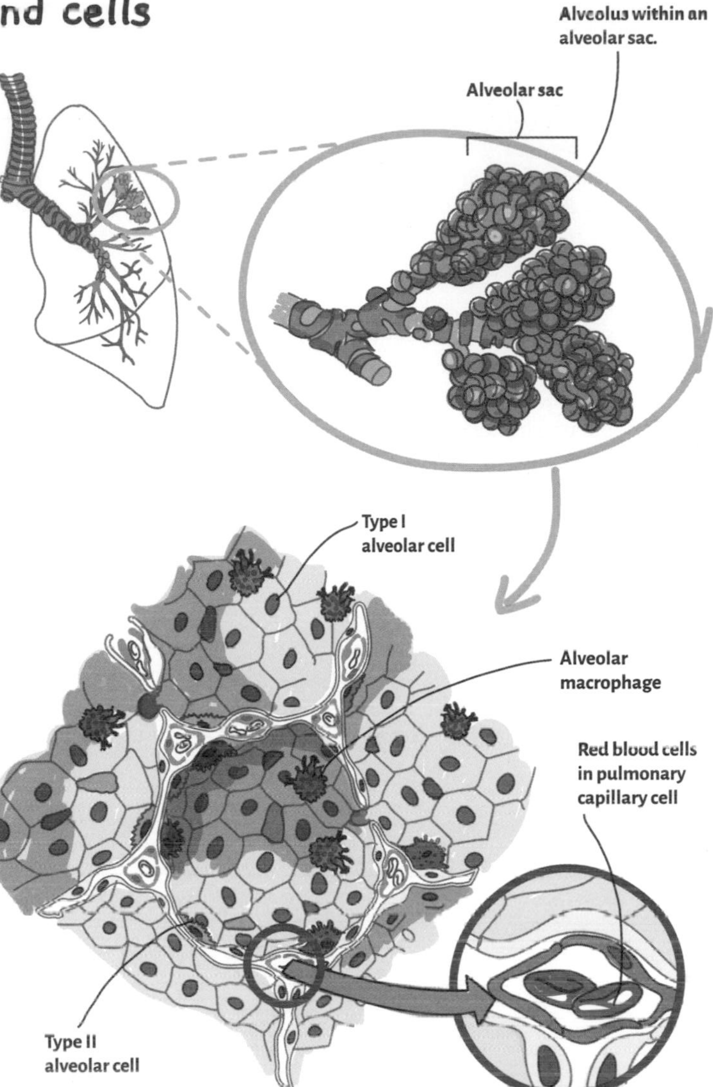

Alveolus within an alveolar sac.
Alveolar sac
Type I alveolar cell
Alveolar macrophage
Red blood cells in pulmonary capillary cell
Type II alveolar cell

ALVEOLI AND GAS EXCHANGE

The primary role of alveoli is gas exchange. The distance between the lumen of an alveolus and the bloodstream is only 1/2,000 mm (1/50,000 inch). This is about 1/15 the thickness of a sheet of paper.

Gas (O_2, CO_2) must diffuse across a respiratory membrane composed of the following:

★Type I alveolar cells
★Fused basement membranes of the alveolus and pulmonary capillary
★Capillary endothelium

O_2 diffuses from the alveolus to the blood; CO_2 diffuses from the blood to the alveolus. Both movements occur down a partial pressure gradient.

ALVEOLI DEVELOPMENT

Alveoli only become functional with our first breath, but they must be prepared in case of premature delivery. No alveoli are present before 24 weeks in utero. The surfactant production only starts at 26 weeks, so independent survival of a premature infant only becomes possible at 26 to 30 weeks. Alveolar walls do not become thin until full term at 40 weeks.

RECAP

NASAL CAVITY AND PARANASAL SINUSES

PARANASAL SINUSES

Are air-filled bone cavities that connect to the nasal cavity.

OLFACTORY EPITHELIUM

Sensory tissue at the top of the nasal cavity. It transmits this olfactory information to the olfactory bulb in the cranial cavity above.

NOSE STRUCTURE

The nose consists of the external nose and the nasal cavity. The external nose is fleshy. The nasal cavity warms and moistens inhaled air to protect the lungs.

RESPIRATORY SYSTEM

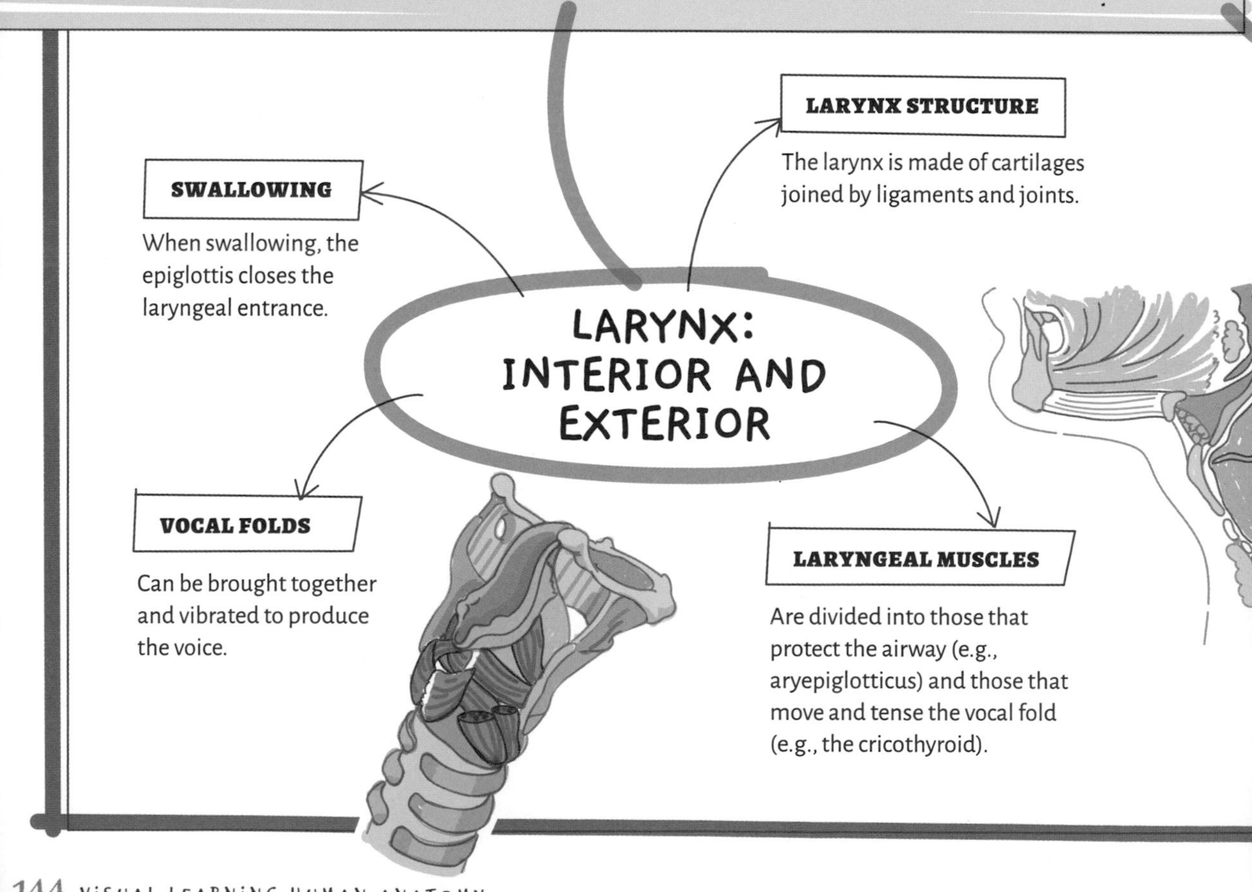

LARYNX: INTERIOR AND EXTERIOR

SWALLOWING

When swallowing, the epiglottis closes the laryngeal entrance.

LARYNX STRUCTURE

The larynx is made of cartilages joined by ligaments and joints.

VOCAL FOLDS

Can be brought together and vibrated to produce the voice.

LARYNGEAL MUSCLES

Are divided into those that protect the airway (e.g., aryepiglotticus) and those that move and tense the vocal fold (e.g., the cricothyroid).

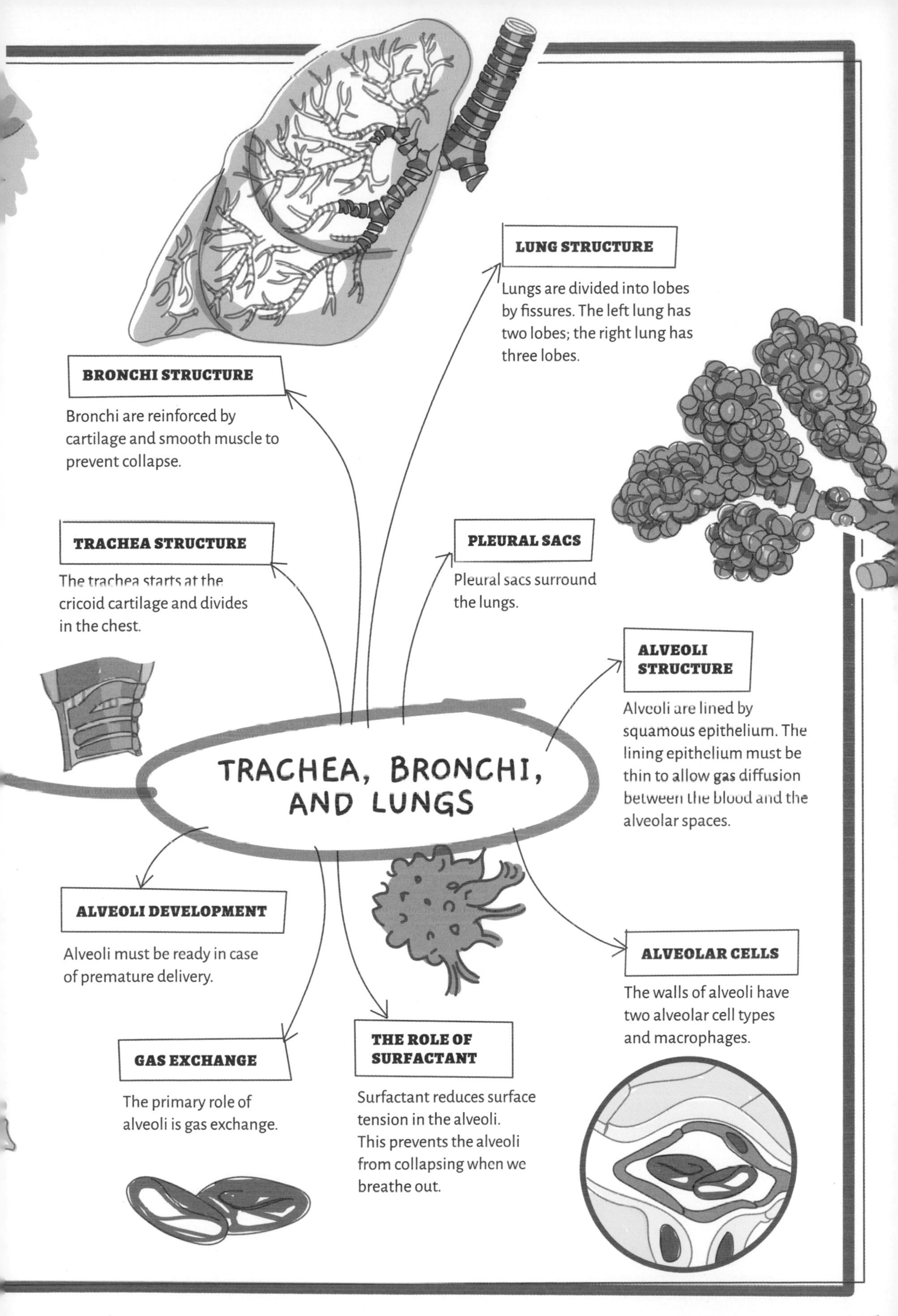
TRACHEA, BRONCHI, AND LUNGS
LUNG STRUCTURE
Lungs are divided into lobes by fissures. The left lung has two lobes; the right lung has three lobes.
BRONCHI STRUCTURE
Bronchi are reinforced by cartilage and smooth muscle to prevent collapse.
TRACHEA STRUCTURE
The trachea starts at the cricoid cartilage and divides in the chest.
PLEURAL SACS
Pleural sacs surround the lungs.
ALVEOLI STRUCTURE
Alveoli are lined by squamous epithelium. The lining epithelium must be thin to allow gas diffusion between the blood and the alveolar spaces.
ALVEOLI DEVELOPMENT
Alveoli must be ready in case of premature delivery.
ALVEOLAR CELLS
The walls of alveoli have two alveolar cell types and macrophages.
GAS EXCHANGE
The primary role of alveoli is gas exchange.
THE ROLE OF SURFACTANT
Surfactant reduces surface tension in the alveoli. This prevents the alveoli from collapsing when we breathe out.

CHAPTER 9

DIGESTIVE SYSTEM

The digestive system must extract nutrients from ingested food as well as protect the body from invaders that we might ingest along with our food. The digestive system includes the alimentary canal (tubular gut) and associated exocrine glands: salivary, liver, and the exocrine part of the pancreas. The alimentary canal also provides a protective environment for the natural gut flora, which supply many vitamins and as much as 10% of nutrients.

The alimentary canal is exposed to ingested pathogens from the external environment, so its walls contain abundant immune system cells (lymphoid follicles). Some of the chemicals used for digestion (e.g., bile salts) are recycled by the liver.

THE ALIMENTARY CANAL

The **alimentary canal** (digestive tract) has four functions: ingestion, digestion, absorption, and excretion or defecation.

The alimentary canal can propel food by swallowing, in the oral cavity and oropharynx, and by rhythmic peristalsis, in the **esophagus** to large intestine.

Absorption of small molecules, such as sugars, amino acids, and fatty acids, is into portal venous channels that flow to the liver. Large fat molecules enter intestinal lymphatics (**lacteals**) that bypass the liver.

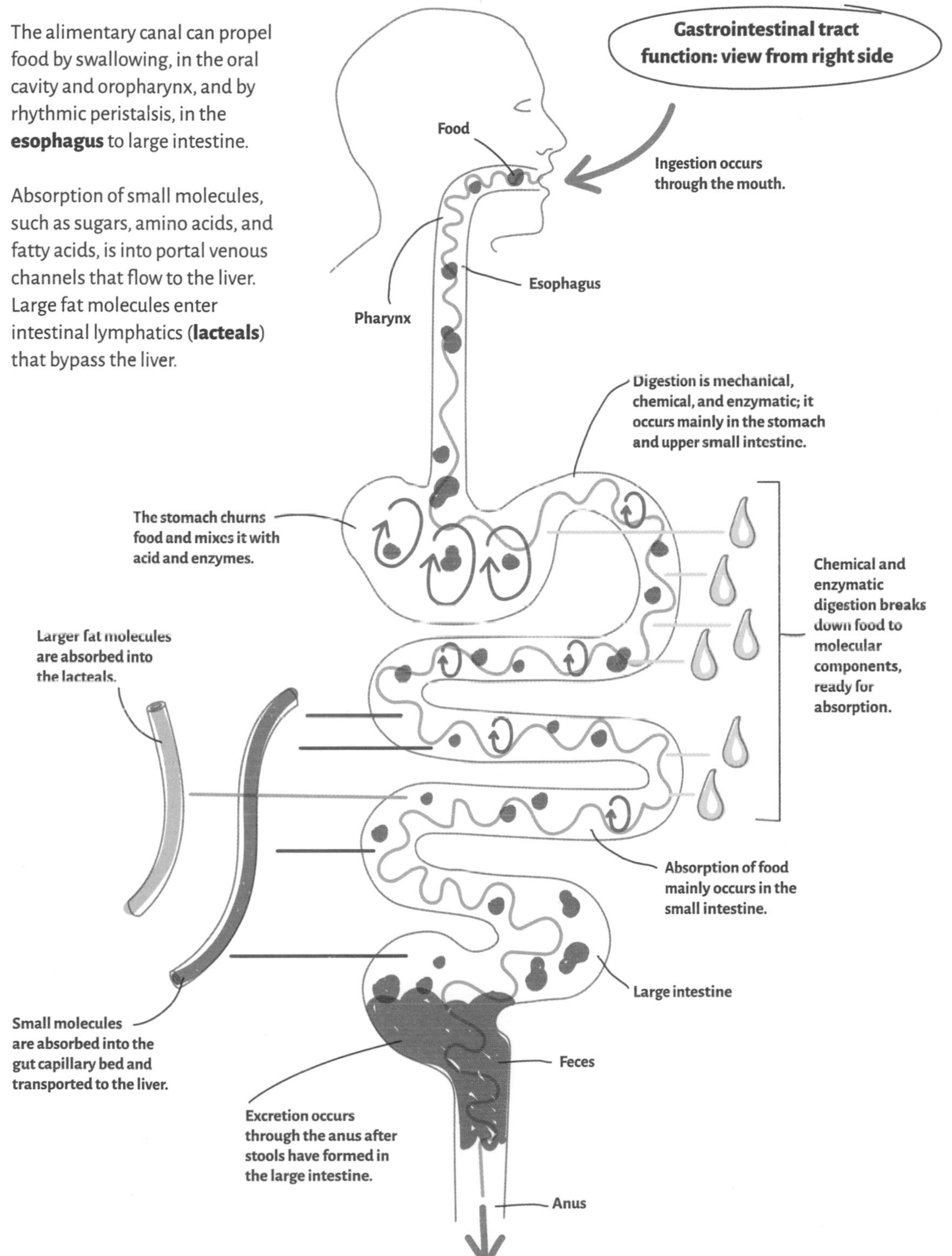

Gastrointestinal tract function: view from right side
Food
Ingestion occurs through the mouth.
Esophagus
Pharynx
Digestion is mechanical, chemical, and enzymatic; it occurs mainly in the stomach and upper small intestine.
The stomach churns food and mixes it with acid and enzymes.
Chemical and enzymatic digestion breaks down food to molecular components, ready for absorption.
Larger fat molecules are absorbed into the lacteals.
Absorption of food mainly occurs in the small intestine.
Large intestine
Small molecules are absorbed into the gut capillary bed and transported to the liver.
Feces
Excretion occurs through the anus after stools have formed in the large intestine.
Anus

SALIVARY GLANDS

Saliva supplies fluid to form masticated food into a bolus (ball of food), provides a fluid seal during sucking, dissolves flavors for taste receptors, and contains starch-digesting enzymes.

There are major and minor salivary glands. The **major salivary glands** are the parotid, submandibular, and sublingual glands; they are macroscopic and drain into the oral cavity. The **minor salivary glands** are microscopic and scattered throughout the oral mucous membranes and the underlying connective tissue.

Major and minor salivary glands

Tongue
Parotid duct
Teeth
Sublingual gland duct openings
Frenulum of the tongue
Sublingual papilla where the submandibular duct opens
Mylohyoid muscle
Submandibular gland
Submandibular duct
Submandibular gland
Superficial part of submandibular gland
The second upper molar adjacent to the papilla
Parotid gland
Deep part of submandibular gland

Sublingual gland

The **sublingual gland** lies under the tongue, in contact with the **sublingual fossa** on the inside of the mandible. The gland has 8 to 20 **sublingual ducts** opening directly into the floor of the mouth or into the submandibular duct.

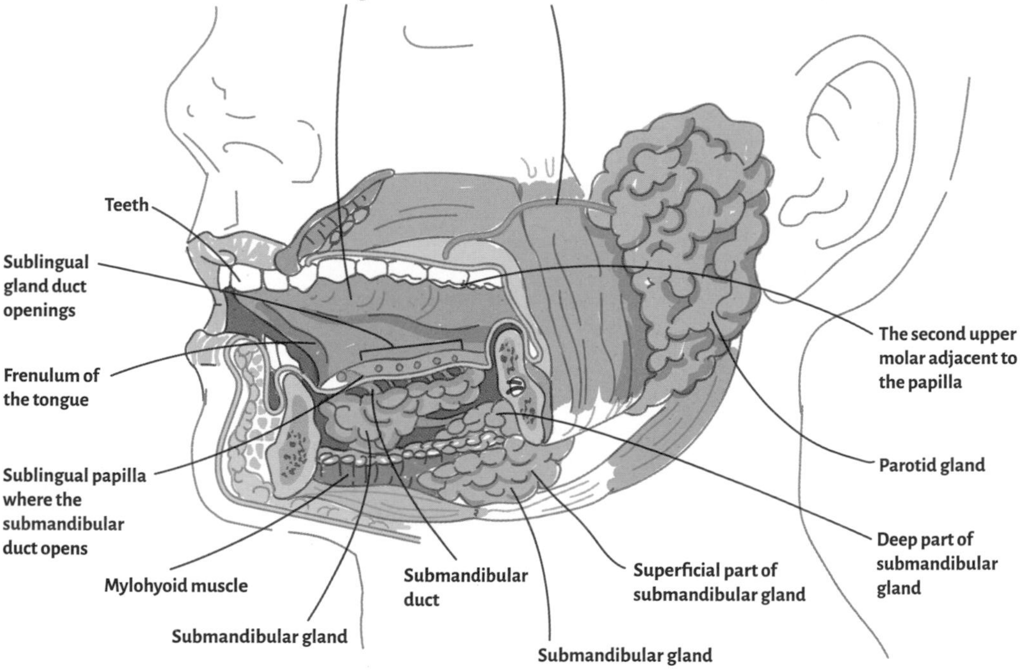

Parotid gland

The **parotid gland** is located anterior and inferior to the external ear. The **parotid duct** carries saliva from the gland to the oral cavity. It ends on a **papilla** near the crown of the second upper molar. This protuberance is easily felt with the tongue tip. The facial nerve and its final branches penetrate the gland.

Submandibular gland

The **submandibular gland** has superficial and deep parts separated by the **mylohyoid muscle**, which forms the floor of the mouth. A **submandibular duct** carries saliva from the gland to the mouth. It is about 5 cm (2 inches) long and opens in the floor of the mouth on the **sublingual papilla** at the base of the **frenulum of the tongue**. The superficial part of the submandibular gland can be felt 2 to 3 cm (about 1 inch) anterior to the mandibular angle.

THE ESOPHAGUS AND THE STOMACH

The esophagus is a muscular tube that carries food to the stomach. The stomach provides a space for the chemical, physical, and enzymatic digestion of protein to begin. It also kills some ingested microorganisms with its acidic environment.

Esophagus structure

The esophagus is approximately 25 cm (12 inches) long and extends from the laryngopharynx to the cardia of the stomach. It begins at the level of the lower border of the cricoid cartilage and passes through the diaphragm at the level of thoracic vertebra 10.

Several structures (inferior constrictor muscle of pharynx, aortic arch, left main bronchus, and diaphragm) compress the esophagus and may slow the passage of food.

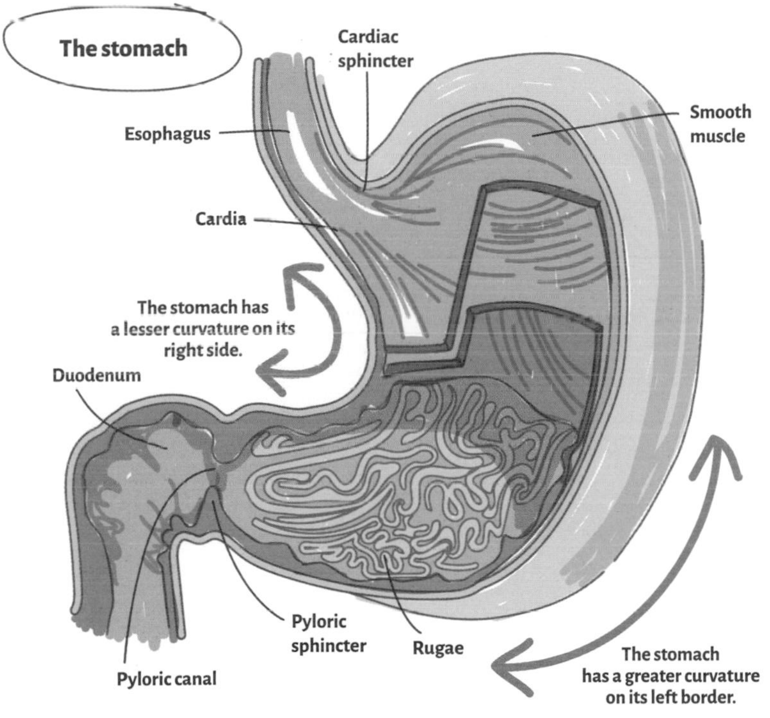

The stomach
Cardiac sphincter
Smooth muscle
Esophagus
Cardia
The stomach has a lesser curvature on its right side.
Duodenum
Pyloric sphincter
Rugae
Pyloric canal
The stomach has a greater curvature on its left border.

Stomach structure

The stomach is a muscular bag located in the upper left abdomen. The esophagus enters the stomach at the **cardia**, and the stomach empties into the first part of the **duodenum** through the **pyloric canal**, which is the outflow from the stomach.

The **cardiac sphincter** is located at the junction of the esophagus and the stomach's opening.

The stomach has two surfaces (anterior and posterior), two borders or curvatures (greater and lesser curvature), and two orifices (cardiac and pyloric). The pyloric orifice is surrounded by the **pyloric sphincter**, which is formed of circular smooth muscle. When empty, the stomach lining forms folds called **rugae**.

Stomach function

The stomach processes food by mechanical, chemical, and enzymatic means. **Mechanical digestion** is due to churning by the three layers of smooth muscle (oblique, longitudinal, and circular) in the stomach wall.

Chemical digestion is due to the hydrochloric acid produced by the parietal or oxyntic cells in the stomach epithelium. **Enzymatic digestion** is by pepsin produced by the chief or zymogenic cells of the stomach epithelium.

THE SMALL AND LARGE INTESTINE

The small intestine is the site of absorption for most important nutrients (sugars, amino acids, nucleic acids, and fats). The large intestine is mainly concerned with absorbing water and minerals to form stool.

Small intestine structure

The **small intestine** consists of the duodenum, jejunum, and ileum.

The mucosa (mucous membrane) of the small intestine, particularly in the duodenum and the jejunum, is arranged in circular folds called **plicae circulares**. Each plica has many fingerlike **villi**, and each epithelial cell has tiny microvilli on its surface. Both villi and circular folds increase the mucosal surface area to assist absorption. All three structures increase the surface area to improve absorption.

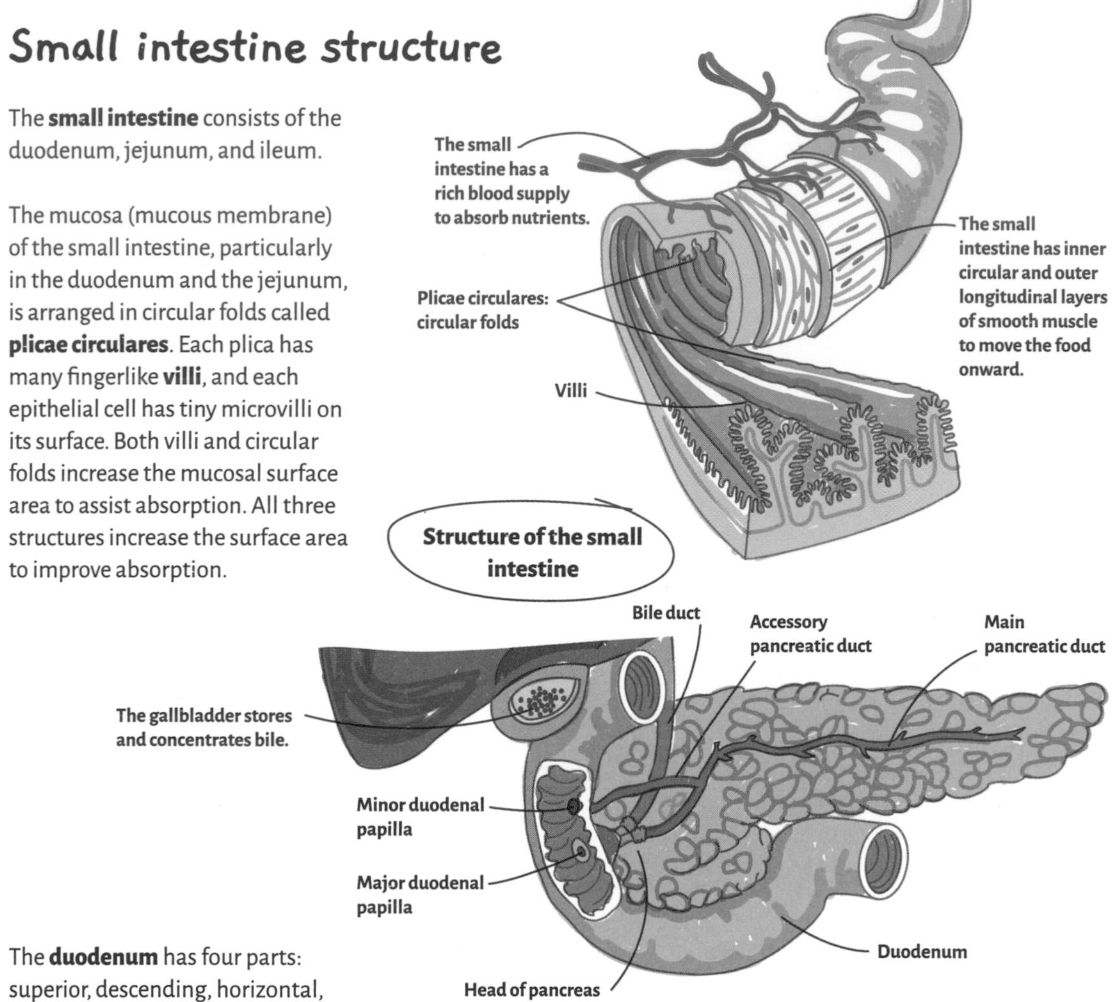
The small intestine has a rich blood supply to absorb nutrients.
Plicae circulares: circular folds
Villi
The small intestine has inner circular and outer longitudinal layers of smooth muscle to move the food onward.
Bile duct
Accessory pancreatic duct
Main pancreatic duct
The gallbladder stores and concentrates bile.
Minor duodenal papilla
Major duodenal papilla
Duodenum
Head of pancreas

Structure of the small intestine

The **duodenum** has four parts: superior, descending, horizontal, and ascending. Two papillae may be seen in the medial wall of the second part of the duodenum.

The **major duodenal papilla** lies 8 to 10 cm (3 to 4 inches) beyond the pylorus and has at its tip the opening of the hepatopancreatic ampulla (of Vater). The ampulla receives the **bile duct** and the **main pancreatic duct**.

The **minor duodenal papilla** lies about 6 to 8 cm (2.5 to 3 inches) beyond the pylorus and has at its tip the opening of the accessory pancreatic duct.

The **jejunum** and **ileum** together are 5 to 8 meters (16 to 26 feet) in length. They are suspended from the posterior abdominal wall by a peritoneal fold called the **mesentery**, which contains the blood, lymph, and nerve supply to the digestive tract between its layers.

The **gallbladder** releases bile to the duodenum when a fatty meal is eaten.

Large intestine structure

The **large intestine** forms a roughly square border enclosing the small intestine. It is 1.5 meters (5 to 6 feet) long in the adult and extends from the terminal ileum (**ileocecal valve**) to the anus. Its parts are the **cecum**, **ascending colon**, **transverse colon**, **descending colon**, **sigmoid colon**, and the **anorectal canal**.

It can be distinguished from the small intestine by the following points:

★ Longitudinal smooth muscle in the large intestine is arranged in three bands (**teniae coli**).

★ The colonic wall is puckered into sacculations (haustra) by the teniae.

★ The large intestine has **epiploic appendages** (fat-filled projections) scattered over the free surface of most of its length.

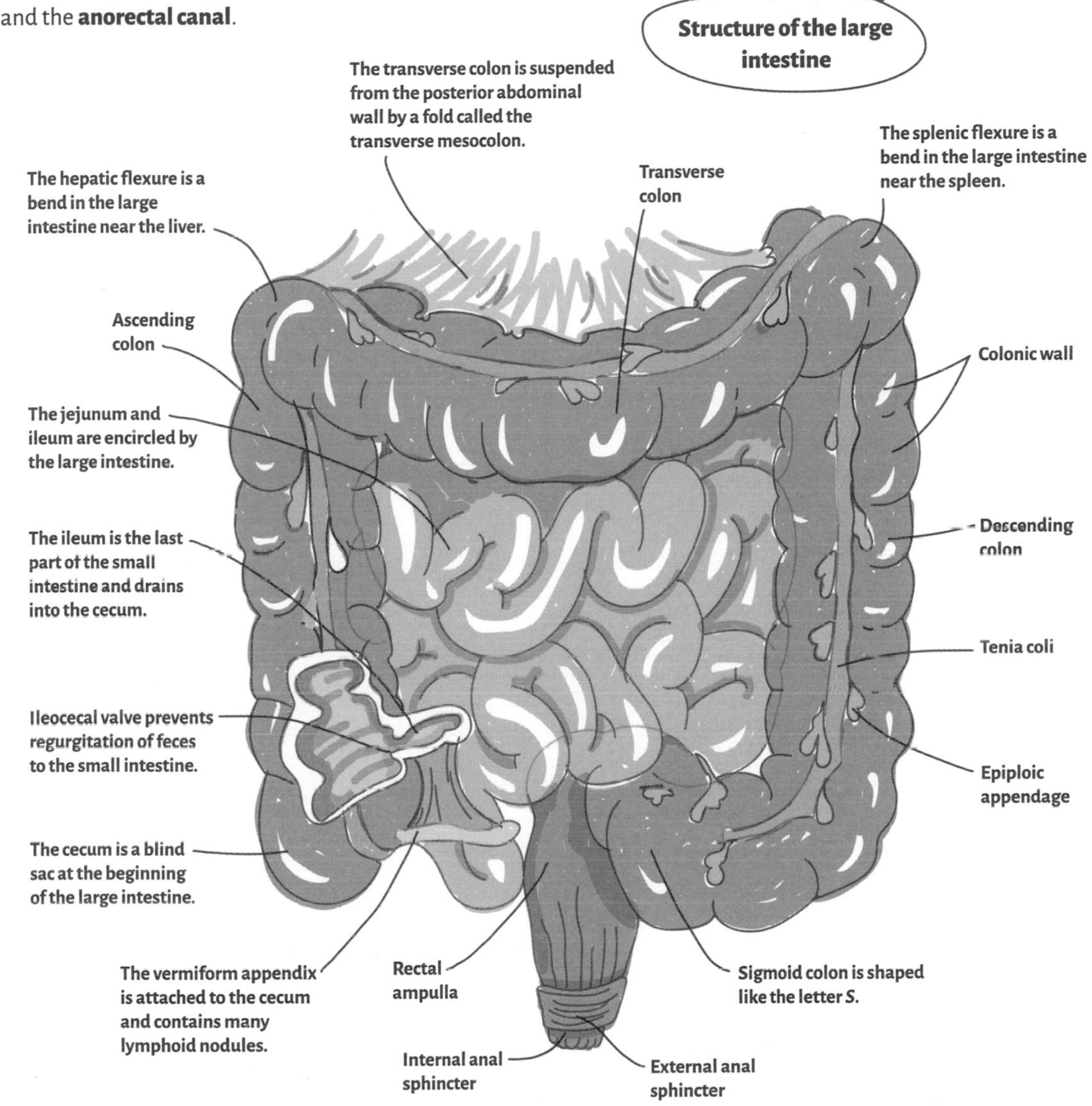
Structure of the large intestine
The transverse colon is suspended from the posterior abdominal wall by a fold called the transverse mesocolon.
Transverse colon
The splenic flexure is a bend in the large intestine near the spleen.
The hepatic flexure is a bend in the large intestine near the liver.
Ascending colon
Colonic wall
The jejunum and ileum are encircled by the large intestine.
Descending colon
The ileum is the last part of the small intestine and drains into the cecum.
Tenia coli
Ileocecal valve prevents regurgitation of feces to the small intestine.
Epiploic appendage
The cecum is a blind sac at the beginning of the large intestine.
The vermiform appendix is attached to the cecum and contains many lymphoid nodules.
Rectal ampulla
Sigmoid colon is shaped like the letter *S*.
Internal anal sphincter
External anal sphincter

Anus

The **anus** is the terminal part of the alimentary canal. It is surrounded by the internal anal sphincter (involuntary smooth muscle) and the external anal sphincter (voluntary skeletal muscle).

The **rectal ampulla** is a dilated region that can hold feces for a short period of time. When feces reach the rectal ampulla above the anus, stretch receptors signal the urge to defecate.

The **external anal sphincter** is a voluntary muscle that controls the movement of feces to the external world. The **internal anal sphincter** is smooth muscle that squeezes the feces into separate masses.

LIVER, GALLBLADDER, AND EXOCRINE PANCREAS

The liver serves many roles, including glycogen storage, plasma protein manufacture, urea production, and bile salt production. The gallbladder stores bile for release on demand. The exocrine pancreas produces enzymes and neutralizing agents.

Liver structure

The liver lies in the upper right quadrant of the abdomen below the right ribs. It consists of two lobes separated by the falciform ligament.

The basic microscopic structure of the liver is multiple hexagonal prisms (**hepatic lobules**).

Each lobule receives intestinal blood from branches of the **portal vein** and oxygenated blood from arteriole branches of the hepatic artery. These branch into **hepatic sinusoids**, which flow between plates of **hepatocytes** (liver cells).

Sinusoids drain into a central vein, and this blood flows to the hepatic veins and the inferior vena cava. Hepatocytes produce bile, which flows by **bile canaliculi** to a **bile duct** branch.

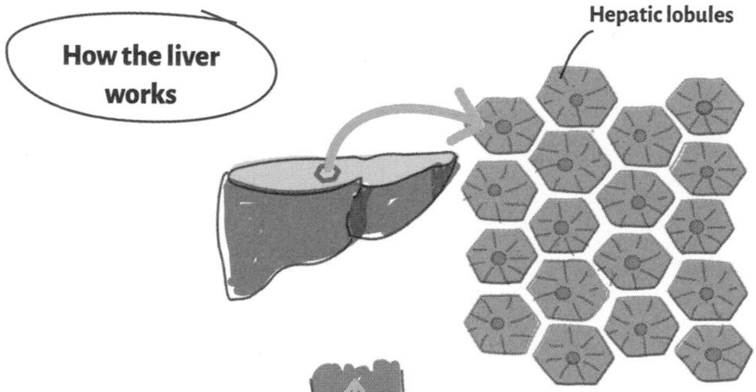

How the liver works

Hepatic lobules

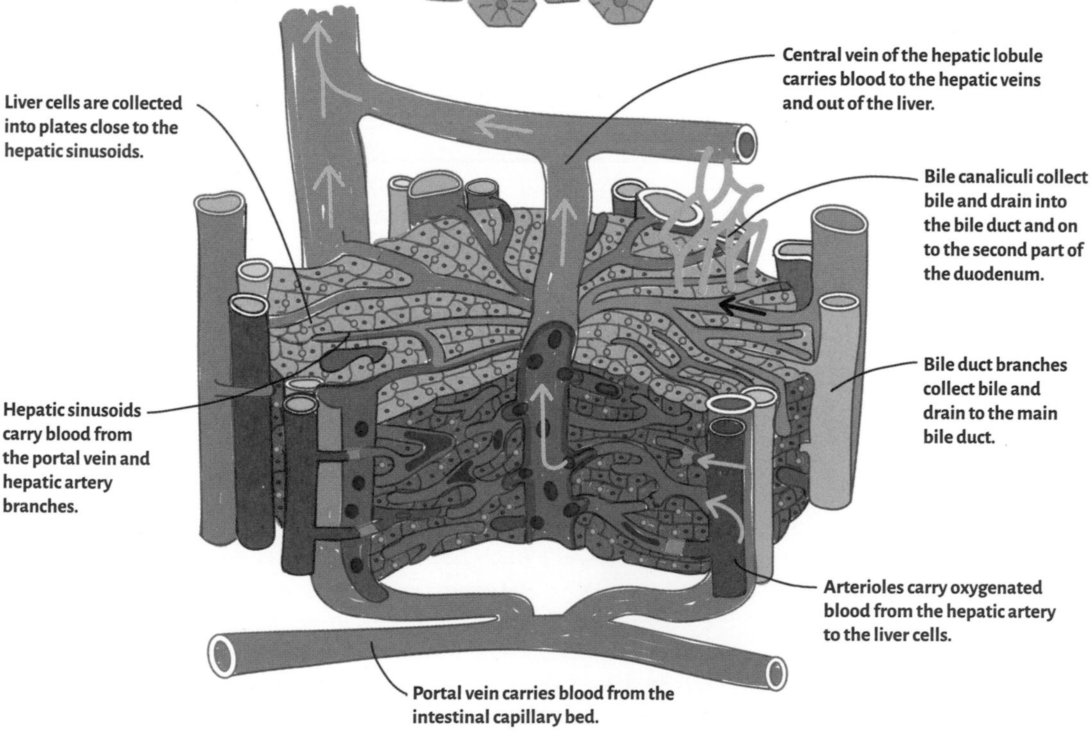

Central vein of the hepatic lobule carries blood to the hepatic veins and out of the liver.

Liver cells are collected into plates close to the hepatic sinusoids.

Bile canaliculi collect bile and drain into the bile duct and on to the second part of the duodenum.

Bile duct branches collect bile and drain to the main bile duct.

Hepatic sinusoids carry blood from the portal vein and hepatic artery branches.

Arterioles carry oxygenated blood from the hepatic artery to the liver cells.

Portal vein carries blood from the intestinal capillary bed.

Liver function

The liver has the following endocrine or metabolic functions:

★ Protein synthesis (albumin, thrombopoietin, and angiotensinogen)
★ Carbohydrate storage as glycogen
★ Fat metabolism
★ Mineral and vitamin storage
★ Production of clotting factors (fibrinogen, factors II, VII, IX, and X)
★ Detoxification of blood from the gut (alcohol, drugs, bacterial and fungal toxins in food)

Hepatocytes also secrete bile for release into the duodenum to emulsify fats and to help in the digestion of fats.

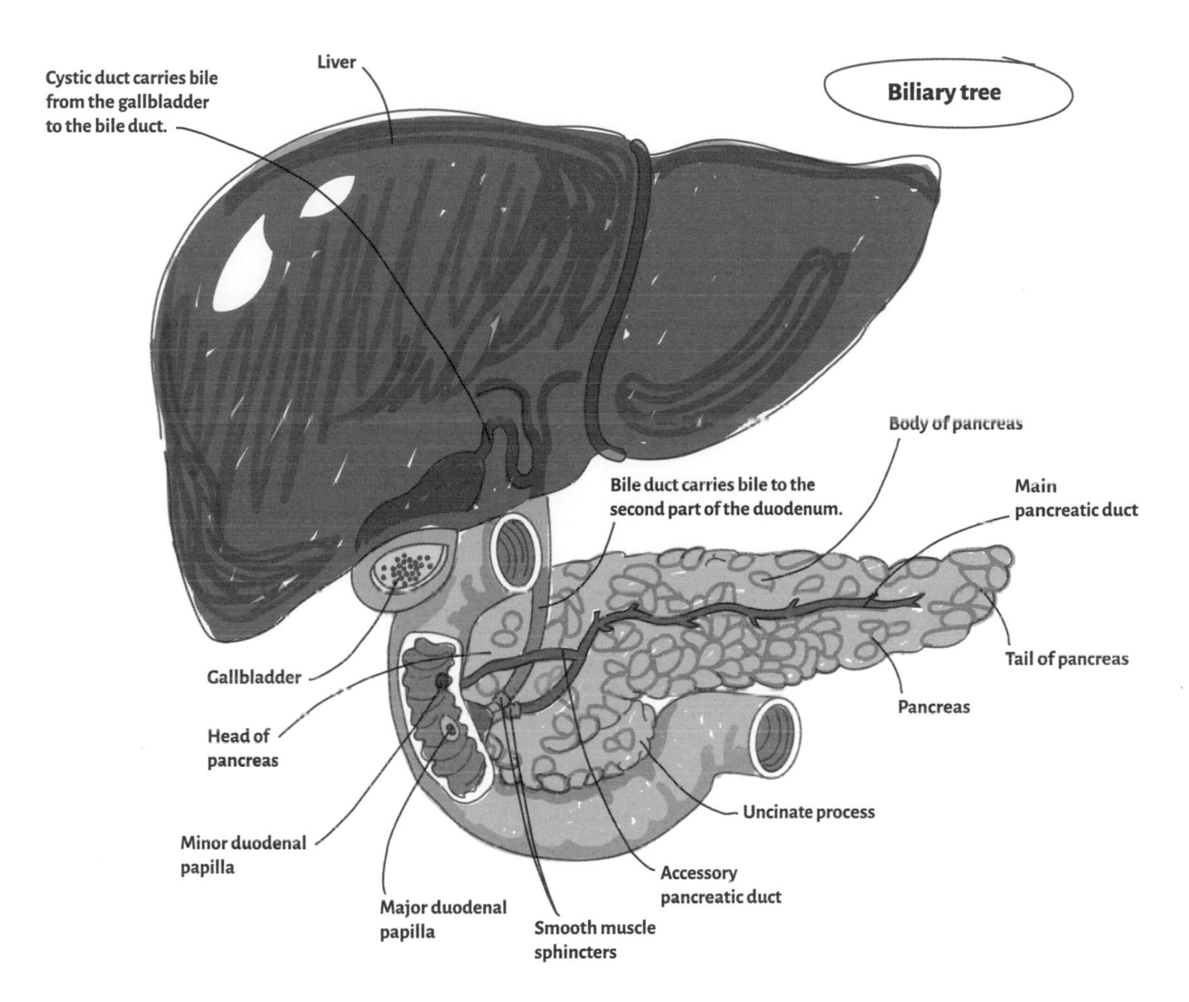
Biliary tree
Liver
Cystic duct carries bile from the gallbladder to the bile duct.
Body of pancreas
Bile duct carries bile to the second part of the duodenum.
Main pancreatic duct
Tail of pancreas
Pancreas
Gallbladder
Head of pancreas
Uncinate process
Minor duodenal papilla
Accessory pancreatic duct
Major duodenal papilla
Smooth muscle sphincters

Exocrine pancreas

The exocrine pancreas consists of a head, neck, body, and tail. Projecting from the lower left part of the head is the uncinate process. The head of the pancreas is encircled by the duodenum, which receives the pancreatic ducts.

Ducts of the pancreas are the main pancreatic duct and the accessory pancreatic duct. **The main pancreatic duct** is joined by the **bile duct** to form an hepatopancreatic ampulla. The ampulla drains into the duodenum at the **major duodenal papilla**. Sphincters control the flow. The **accessory pancreatic duct** drains into the **minor duodenal papilla** in the duodenum.

Biliary tree

Bile salts from the liver are stored in the **gallbladder** and released on demand to emulsify fats in food.

Left and right hepatic ducts combine to form a common hepatic duct, which joins with the **cystic duct** from the gallbladder to form the bile duct.

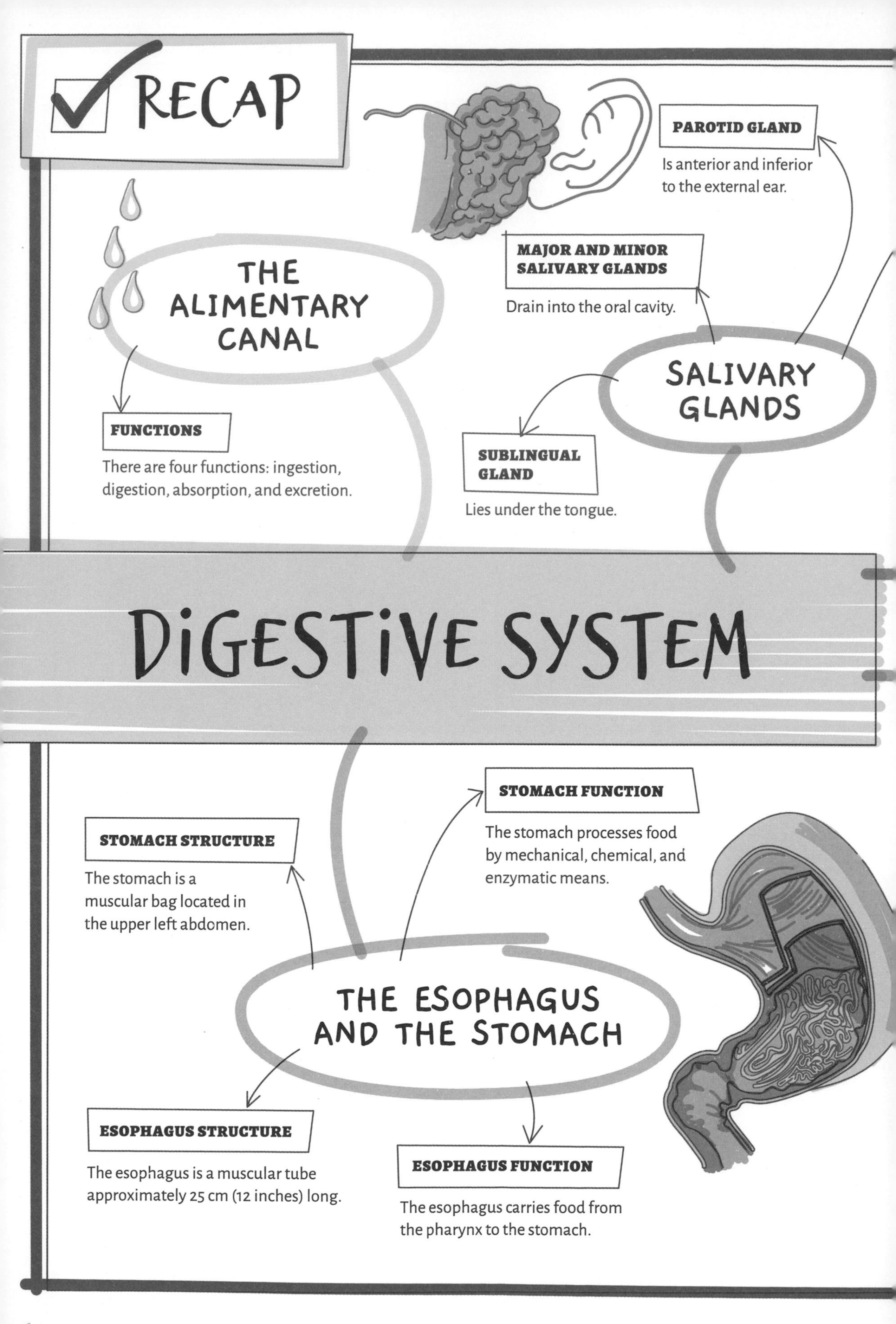
RECAP
THE ALIMENTARY CANAL
FUNCTIONS
There are four functions: ingestion, digestion, absorption, and excretion.
SALIVARY GLANDS
PAROTID GLAND
Is anterior and inferior to the external ear.
MAJOR AND MINOR SALIVARY GLANDS
Drain into the oral cavity.
SUBLINGUAL GLAND
Lies under the tongue.
DIGESTIVE SYSTEM
THE ESOPHAGUS AND THE STOMACH
STOMACH STRUCTURE
The stomach is a muscular bag located in the upper left abdomen.
STOMACH FUNCTION
The stomach processes food by mechanical, chemical, and enzymatic means.
ESOPHAGUS STRUCTURE
The esophagus is a muscular tube approximately 25 cm (12 inches) long.
ESOPHAGUS FUNCTION
The esophagus carries food from the pharynx to the stomach.

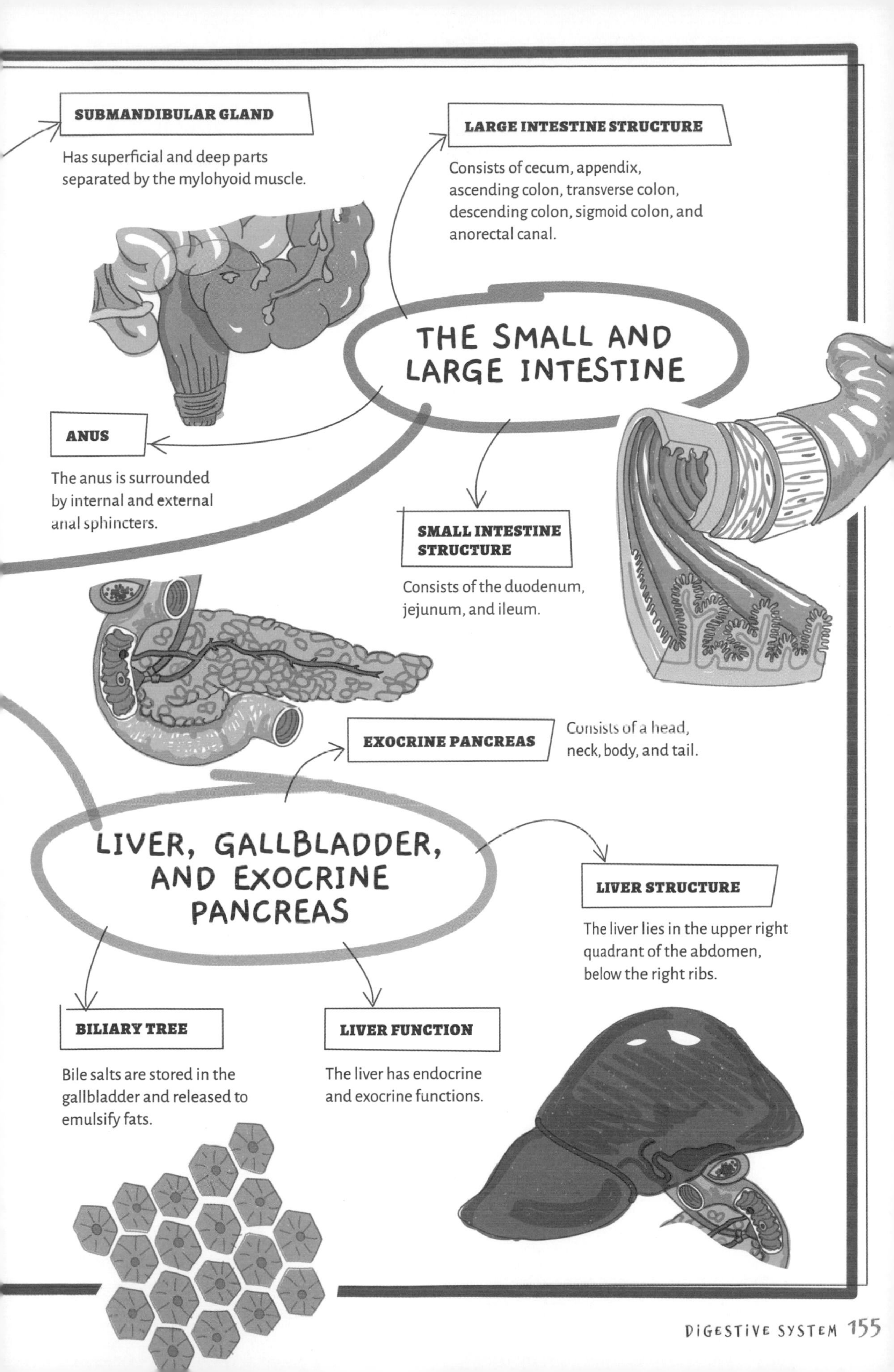
SUBMANDIBULAR GLAND
Has superficial and deep parts separated by the mylohyoid muscle.
LARGE INTESTINE STRUCTURE
Consists of cecum, appendix, ascending colon, transverse colon, descending colon, sigmoid colon, and anorectal canal.
THE SMALL AND LARGE INTESTINE
ANUS
The anus is surrounded by internal and external anal sphincters.
SMALL INTESTINE STRUCTURE
Consists of the duodenum, jejunum, and ileum.
EXOCRINE PANCREAS
Consists of a head, neck, body, and tail.
LIVER, GALLBLADDER, AND EXOCRINE PANCREAS
LIVER STRUCTURE
The liver lies in the upper right quadrant of the abdomen, below the right ribs.
BILIARY TREE
Bile salts are stored in the gallbladder and released to emulsify fats.
LIVER FUNCTION
The liver has endocrine and exocrine functions.

CHAPTER 10

URINARY SYSTEM

The purpose of the urinary system is to drain urine and other waste from the body. The urinary tract consists of two kidneys on the posterior abdominal wall, two ureters, a midline urinary bladder, and a midline urethra. The kidneys filter blood plasma, but immediately return the bulk of the fluid to the bloodstream.

The weight of the two kidneys declines by 30% from young adulthood to old age, and renal function drops by half. Urine is channeled by the ureters to the urinary bladder and from there to the external environment by the urethra.

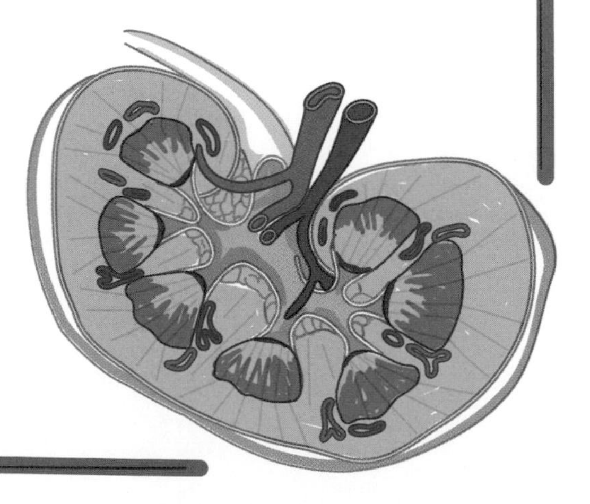

THE URINARY TRACT

Although the primary role of the kidneys is to extract nitrogenous waste from the blood, they perform a multitude of other tasks.

Functions of the urinary tract

The functions of the urinary tract include the following:

★ Excretion of nitrogenous waste, drugs, bilirubin, creatinine, uric acid, and toxins
★ Regulation of blood ionic composition (sodium, potassium, and chloride)
★ Regulation of blood pH (acid-base balance)
★ Regulation of blood volume
★ Regulation of blood pressure
★ Regulation of blood osmolarity (dissolved particles per volume)
★ Production of hormones for calcium metabolism
★ Production of hormones for red blood cell production
★ Regulation of blood glucose

The system that produces and excretes urine

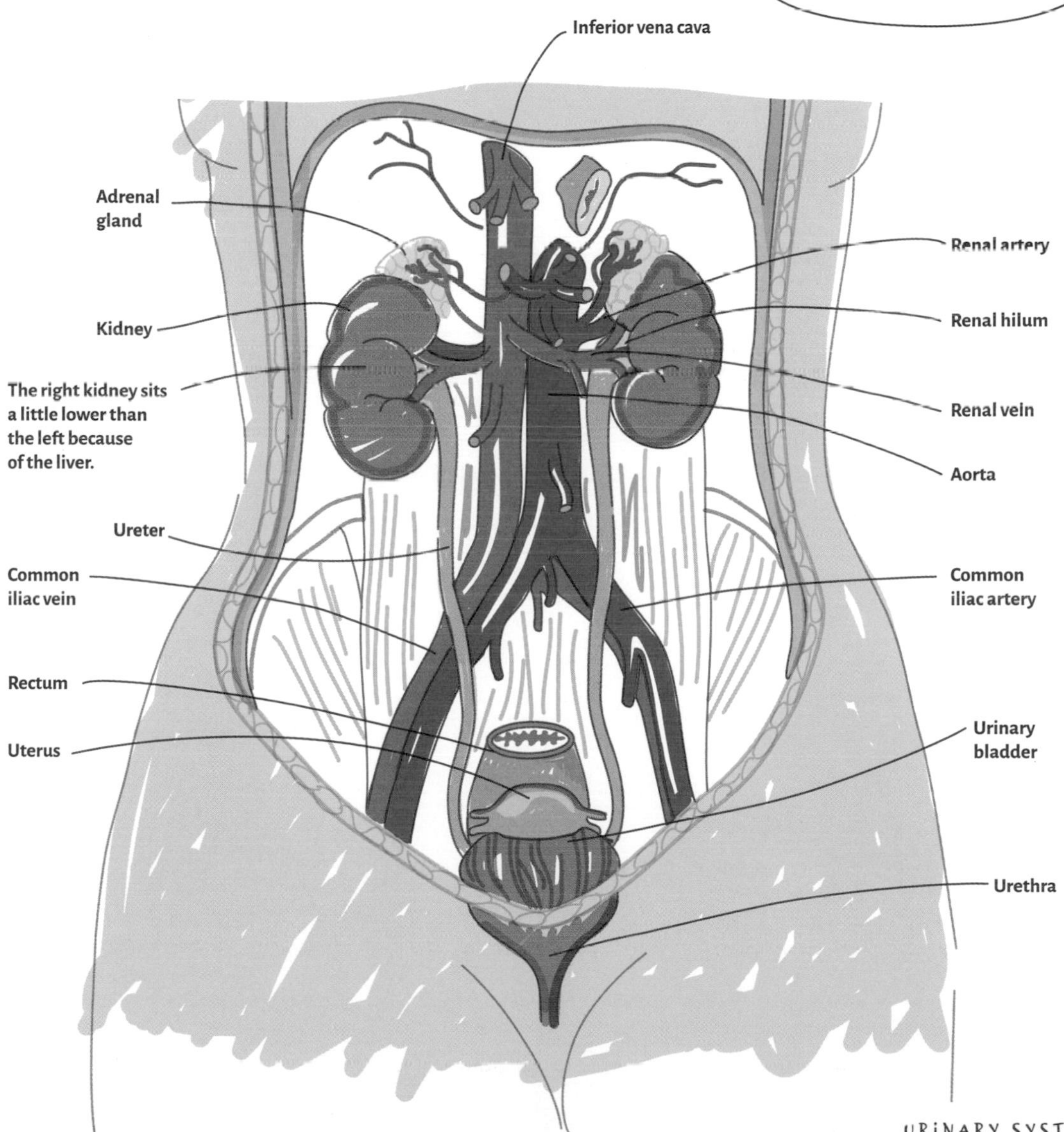

Inferior vena cava
Adrenal gland
Kidney
The right kidney sits a little lower than the left because of the liver.
Ureter
Common iliac vein
Rectum
Uterus
Renal artery
Renal hilum
Renal vein
Aorta
Common iliac artery
Urinary bladder
Urethra

THE KiDNEYS

The two kidneys lie on the posterior abdominal wall behind the peritoneal cavity. They have a rich blood supply (25% of cardiac output) and remove between 12 and 20 grams (0.4 to 0.7 ounce) of urea from the blood each day.

Kidney structure

Each kidney is surrounded by a **fibrous capsule**. The interior of each kidney is divided into outer cortex and inner medulla.

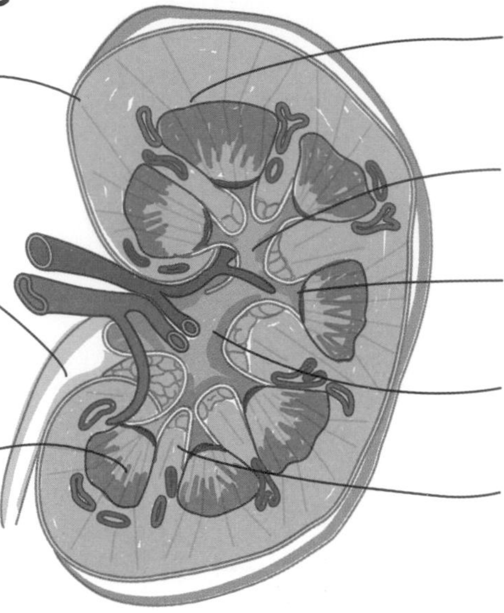

Ureter: Receives urine from the renal pelvis, leaves the renal hilum, and descends to the bladder

Medulla: Contains papillae of pyramids with nephron loops (of Henle)

Cortex: Contains renal corpuscles, and proximal and distal convoluted tubules

Major calyx: Receives urine from the minor calyces

Minor calyx: Receives urine from the papillae of pyramids

Renal pelvis: Receives urine from the major calyces

Renal column: Cortical tissue extends into the medulla

Kidney function

The functional unit of the kidney is the nephron, of which there are about 1 million in each kidney.

The three functions of the kidney are:

1. **Filtration** of plasma and dissolved substances (in the glomeruli of the renal corpuscle)
2. **Tubular reabsorption** of water and of useful dissolved substances (in the renal tubules)

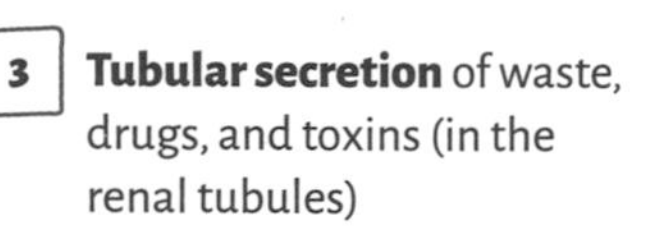

3. **Tubular secretion** of waste, drugs, and toxins (in the renal tubules)

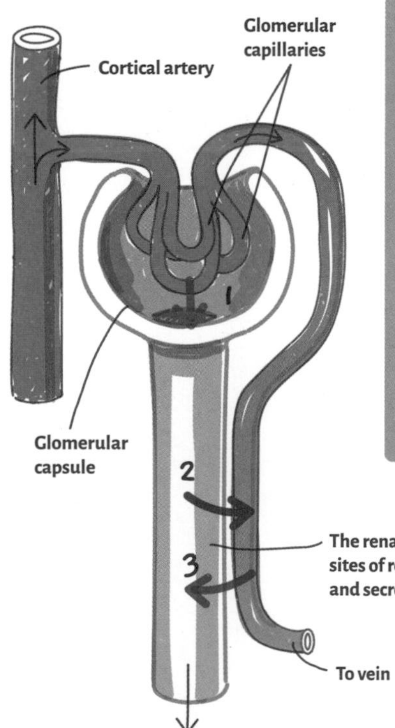
Cortical artery
Glomerular capillaries
Glomerular capsule
2
3
The renal tubules are sites of reabsorption and secretion.
To vein

GLOMERULAR FILTRATE

The daily volume of glomerular filtrate is 150 liters (33 gallons) for women and 180 liters (40 gallons) for men. Tubular reabsorption returns more than 99% of the filtrate to the bloodstream, so only 1% (1 to 2 liters; 0.25 to 0.5 gallons) is excreted as urine.

Glomerular and tubular structure

Each **renal corpuscle** consists of a **glomerular (Bowman's) capsule** with a **capsular space** and a **glomerular tuft of capillaries** covered by podocytes. Plasma fluid is filtered between podocyte processes into the capsular space and on into the proximal convoluted tubule.

Glomerular filtrate passes (in sequence) through the proximal convoluted tubule, nephron loop, distal convoluted tubule, collecting duct, and papillary duct, to emerge at the summit of the renal papilla.

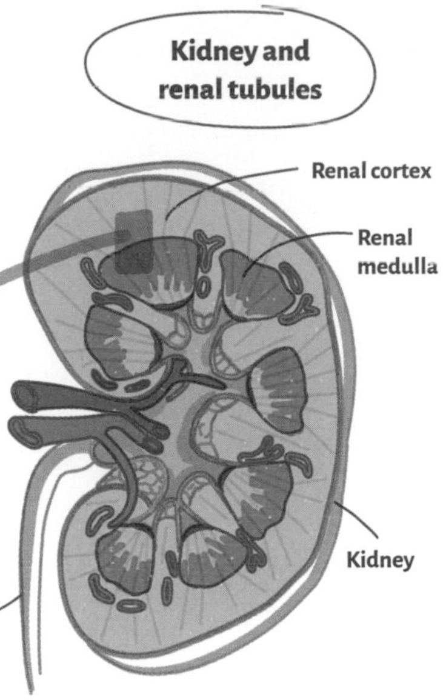

Kidney and renal tubules

Renal cortex

Renal medulla

Kidney

Ureter

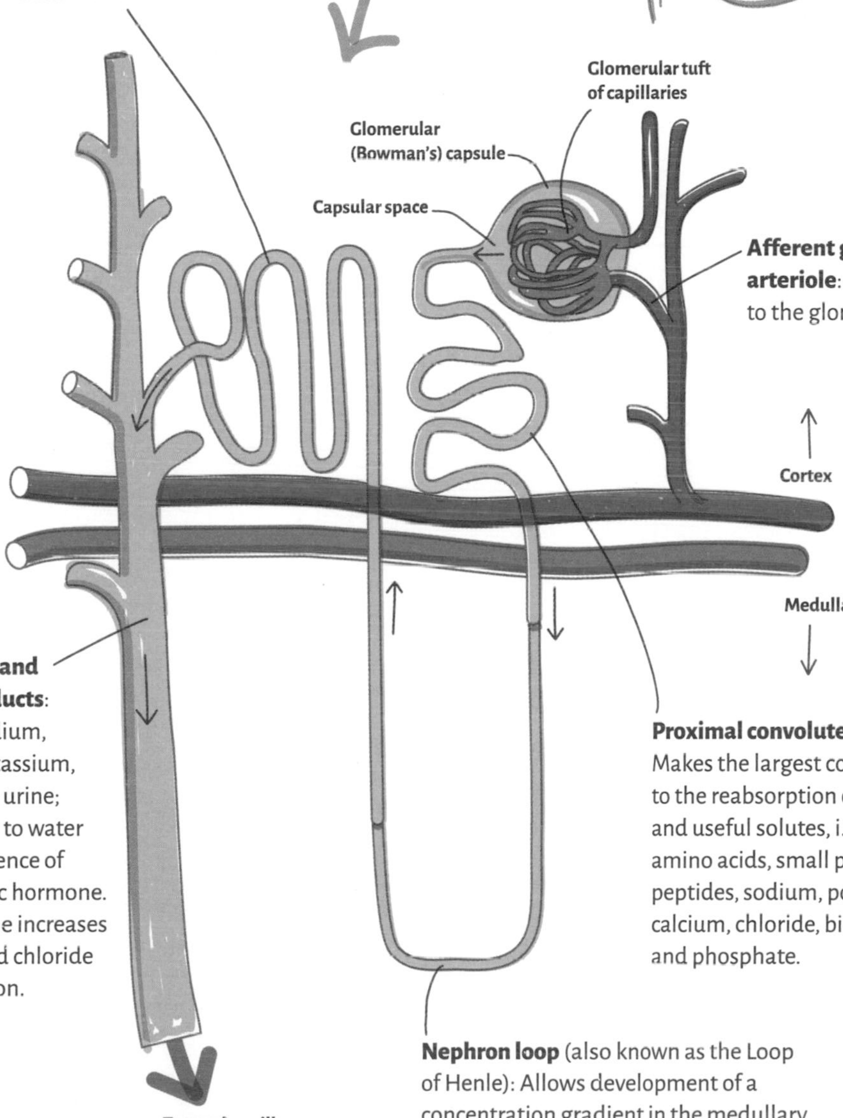

Distal convoluted tubule: For secretion of drugs. It is permeable to water in the presence of antidiuretic hormone.

Glomerular tuft of capillaries

Glomerular (Bowman's) capsule

Capsular space

Afferent glomerular arteriole: Carries blood to the glomerular tuft.

Cortex

Medulla

Collecting and papillary ducts: Absorb sodium, excrete potassium, and collect urine; permeable to water in the presence of antidiuretic hormone. Aldosterone increases sodium and chloride reabsorption.

Proximal convoluted tubule: Makes the largest contribution to the reabsorption of water and useful solutes, i.e., glucose, amino acids, small proteins and peptides, sodium, potassium, calcium, chloride, bicarbonate, and phosphate.

To renal papilla and minor calyx

Nephron loop (also known as the Loop of Henle): Allows development of a concentration gradient in the medullary pyramids; it absorbs 15% of filtered water.

URETER, URINARY BLADDER, AND URETHRA

Urine is carried down the posterior abdominal wall in the paired ureters, to be stored in the muscular urinary bladder of the pelvis, before it is discharged into the external environment by the midline urethra.

Ureters

The **ureters** are paired tubes that carry urine from the renal pelvis to the urinary bladder. They lie anterior to the psoas major muscle before crossing the sacroiliac joint to enter the pelvis.

Urinary bladder

The ureters enter the urinary bladder at the upper corners of the trigone. The bladder is a muscular bag that can expand to 1 liter (0.22 gallon) volume and can empty to no volume.

The **bladder wall** in both sexes is made of smooth muscle (the **detrusor**) that can contract to expel urine. This wall is lined with transitional epithelium that can tolerate extreme stretching and relaxation. The urethra drains the bladder neck.

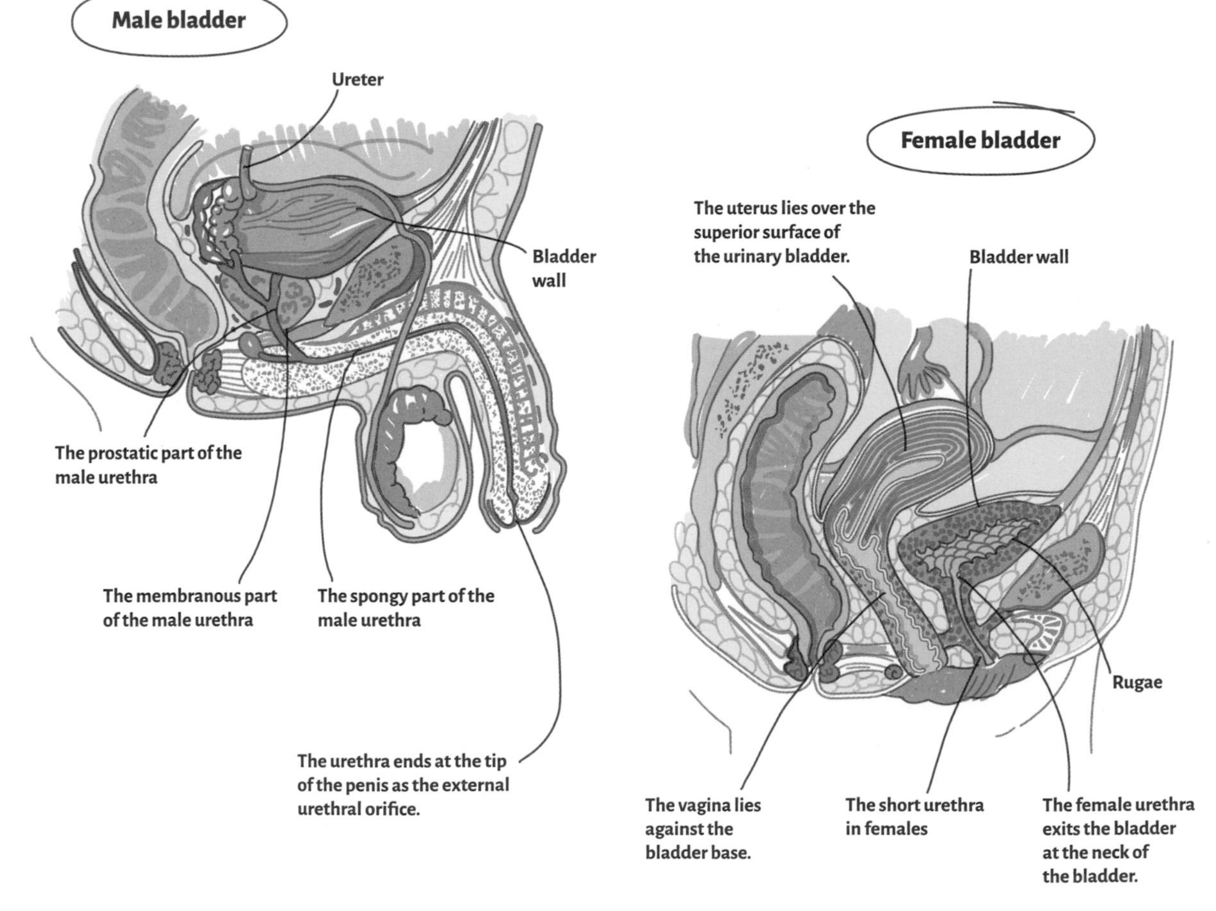

Male bladder
Ureter
Bladder wall
The prostatic part of the male urethra
The membranous part of the male urethra
The spongy part of the male urethra
The urethra ends at the tip of the penis as the external urethral orifice.
Female bladder
The uterus lies over the superior surface of the urinary bladder.
Bladder wall
Rugae
The vagina lies against the bladder base.
The short urethra in females
The female urethra exits the bladder at the neck of the bladder.

Urethra

The urethra extends from the internal to the external urethral orifice. An involuntary **internal urethral sphincter** surrounds the **internal orifice**.

The urethra in both sexes passes through the voluntary **external urethral sphincter** of the muscular **urogenital diaphragm**. The urethra is very different in length in the two sexes.

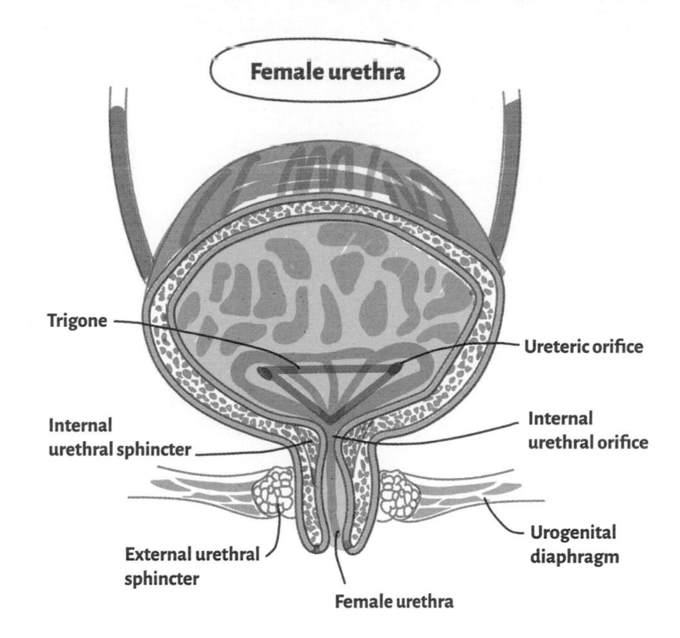
Female urethra
Trigone
Ureteric orifice
Internal urethral sphincter
Internal urethral orifice
Urogenital diaphragm
External urethral sphincter
Female urethra

The **female urethra** is only 4 cm (1.5 inches) long and runs from the neck of the bladder to the space between the labia minora. It passes through the voluntary external urethral sphincter of the muscular urogenital diaphragm. The short length of the female urethra makes girls and women vulnerable to urinary tract infection.

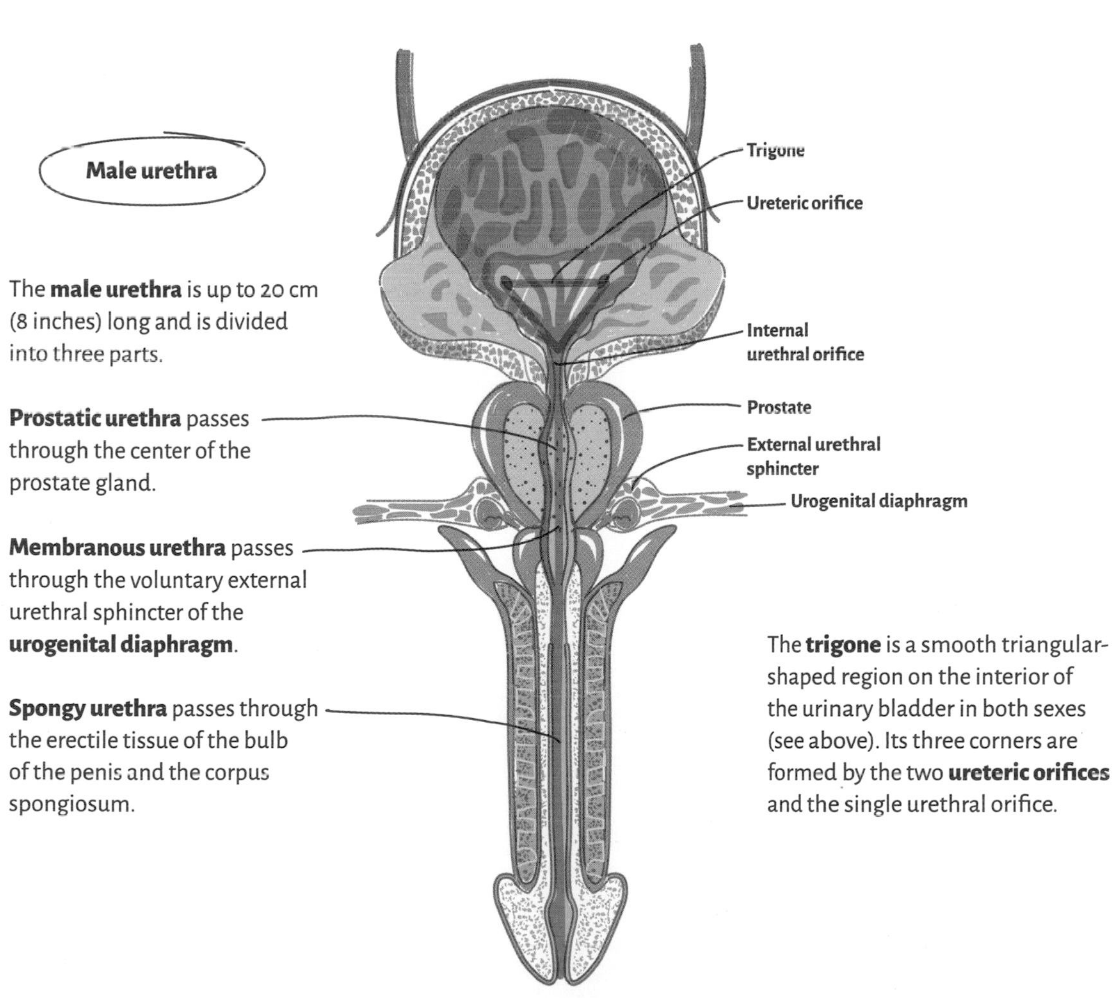
Male urethra
Trigone
Ureteric orifice
Internal urethral orifice
Prostate
External urethral sphincter
Urogenital diaphragm

The **male urethra** is up to 20 cm (8 inches) long and is divided into three parts.

Prostatic urethra passes through the center of the prostate gland.

Membranous urethra passes through the voluntary external urethral sphincter of the **urogenital diaphragm**.

Spongy urethra passes through the erectile tissue of the bulb of the penis and the corpus spongiosum.

The **trigone** is a smooth triangular-shaped region on the interior of the urinary bladder in both sexes (see above). Its three corners are formed by the two **ureteric orifices** and the single urethral orifice.

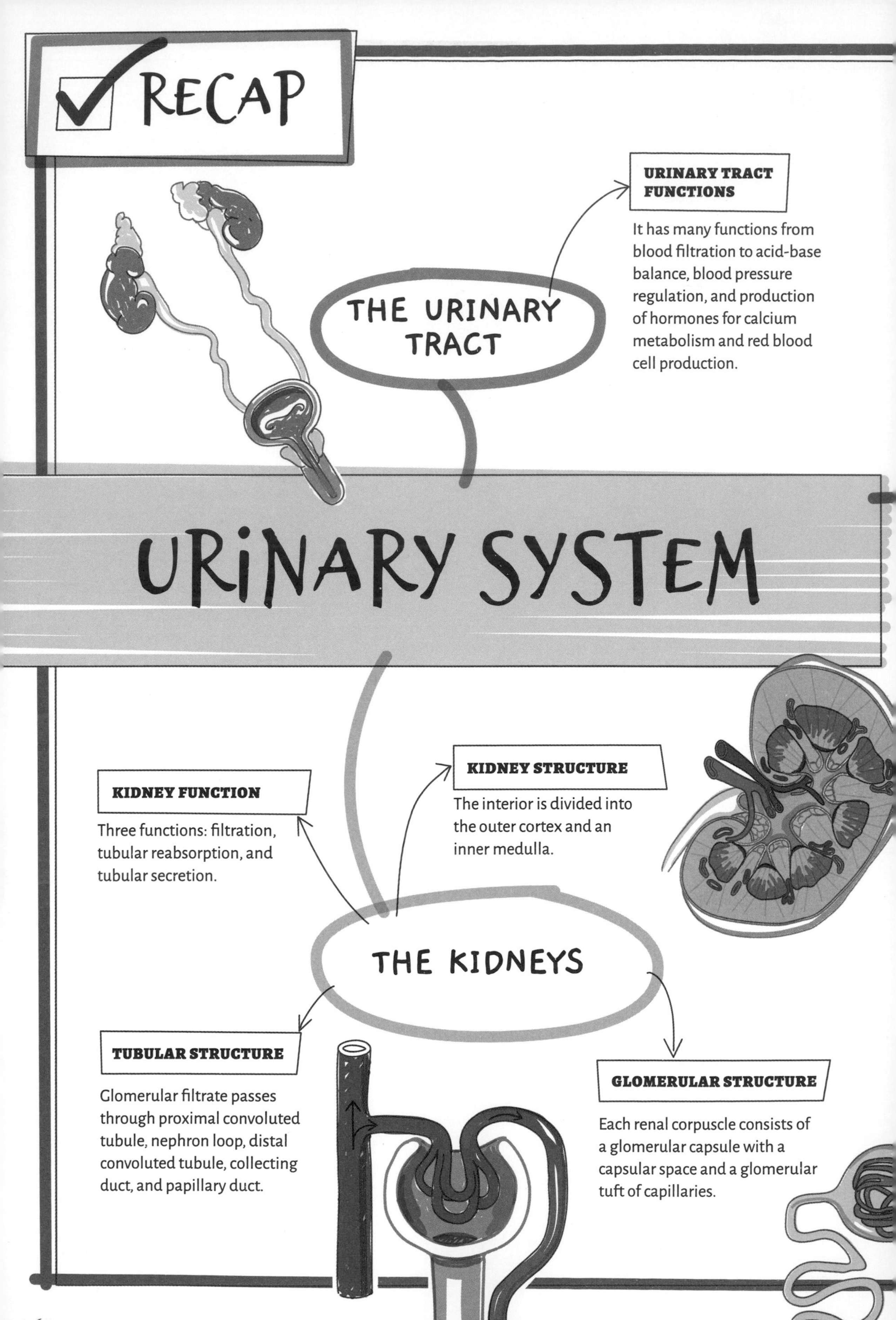
RECAP
THE URINARY TRACT
URINARY TRACT FUNCTIONS
It has many functions from blood filtration to acid-base balance, blood pressure regulation, and production of hormones for calcium metabolism and red blood cell production.
URINARY SYSTEM
THE KIDNEYS
KIDNEY FUNCTION
Three functions: filtration, tubular reabsorption, and tubular secretion.
KIDNEY STRUCTURE
The interior is divided into the outer cortex and an inner medulla.
TUBULAR STRUCTURE
Glomerular filtrate passes through proximal convoluted tubule, nephron loop, distal convoluted tubule, collecting duct, and papillary duct.
GLOMERULAR STRUCTURE
Each renal corpuscle consists of a glomerular capsule with a capsular space and a glomerular tuft of capillaries.

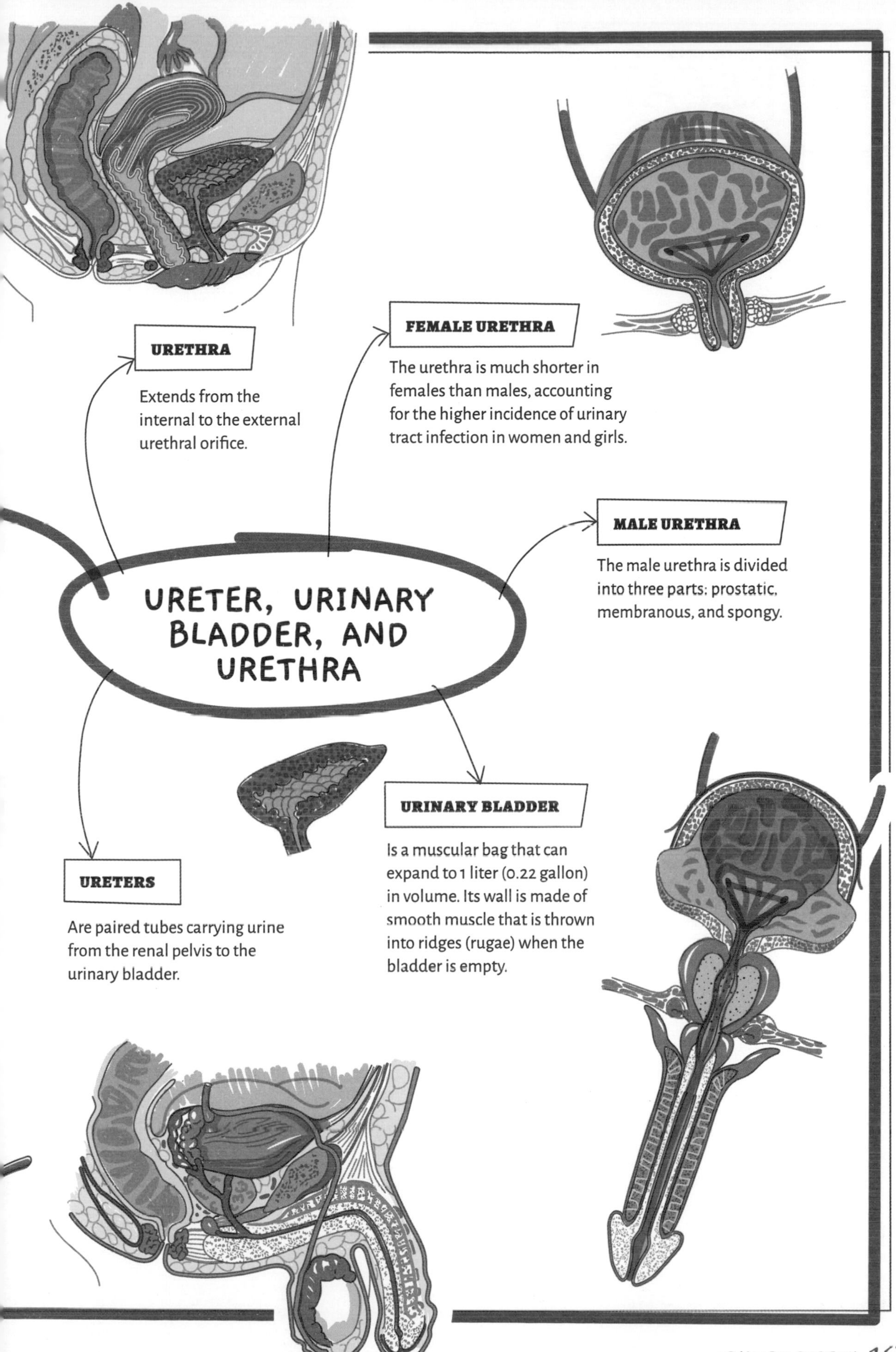
URETHRA
Extends from the internal to the external urethral orifice.
FEMALE URETHRA
The urethra is much shorter in females than males, accounting for the higher incidence of urinary tract infection in women and girls.
MALE URETHRA
The male urethra is divided into three parts: prostatic, membranous, and spongy.
URETER, URINARY BLADDER, AND URETHRA
URINARY BLADDER
Is a muscular bag that can expand to 1 liter (0.22 gallon) in volume. Its wall is made of smooth muscle that is thrown into ridges (rugae) when the bladder is empty.
URETERS
Are paired tubes carrying urine from the renal pelvis to the urinary bladder.

CHAPTER 11

REPRODUCTIVE SYSTEM

Sexual reproduction requires the union of sex cells (gametes) in a process called fertilization. The gonads—testes and ovaries—produce gametes and some of the steroid sex hormones that produce secondary sexual characteristics, such as breast fat, pubic and axillary hair, and in males facial hair, deepening voice, and muscle mass increase. The gonadal hormones estrogen, progesterone, and testosterone also regulate sexual function.

The reproductive tracts of both sexes have tubular structures to carry gametes and embryos, as well as accessory glandular structures to support the gametes and conceptus.

EARLY SEX CELLS

The cells that will produce our offspring are being put aside and prepared even before we are born. The number of ova (female sex cells) that a woman will produce is determined before birth.

The primordial germ cells that will give rise to gametes are sequestered from other cells of the body very early in embryonic development. These cells reside in the gonadal ridges, which form in the fifth week of pregnancy alongside the primitive intermediate kidney. So, initially, the gonads of both sexes are found on the upper posterior abdominal wall.

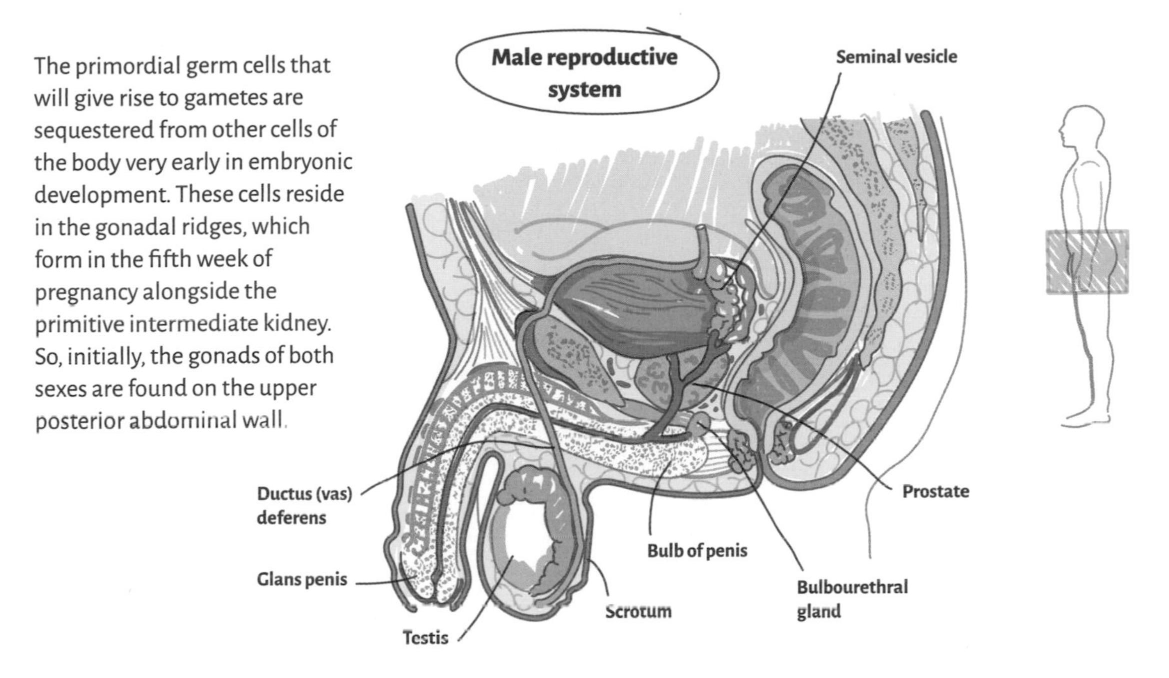

Male reproductive system
Seminal vesicle
Prostate
Ductus (vas) deferens
Glans penis
Bulb of penis
Bulbourethral gland
Scrotum
Testis

Males have an X and Y chromosome, whereas females have two X chromosomes. Male development is initiated by a Y chromosome gene called **SRY** (**S**ex-determining **R**egion of the **Y** chromosome).

The male gonad (**testis**) will usually descend to the **scrotum** during late pregnancy, whereas the female gonad (**ovary**) descends only as far as the lateral pelvic wall.

External genitalia are similar in the two sexes until about eight weeks in utero development. From the tenth week in utero, the genital tubercle develops into a **penis** in the male and a **clitoris** in the female.

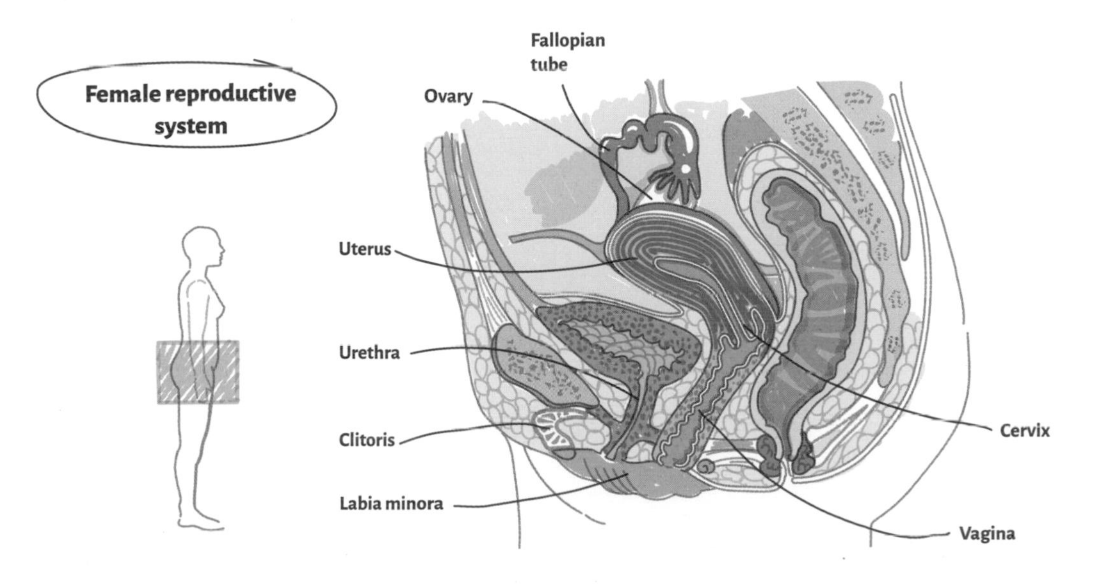

Female reproductive system
Fallopian tube
Ovary
Uterus
Urethra
Clitoris
Labia minora
Cervix
Vagina

MALE REPRODUCTIVE SYSTEM

The male reproductive system includes the testes to produce spermatozoa, ducts to carry sperm, accessory glands for production of the semen, and the penis to insert semen into the female reproductive tract.

Testis and epididymis

The testis and epididymis lie in the scrotal sac suspended below the abdomen to keep them cool.

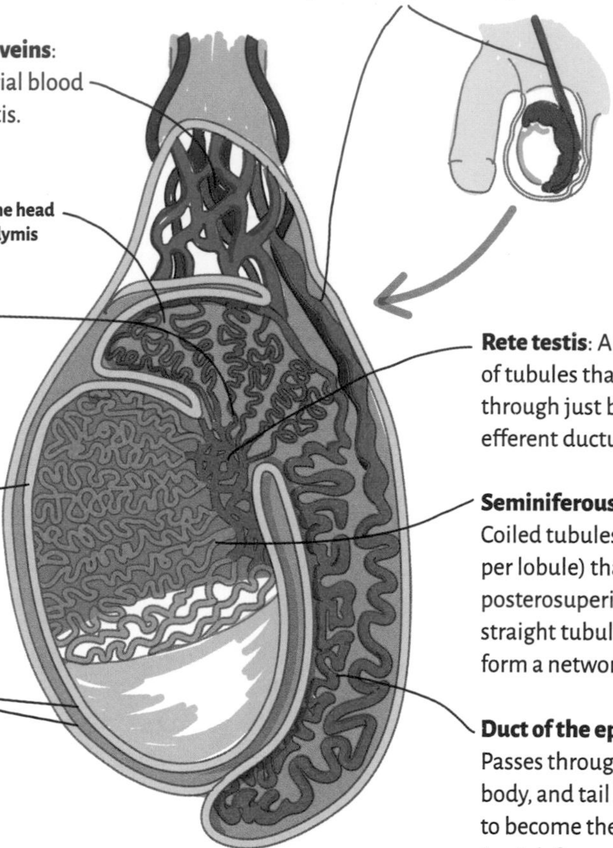

Ductus or vas deferens: Carries sperm from the epididymis to the region of the bladder base. The ductus can be cut (vasectomy) just above the testis to prevent sperm cells being ejaculated (see below).

Pampiniform plexus of veins: Cools the testicular arterial blood before it reaches the testis.

Tubules of the head of the epididymis

Efferent ductules of testis: Carry sperm from the rete testis to the tubules of the head of the epididymis, the site of sperm maturation.

Rete testis: A network of tubules that sperm pass through just before the efferent ductules.

Tunica albuginea: Dense fibrous capsule that surrounds the testis. It sends septa into the interior of the testis dividing the organ into lobules.

Seminiferous tubules: Coiled tubules (up to four per lobule) that converge posterosuperiorly to become straight tubules that in turn form a network (rete testis).

Tunica vaginalis: Double layered sac that partially surrounds the testis and epididymis. Its cavity contains a thin film of fluid.

Duct of the epididymis: Passes through the head, body, and tail of the epididymis to become the ductus (vas) deferens.

DUCTUS (VAS) DEFERENS

The **ductus deferens** is 45 cm (18 inches) long and carries spermatozoa from the tail of the epididymis to the region of the bladder base, where each ductus is dilated as an ampulla. Each ductus ascends in the spermatic cord, which also carries vessels and nerves for the testis and epididymis. The ductus passes through the inguinal canal to enter the abdomen. Each ductus joins with the duct of the seminal vesicle to form an ejaculatory duct, which opens into the prostatic urethra.

Penis

The **penis** is an erectile organ that becomes engorged with blood and rigid during sexual arousal. It has three erectile bodies and is divided into a root attached to the pelvis and a free-hanging body.

The paired **crura** (sing. crus) of the penile root extend into the penile body as the **corpora cavernosa** (sing. corpus cavernosum).

The single **bulb of the penis** in the penile root extends into the body as the **corpus spongiosum**. The **spongy urethra** passes through the center of the bulb, and the corpus spongiosum, which expands distally as the **glans**, is covered in uncircumcised males by a **prepuce** (foreskin).

Accessory glands

The **prostate** surrounds the **prostatic urethra** and has the paired ejaculatory ducts passing through it. Prostatic secretions include the following:

★Citric acid, which is used as an energy source by sperm
★Proteolytic enzymes, which break down the clotting proteins from the seminal vesicles
★Seminalplasmin, which is an antibiotic that destroys bacteria

The **seminal vesicles** lie posterior to the bladder base. Secretions are alkaline and include:

★Fructose, which is used for sperm energy
★Prostaglandins, which promote sperm motility and assist in sperm transport
★Clotting proteins, which coagulate semen after ejaculation and hold sperm against the cervix

Bulbourethral glands drain into the intrabulbar fossa of the spongy urethra. During sexual arousal, they produce an alkaline mucoid secretion that clears the urethra and contributes to lubrication during intercourse.

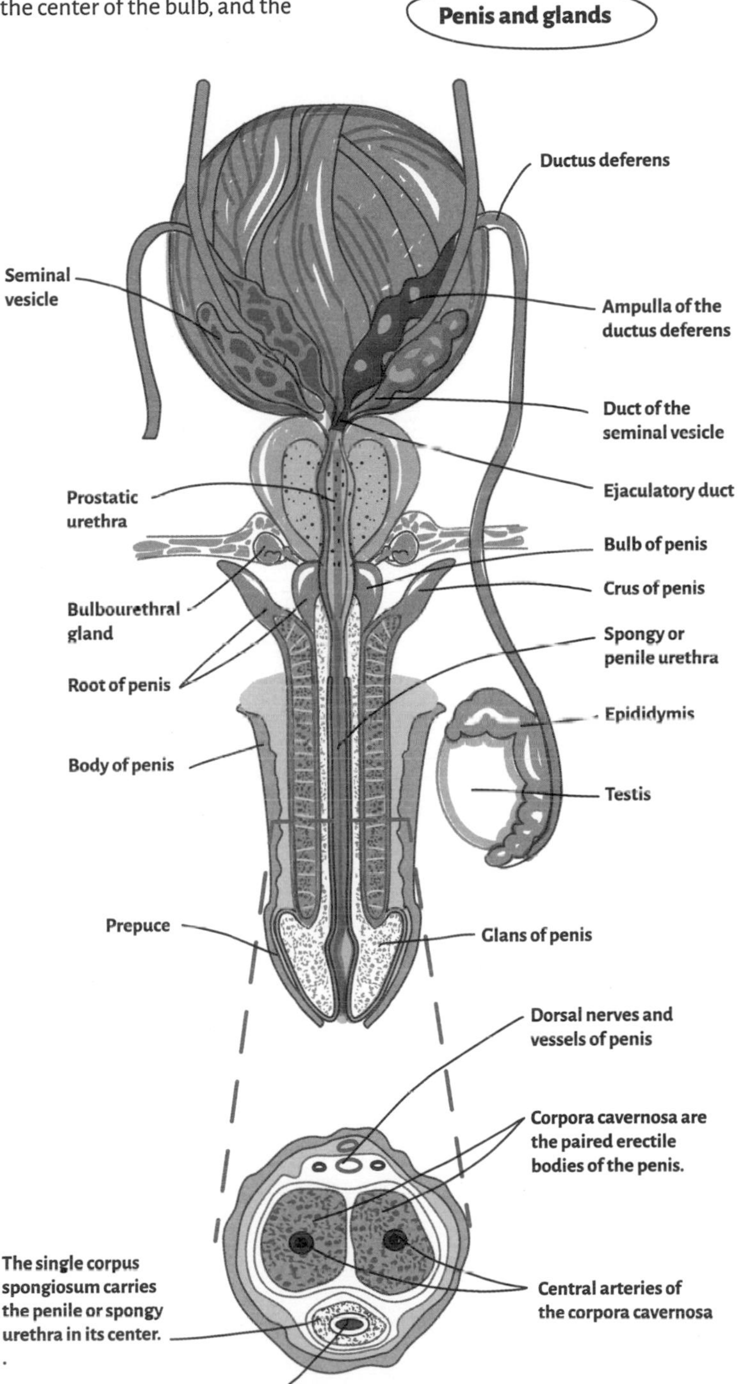
Penis and glands
Ductus deferens
Seminal vesicle
Ampulla of the ductus deferens
Duct of the seminal vesicle
Ejaculatory duct
Prostatic urethra
Bulb of penis
Crus of penis
Bulbourethral gland
Spongy or penile urethra
Root of penis
Epididymis
Body of penis
Testis
Prepuce
Glans of penis
Dorsal nerves and vessels of penis
Corpora cavernosa are the paired erectile bodies of the penis.
The single corpus spongiosum carries the penile or spongy urethra in its center.
Central arteries of the corpora cavernosa
Spongy or penile urethra carries both urine and semen.

Testis and spermatogenesis

The **testes** not only produce spermatozoa (sperm cells) but also the androgen testosterone by the **interstitial (Leydig) cells**. The testes must be kept at a temperature of 34° to 35°C (93.2° to 95°F) for optimal spermatogenesis and are suspended below the body.

The paired testes are suspended in the scrotum to maintain optimal temperature. The testes may be elevated in cold weather by the smooth dartos muscles of the scrotum, or by the skeletal cremaster muscle that attaches to the spermatic cord, bringing the testes closer to the warm abdomen. Hot conditions cause both muscles to relax, hanging the testes further from the heat of the trunk.

Blood to the testis from the abdominal cavity also needs to be cooled. The warm blood of the testicular artery is cooled by heat transfer to the cooler returning blood of testicular veins that form a pampiniform plexus around the artery.

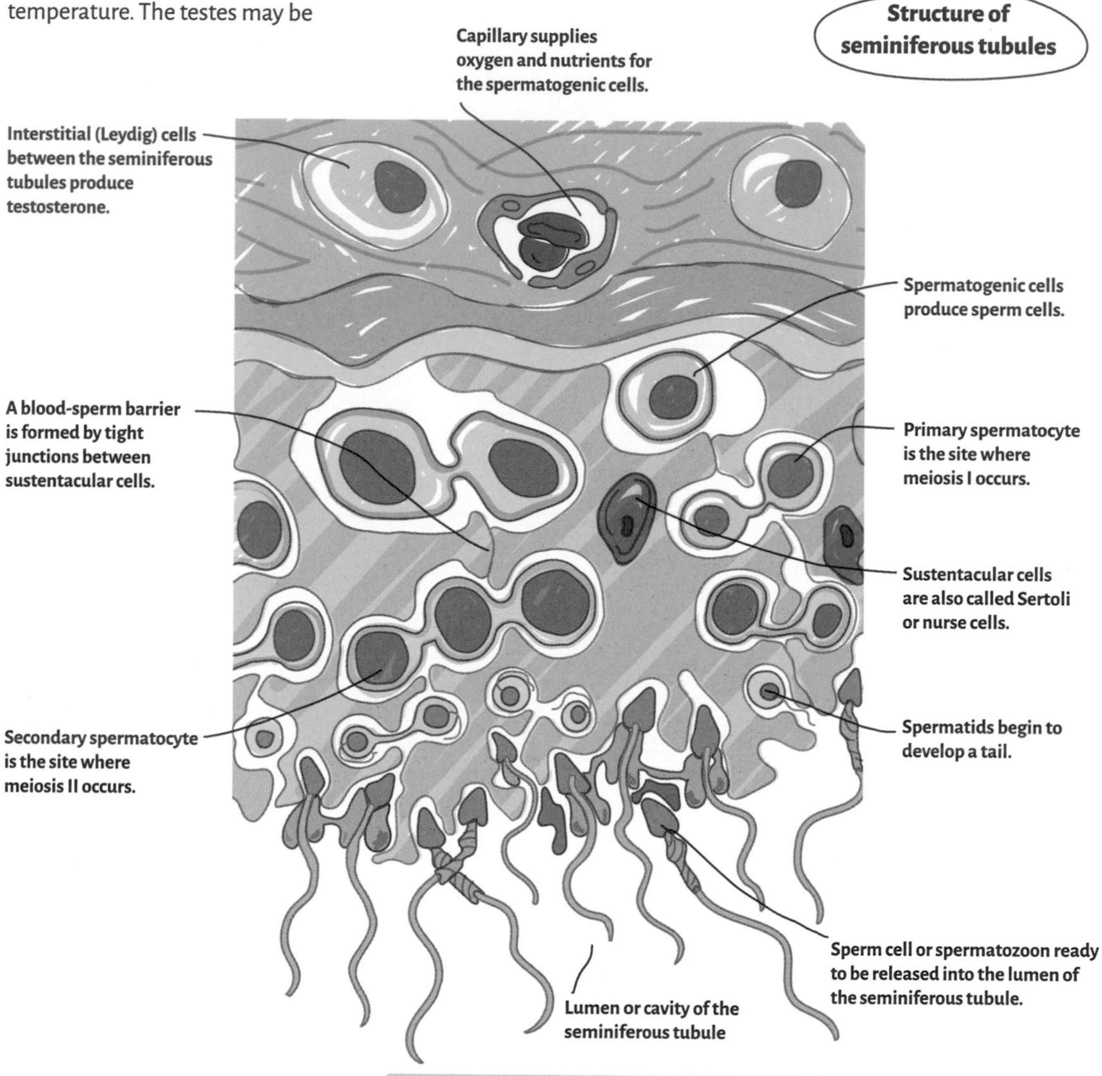
Structure of seminiferous tubules
Capillary supplies oxygen and nutrients for the spermatogenic cells.
Interstitial (Leydig) cells between the seminiferous tubules produce testosterone.
Spermatogenic cells produce sperm cells.
A blood-sperm barrier is formed by tight junctions between sustentacular cells.
Primary spermatocyte is the site where meiosis I occurs.
Sustentacular cells are also called Sertoli or nurse cells.
Secondary spermatocyte is the site where meiosis II occurs.
Spermatids begin to develop a tail.
Sperm cell or spermatozoon ready to be released into the lumen of the seminiferous tubule.
Lumen or cavity of the seminiferous tubule

Spermatogenesis is the process by which the seminiferous tubules produce sperm. This requires meiosis to produce the gametes and formation of sperm cells (spermiogenesis).

MICROSCOPIC STRUCTURE OF THE TESTIS

Spermatogonia develop from primordial germ cells but remain dormant until puberty, when spermatogonia give rise (in sequence) to primary spermatocytes, secondary spermatocytes, spermatids, and sperm cells.

FEMALE REPRODUCTIVE SYSTEM

The female reproductive system includes two ovaries (for gamete production), Fallopian (uterine) tubes, the uterus, the vagina, and external genitalia. The female reproductive tract communicates with the interior of the abdominal cavity through the abdominal ostium of the tube.

Ovaries

There are two ovaries that lie against the lateral pelvic wall and are supplied by vessels descending from the upper abdomen in the **suspensory ligament of the ovary**. They are attached to the side of the uterus by the **ovarian ligament**. The suspensory ligament carries the ovarian artery, vein, nerve, and lymphatic plexus.

Frontal view of the female reproductive system

Suspensory ligament of the ovary
Cone-shaped infundibulum
Ampulla
Fallopian tubes
Isthmus of the Fallopian tube
Fundus
Fimbriae at edge of infundibulum
Ovary
Endometrium
Ovarian ligament
Blood vessels supply the uterus and the Fallopian tube.
Myometrium
Perimetrium or serosa
The internal os of the cervix opens into the uterine cavity.
Cervix
Isthmus of the uterus
The external os of the cervix opens into the vagina.
Vagina

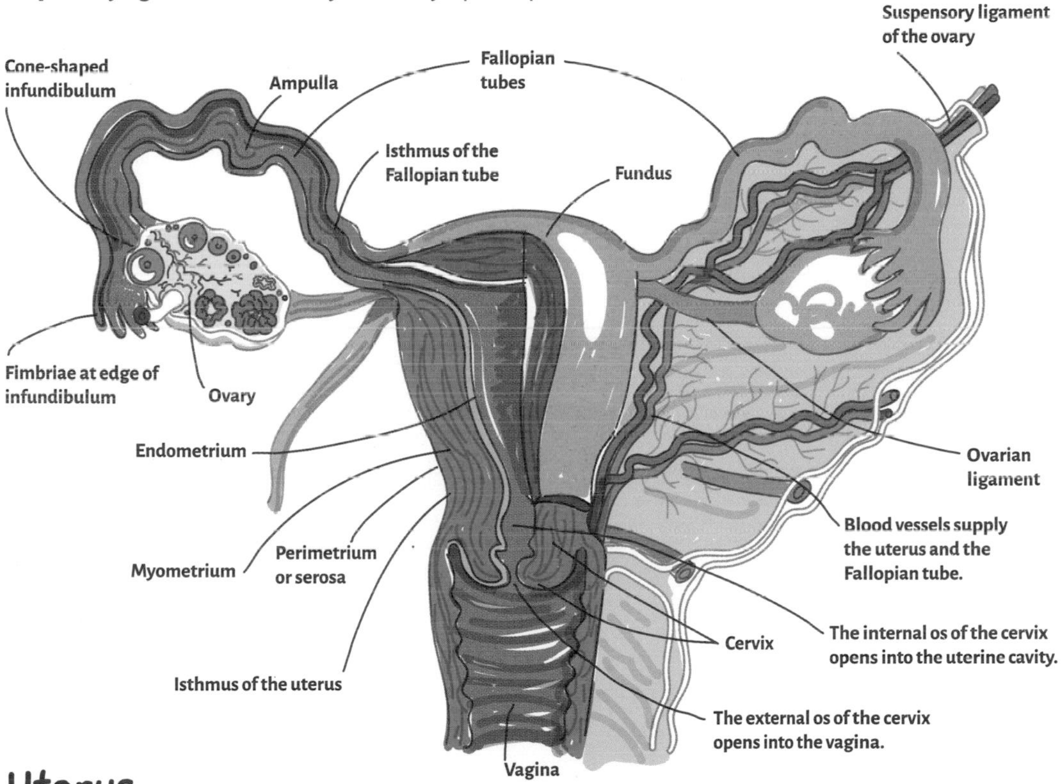

Uterus

The **uterus** is the organ of gestation (fetal development) and parturition (birth). It has a three-layered structure with a lining **endometrium**, a smooth muscle **myometrium**, and an external **perimetrium** or **serosa**.

The smooth muscle of the myometrium is arranged in three directions (longitudinal, oblique, and circular), so that the uterus can produce uniform, coordinated contractions to push the fetus out during birth. The uterus is a pear-shaped organ with a **fundus** above the attachments of the Fallopian tubes (see page 170), a body, and a **cervix** opening into the **vagina**. The **isthmus of the uterus** is the narrow region where the uterine body meets the cervix.

UTERINE CAVITY

The cervix has a canal opening into the uterine body (**internal os**) and to the vaginal lumen (**external os**). The mucosa of the cervical canal has mucosal folds that lock together to prevent microbial penetration but can open around ovulation to permit sperm to enter, and during menstruation to allow blood and sloughed endometrium to exit.

The uterine cavity is continuous with the cavity of the Fallopian tube and the cervical canal, so that sperm can ascend from the vagina to the Fallopian tube and ovary.

FALLOPIAN TUBES

The paired **Fallopian tubes** carry sperm up to the ovary and the ovum down toward the uterine body.

The Fallopian tubes have four regions: **intramural**, **isthmus**, **ampulla** (where fertilization commonly occurs), and the cone-shaped **infundibulum**.

The infundibulum rim has fingerlike projections called **fimbriae**, which embrace the ovary, and a central opening called the **abdominal ostium** through which sperm and the ovum can pass (see page 169).

Vagina

The **vagina** is a muscular tube that receives the erect penis for transfer of semen. It is also an important component of the birth canal (see page 173). The vagina can expand to 10 cm (4 inches) in diameter during childbirth and later can return to its normal size.

When it is empty, the vaginal lumen (vaginal canal) has an H-shaped outline, with anterior and posterior platforms.

The uterine cervix opens into the anterior wall of the upper vagina as the **vaginal cervix**. The **fornix** is a recess in the vagina that surrounds the vaginal cervix.

The opening of the lower vagina is called the **vaginal vestibule**, and it is located between the two **labia minora**.

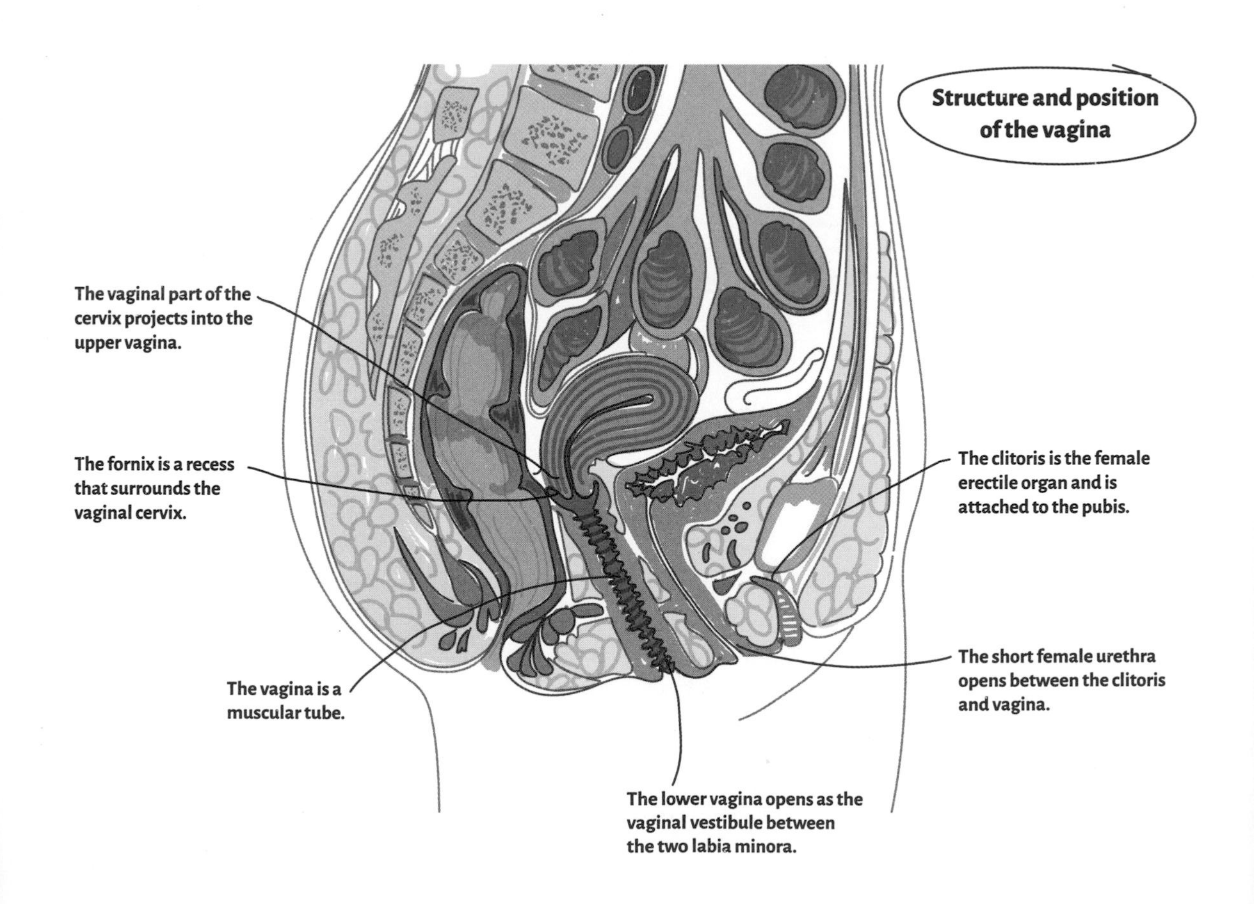
Structure and position of the vagina
The vaginal part of the cervix projects into the upper vagina.
The fornix is a recess that surrounds the vaginal cervix.
The vagina is a muscular tube.
The lower vagina opens as the vaginal vestibule between the two labia minora.
The clitoris is the female erectile organ and is attached to the pubis.
The short female urethra opens between the clitoris and vagina.

The ovary and oogenesis

The **ovary** not only produces the ovum in a process called oogenesis but also produces estrogen and progesterone for secondary sexual characteristics, and for recurring changes during the menstrual cycle and pregnancy.

The ovary lies against the lateral pelvic wall between the ureter and the external iliac vein.

The Fallopian tube arches over the ovary's superior pole, and the fimbriae of the Fallopian tube infundibulum embrace the surface of the ovary, ready to receive the secondary oocyte. The ovary is covered by a **germinal epithelium**, with an underlying fibrous layer called the **tunica albuginea**. It is divided into an outer **cortex** and an inner **medulla**.

The cortex contains **ovarian follicles** (oocytes, in various stages of development, plus surrounding cells) surrounded by connective tissue. When the surrounding cells form a single layer, they are called **follicular cells**. The medulla consists of connective tissue and vascular branches.

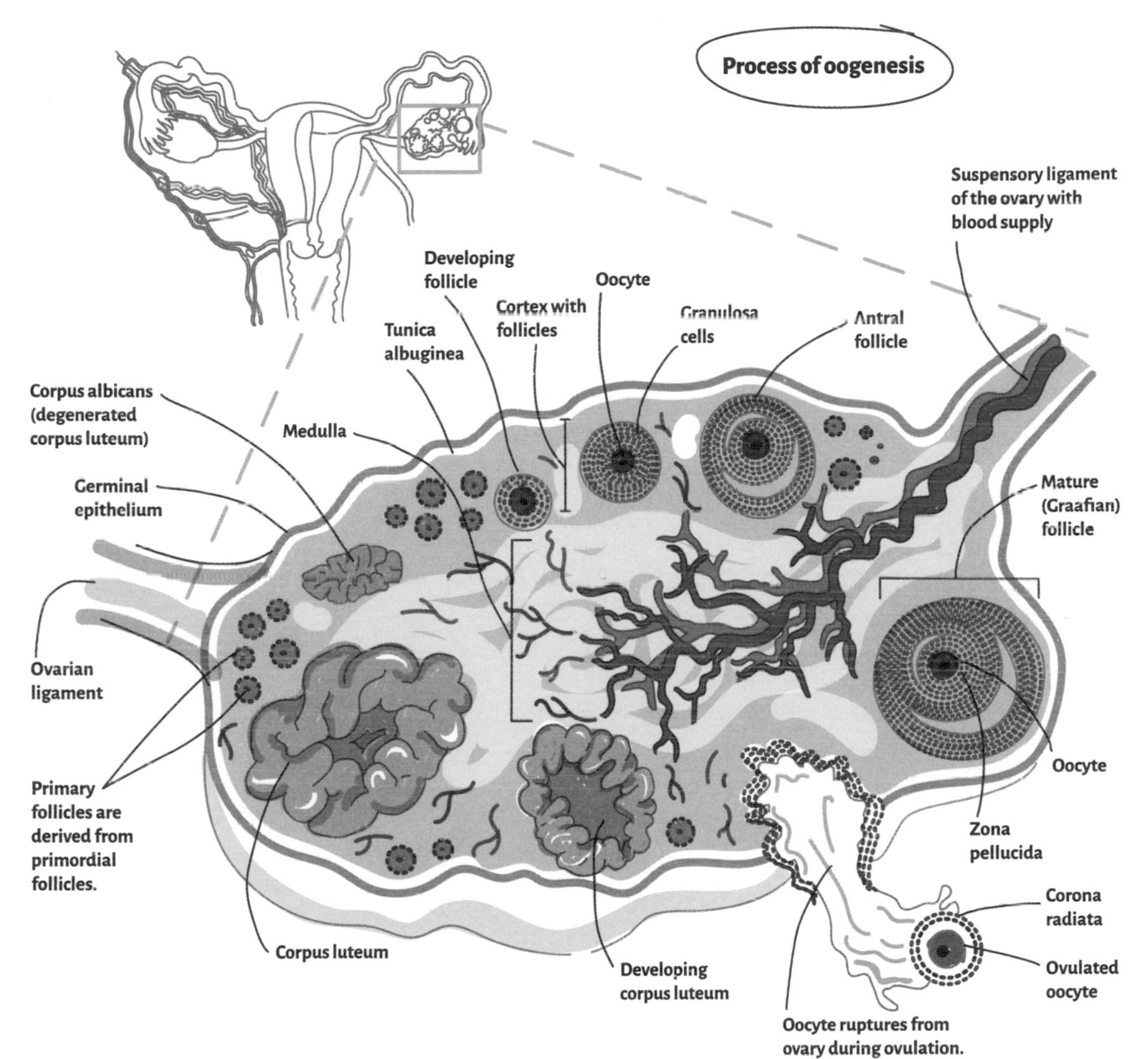

Process of oogenesis
Suspensory ligament of the ovary with blood supply
Developing follicle
Oocyte
Cortex with follicles
Tunica albuginea
Granulosa cells
Antral follicle
Corpus albicans (degenerated corpus luteum)
Medulla
Germinal epithelium
Mature (Graafian) follicle
Ovarian ligament
Primary follicles are derived from primordial follicles.
Oocyte
Zona pellucida
Corona radiata
Ovulated oocyte
Corpus luteum
Developing corpus luteum
Oocyte ruptures from ovary during ovulation.

Oogenesis is the process of the formation of gametes in the ovary. The initial stages of oogenesis begin in the ovary during fetal life, when germ cells differentiate into oogonia. Most oogonia will degenerate, but some develop into **primary oocytes** that enter prophase of meiosis I during fetal life, but develop no further. Each primary oocyte and surrounding follicular cells is called a **primordial follicle**.

Follicles

After puberty and until menopause, monthly exposure to FSH and LH from the anterior pituitary (see below) stimulates primordial follicles to develop into **primary follicles** (an **oocyte** surrounded by **granulosa cells**).

Primary follicles develop into **secondary follicles**, when granulosa cells accumulate follicular fluid to form an antral follicle. The innermost layer of granulosa cells is attached to the zona pellucida of the oocyte as the **corona radiata**.

One or two secondary follicles each month turn into **mature** (**Graafian**) **follicles** when meiosis I is completed and meiosis II begins and proceeds to metaphase. At ovulation, the mature follicle ruptures, releasing the secondary oocyte into the peritoneal cavity.

After ovulation, the remains of the follicle develop into a **corpus luteum**. If pregnancy does not occur, the corpus luteum degenerates after two weeks to become a corpus albicans.

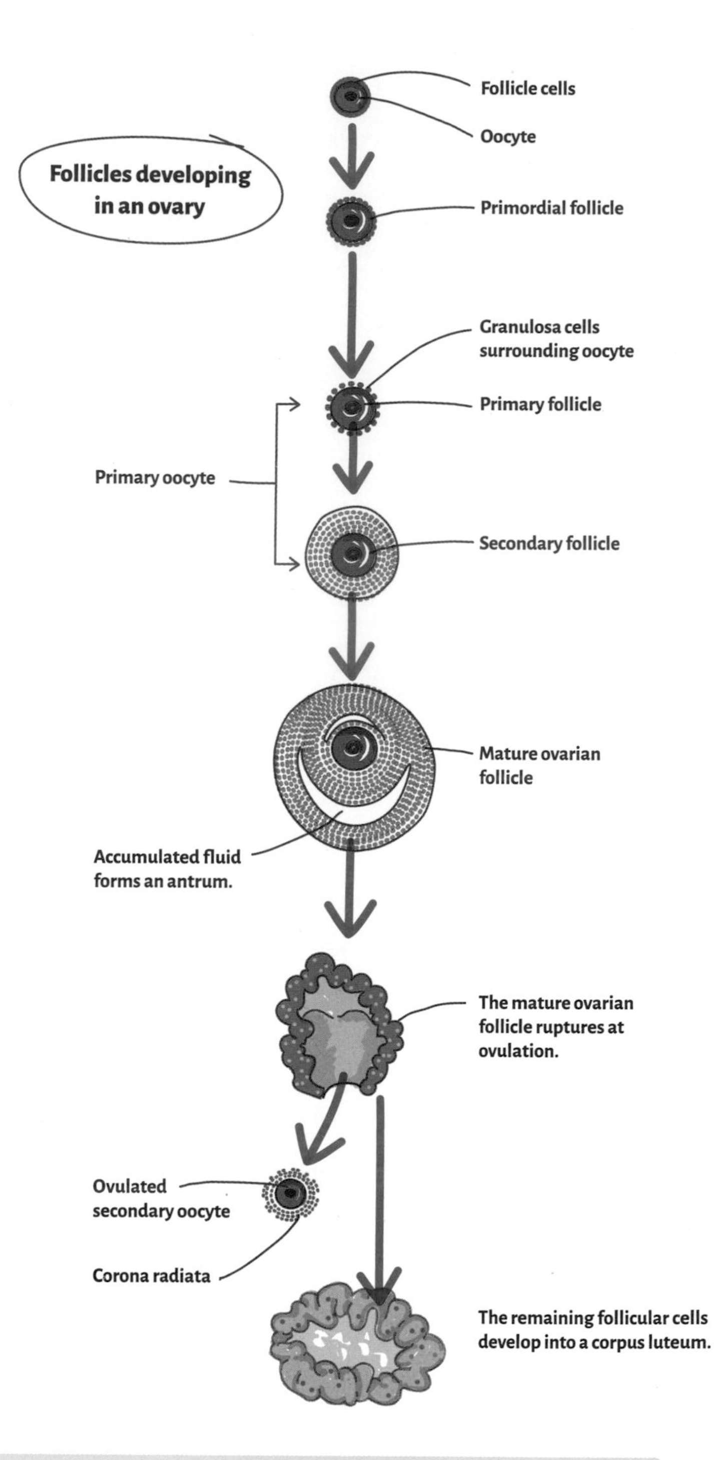
Follicles developing in an ovary
Follicle cells
Oocyte
Primordial follicle
Granulosa cells surrounding oocyte
Primary follicle
Primary oocyte
Secondary follicle
Mature ovarian follicle
Accumulated fluid forms an antrum.
The mature ovarian follicle ruptures at ovulation.
Ovulated secondary oocyte
Corona radiata
The remaining follicular cells develop into a corpus luteum.

THE OVARY IN PREGNANCY AND AFTER MENOPAUSE

If pregnancy occurs, the corpus luteum does not degenerate, and it produces progesterone, estrogen, relaxin, and inhibin to support the early pregnancy and to prepare for parturition (birth).

After the end of reproductive life, called the **menopause**, the ovaries become less responsive to hormonal stimulation and produce less estrogen. The ovaries eventually atrophy (waste away).

Birth

During childbirth (parturition), the smooth muscle of the uterine wall contracts rhythmically to expel the fetus and its placenta. Pacemaker cells near the attachments of the Fallopian tubes set a regular rhythm, called **contractions**, under the influence of the hormone oxytocin from the posterior pituitary.

The fetus in the birth canal

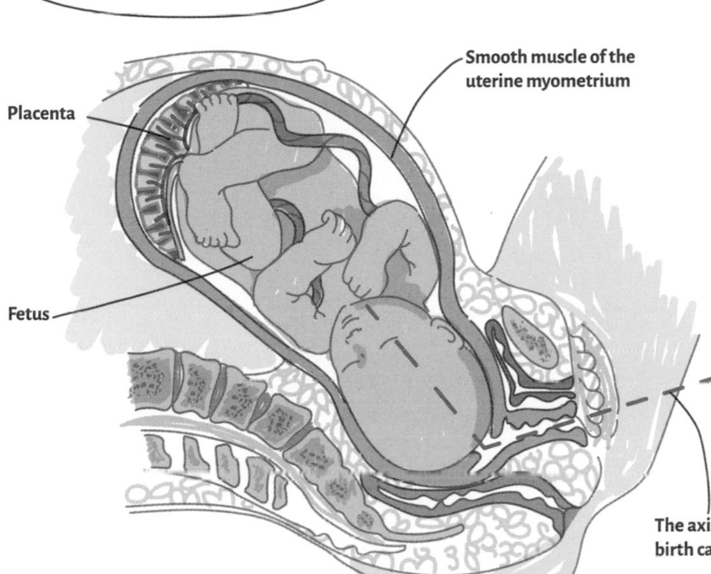
Placenta
Smooth muscle of the uterine myometrium
Fetus
The axis of the birth canal

The **placenta** supplies the **fetus** with oxygen and nutrients and is an important endocrine organ. The placenta is actually derived from the embryo and is expelled during the third stage of labor.

The **smooth muscle of the uterine myometrium** sets up a rhythmic contraction during the first stage of labor. This dilates the cervix, so that the fetus can be delivered during the second stage.

The **birth canal** is the path taken by the fetus during birth. It runs from the axis of the uterine body to the vaginal axis.

External genitalia

The female external genitalia are found in the **perineum** (the space between the thighs). Most of the female perineum is taken up by paired fleshy **labia majora** (sing. labium majus), separated by a midline pudendal cleft.

Where the labia majora meet anteriorly, there is a midline fatty elevation called the **mons pubis** over the pubic bones and pubic symphysis. The labia majora and mons pubis are pigmented and covered by pubic hair after puberty.

Female external genitalia

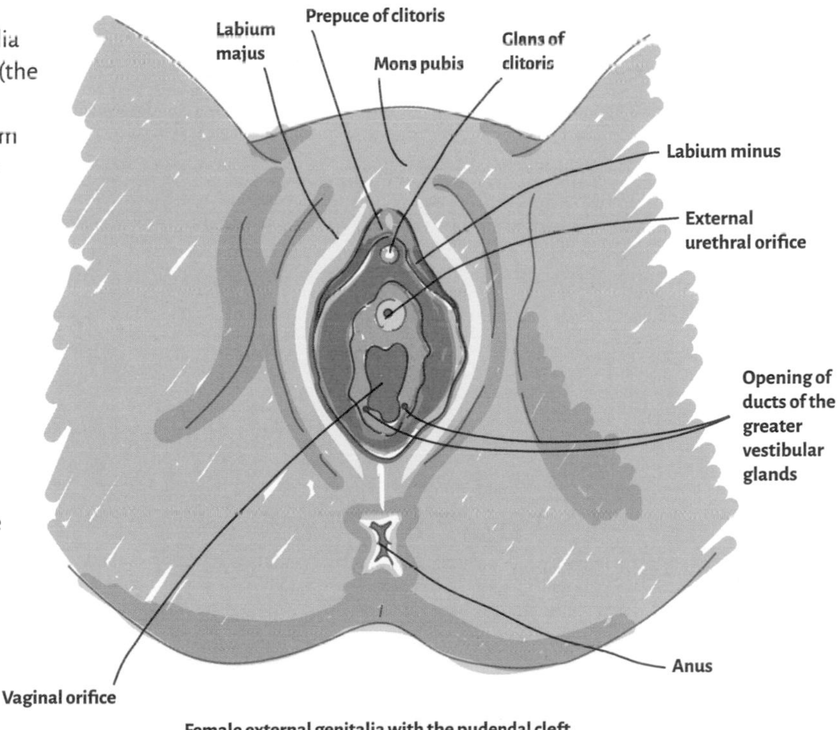
Labium majus
Prepuce of clitoris
Mons pubis
Glans of clitoris
Labium minus
External urethral orifice
Opening of ducts of the greater vestibular glands
Anus
Vaginal orifice

Female external genitalia with the pudendal cleft opened by separating the labia majora

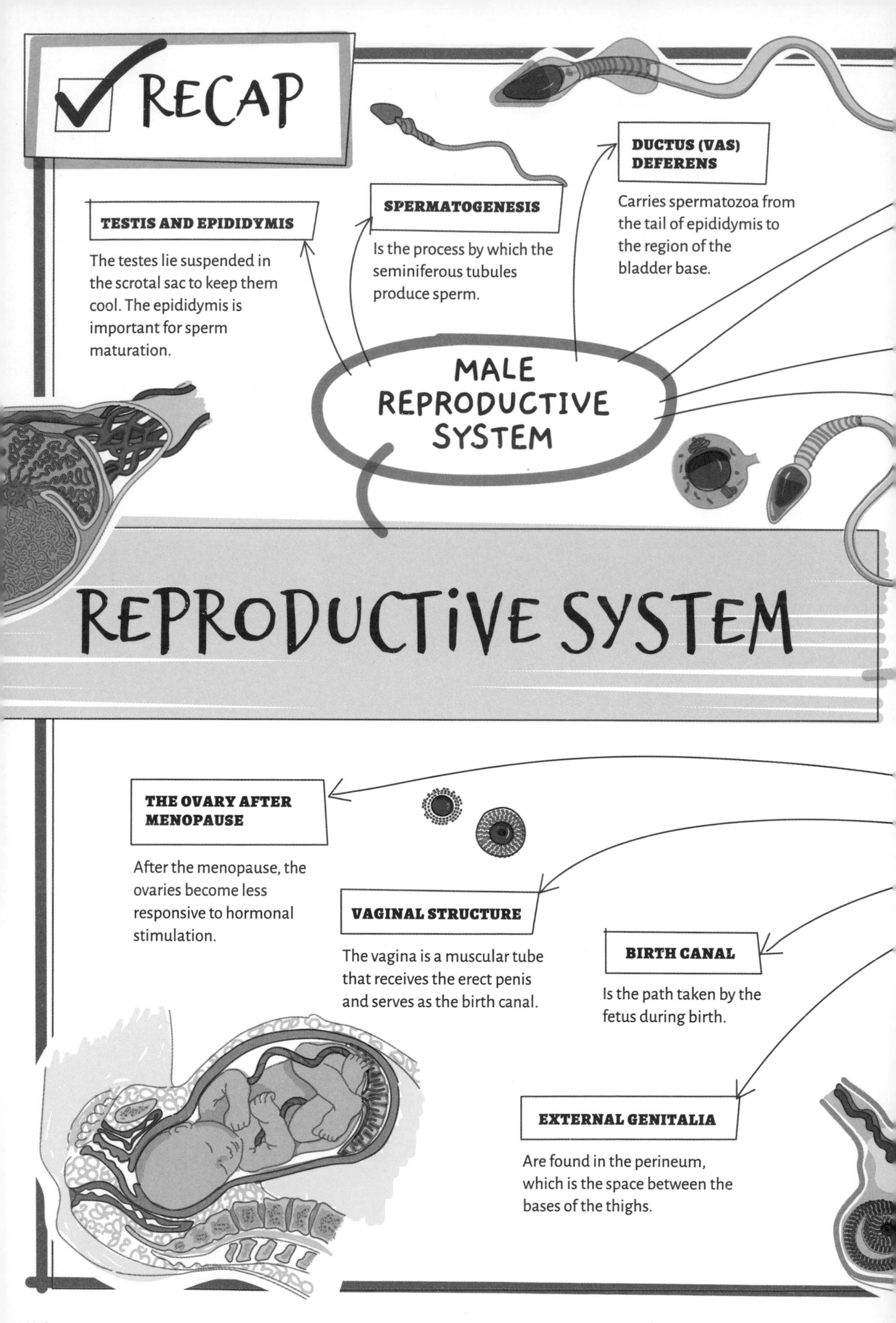

RECAP

REPRODUCTIVE SYSTEM

MALE REPRODUCTIVE SYSTEM

TESTIS AND EPIDIDYMIS

The testes lie suspended in the scrotal sac to keep them cool. The epididymis is important for sperm maturation.

SPERMATOGENESIS

Is the process by which the seminiferous tubules produce sperm.

DUCTUS (VAS) DEFERENS

Carries spermatozoa from the tail of epididymis to the region of the bladder base.

THE OVARY AFTER MENOPAUSE

After the menopause, the ovaries become less responsive to hormonal stimulation.

VAGINAL STRUCTURE

The vagina is a muscular tube that receives the erect penis and serves as the birth canal.

BIRTH CANAL

Is the path taken by the fetus during birth.

EXTERNAL GENITALIA

Are found in the perineum, which is the space between the bases of the thighs.

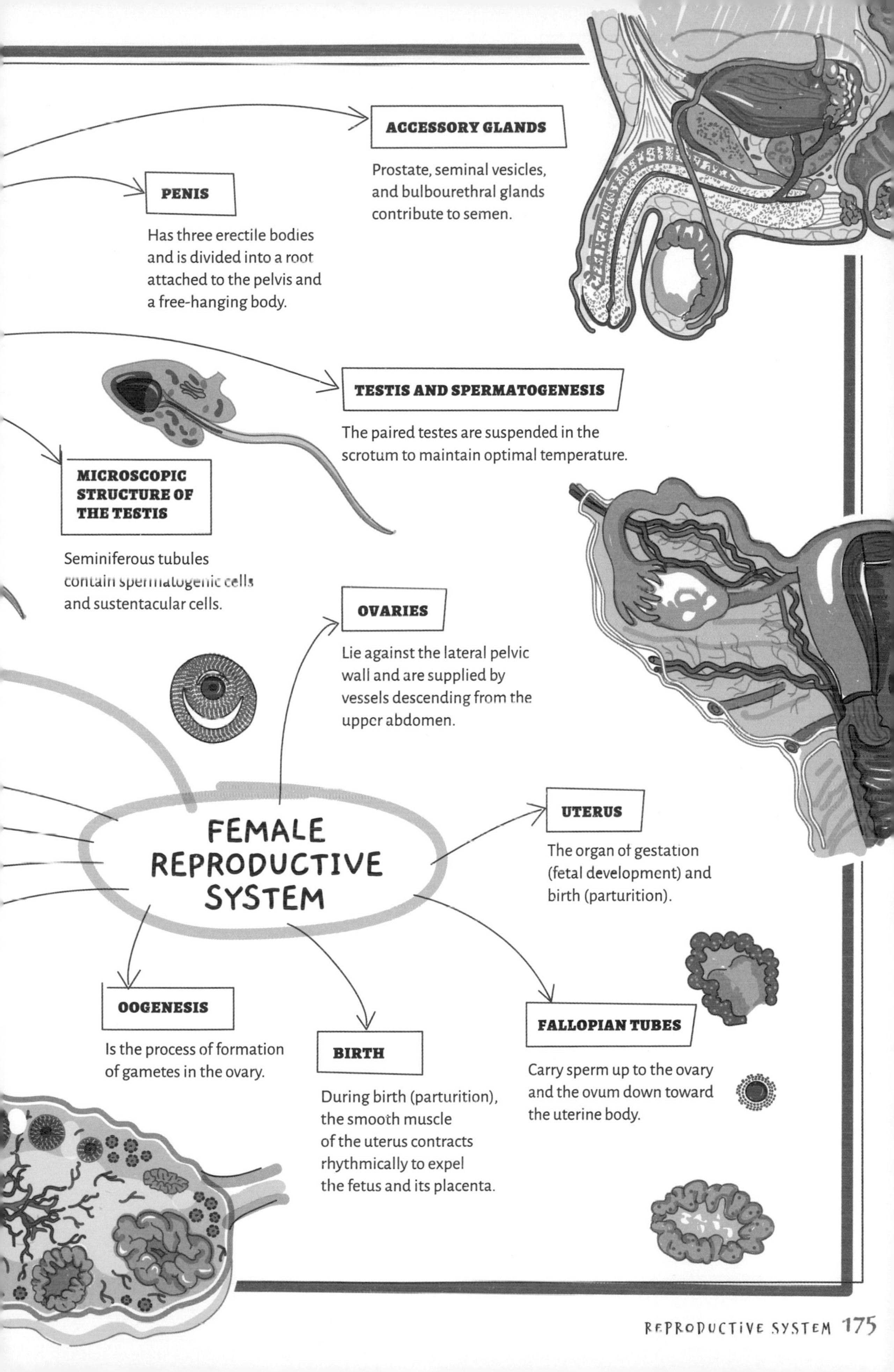

ACCESSORY GLANDS

Prostate, seminal vesicles, and bulbourethral glands contribute to semen.

PENIS

Has three erectile bodies and is divided into a root attached to the pelvis and a free-hanging body.

TESTIS AND SPERMATOGENESIS

The paired testes are suspended in the scrotum to maintain optimal temperature.

MICROSCOPIC STRUCTURE OF THE TESTIS

Seminiferous tubules contain spermatogenic cells and sustentacular cells.

OVARIES

Lie against the lateral pelvic wall and are supplied by vessels descending from the upper abdomen.

FEMALE REPRODUCTIVE SYSTEM

UTERUS

The organ of gestation (fetal development) and birth (parturition).

OOGENESIS

Is the process of formation of gametes in the ovary.

BIRTH

During birth (parturition), the smooth muscle of the uterus contracts rhythmically to expel the fetus and its placenta.

FALLOPIAN TUBES

Carry sperm up to the ovary and the ovum down toward the uterine body.

CHAPTER 12

ENDOCRINE SYSTEM

Endocrine glands secrete their products (hormones) to the bloodstream or body cavity in contrast to exocrine glands, which secrete to an epithelial surface. Hormones may be either peptides (e.g., insulin) or steroids, such as estrogen and progesterone. Peptide hormones exert their actions by locking into a receptor molecule on the cell surface, which then initiates changes in the cytoplasm. Steroids are transported into the cell to attach to chaperone molecules in the cytoplasm before entering the nucleus of the cell, where they alter cell activity.

ENDOCRINE GLANDS

The **endocrine glands** are paired or midline structures scattered through the head, neck, and trunk. They have rich blood supplies.

Feedback control systems

The endocrine system is regulated by negative feedback loops, where production of a hormone by a gland leads to blood levels of the hormone or a change in body state that turn off the stimulus to make the hormone.

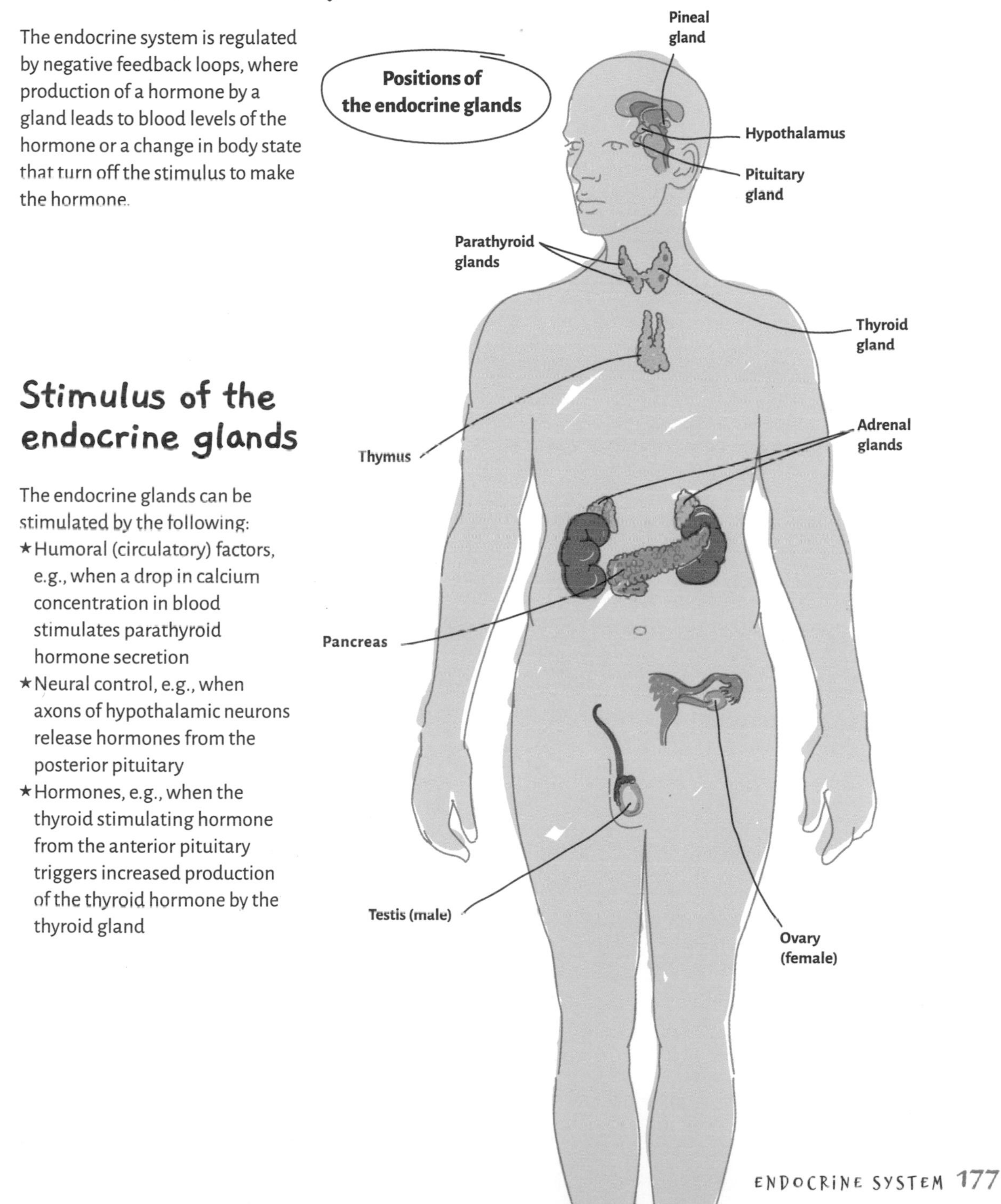
Positions of the endocrine glands
Pineal gland
Hypothalamus
Pituitary gland
Parathyroid glands
Thyroid gland
Thymus
Adrenal glands
Pancreas
Testis (male)
Ovary (female)

Stimulus of the endocrine glands

The endocrine glands can be stimulated by the following:

* Humoral (circulatory) factors, e.g., when a drop in calcium concentration in blood stimulates parathyroid hormone secretion
* Neural control, e.g., when axons of hypothalamic neurons release hormones from the posterior pituitary
* Hormones, e.g., when the thyroid stimulating hormone from the anterior pituitary triggers increased production of the thyroid hormone by the thyroid gland

ANTERIOR PITUITARY GLAND AND ITS HORMONES

The pituitary gland develops from a bulge of the roof of the primitive oral cavity (Rathke's pouch) and a projection from the brain (neurohypophysis), so it has both epithelial and neural origins.

Position and anatomy of the pituitary

The pituitary (hypophysis) gland lies in the pituitary fossa of the sphenoid bone, immediately inferior to the **hypothalamus** of the brain. The pituitary is connected to the hypothalamus by the **pituitary stalk** or **infundibulum** and receives a rich vascular supply.

The pituitary gland is divided into anterior and posterior parts, with the anterior part making up 75% of the gland mass. The anterior pituitary is in turn divided into a bulbous **pars distalis** and a tubular **pars tuberalis** (as a covering around the infundibulum).

Hypothalamus and pituitary

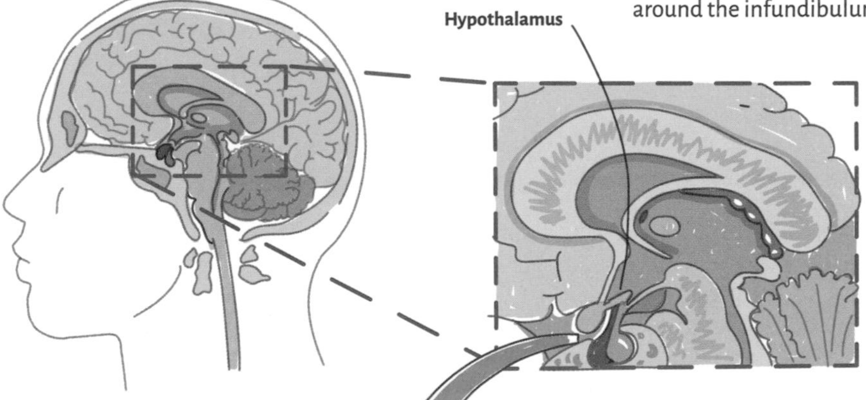

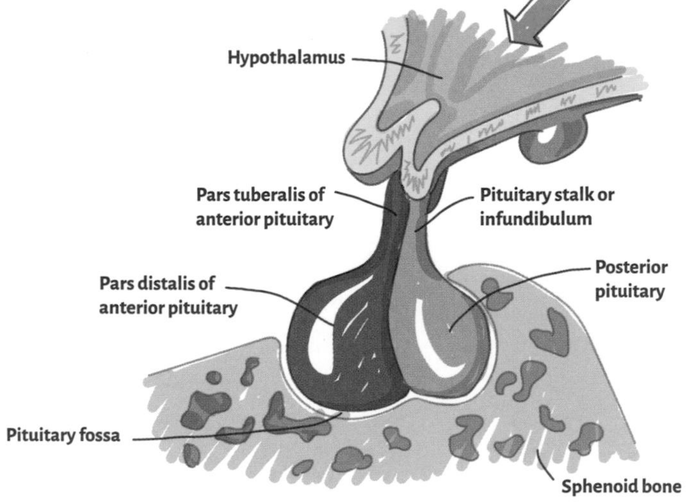

Hypothalamus
Hypothalamus
Pars tuberalis of anterior pituitary
Pituitary stalk or infundibulum
Pars distalis of anterior pituitary
Posterior pituitary
Pituitary fossa
Sphenoid bone

CELL TYPES

The cells of the anterior pituitary are grouped according to the way they stain with dyes.

- ★ **Basophils** are 10% of cells; they make TSH, FSH, LH, and ACTH.
- ★ **Acidophils** are 40% of cells; they make GH and PRL.
- ★ **Chromophobes** are 50% of cells; they do not secrete hormones. They may be basophils or acidophils that have already released their hormones.

Control by the hypothalamus

The anterior pituitary is regulated by releasing or inhibiting hormones or factors that pass from the hypothalamus to the anterior pituitary through the **hypophyseal** portal system of vessels.

Neurosecretory cells in the hypothalamus discharge releasing or inhibitory hormones from their axon terminals into the portal system capillaries to flow down the infundibulum to the anterior pituitary.

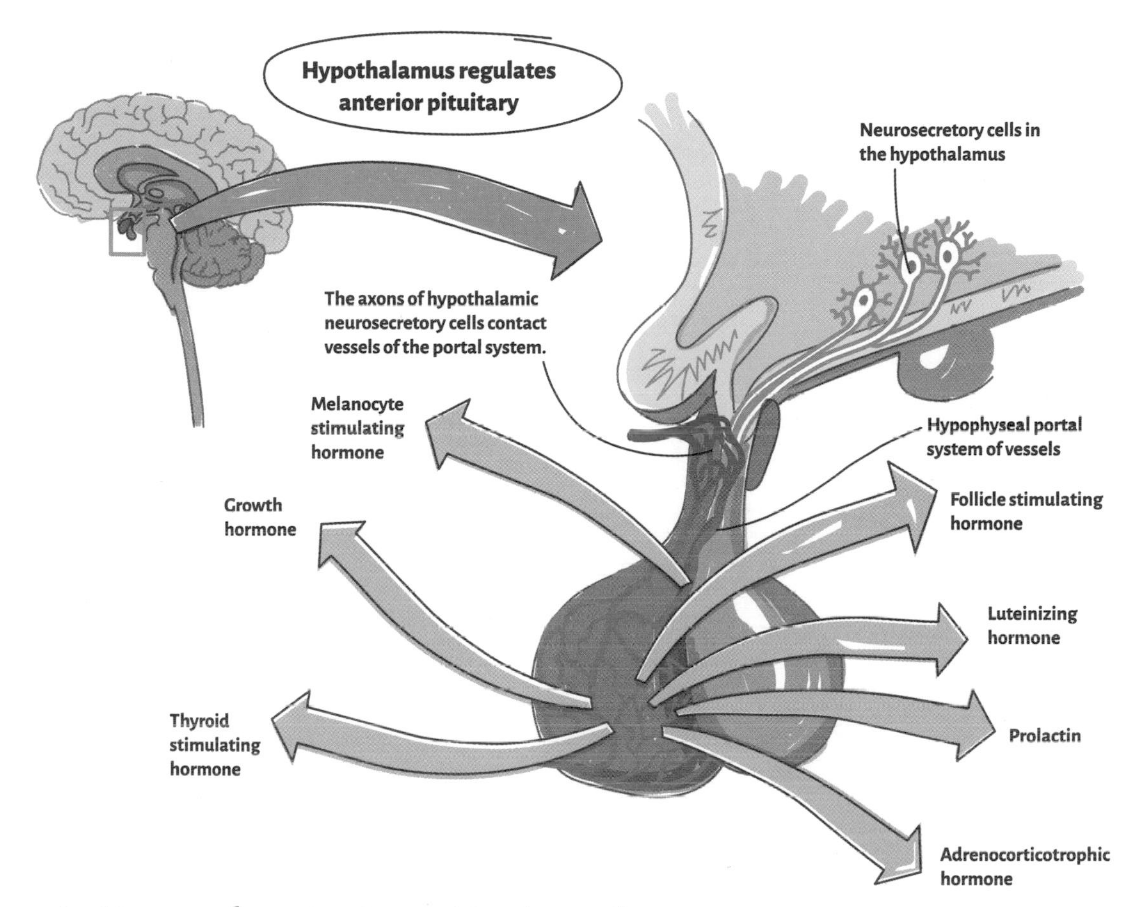
Hypothalamus regulates anterior pituitary
Neurosecretory cells in the hypothalamus
The axons of hypothalamic neurosecretory cells contact vessels of the portal system.
Melanocyte stimulating hormone
Hypophyseal portal system of vessels
Growth hormone
Follicle stimulating hormone
Luteinizing hormone
Thyroid stimulating hormone
Prolactin
Adrenocorticotrophic hormone

Actions of anterior pituitary hormones

Growth hormone (somatotropin, GH) stimulates tissues to produce insulin-like growth factors (IGFs), which stimulate body growth and regulate metabolism.

Thyroid stimulating hormone (thyrotropin, TSH) controls the secretions of the thyroid gland.

Follicle stimulating hormone (FSH) in females stimulates the development of ovarian follicles and induces ovarian secretion of estrogen. In males, FSH stimulates sperm production.

Luteinizing hormone (LH) in females stimulates secretion of estrogen and progesterone by the ovary and the formation of corpus luteum. In males, LH stimulates the interstitial cells of the testis to produce testosterone.

Prolactin (PRL) prepares the breast for milk production.

Adrenocorticotrophic hormone (ACTH) stimulates secretion of the glucocorticoids from the adrenal cortex.

Melanocyte stimulating hormone (MSH) is made in the part of the pituitary between the anterior and the posterior lobes (pars intermedia). Its role in humans is uncertain, but it may cause skin darkening.

POSTERIOR PITUITARY GLAND AND ITS HORMONES

The posterior pituitary (neurohypophysis) gland is derived from a projection from the embryonic hypothalamus and retains axonal connections with neuron groups in the brain.

Posterior pituitary structure

The posterior pituitary is also called the pars nervosa of the pituitary, reflecting its neural origin. The posterior pituitary does not synthesize hormones itself but contains hormone-secreting axons that descend from neurons in the hypothalamus.

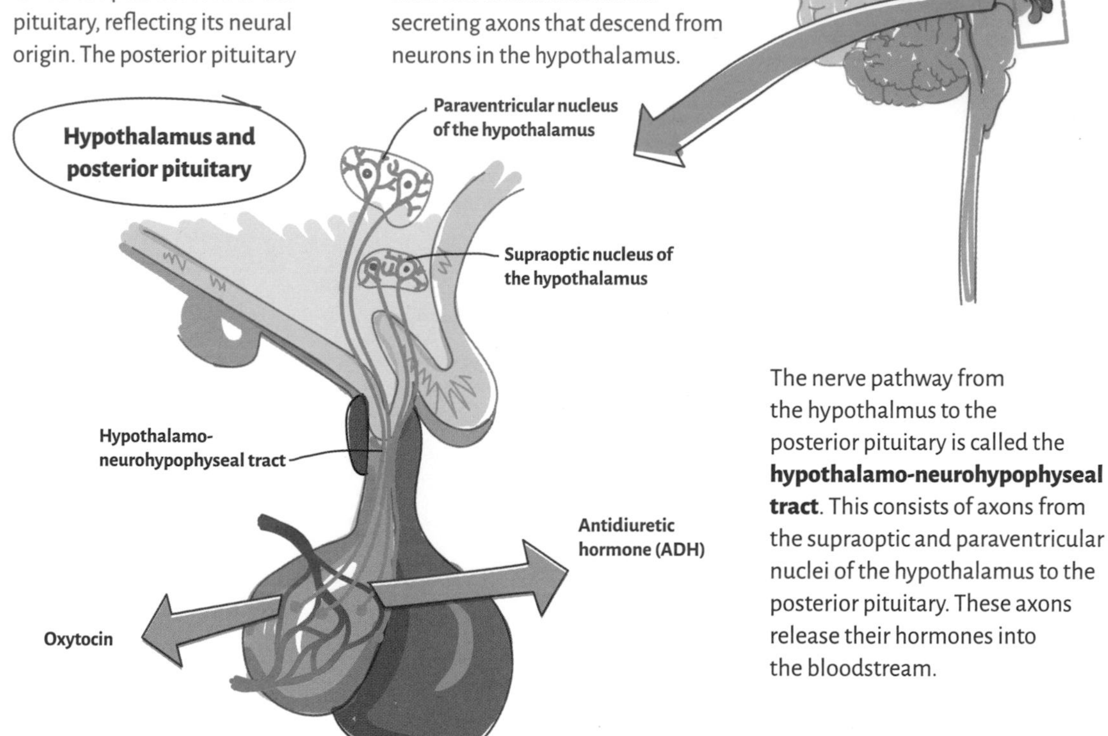
Hypothalamus and posterior pituitary
Paraventricular nucleus of the hypothalamus
Supraoptic nucleus of the hypothalamus
Hypothalamo-neurohypophyseal tract
Antidiuretic hormone (ADH)
Oxytocin

The nerve pathway from the hypothalmus to the posterior pituitary is called the **hypothalamo-neurohypophyseal tract**. This consists of axons from the supraoptic and paraventricular nuclei of the hypothalamus to the posterior pituitary. These axons release their hormones into the bloodstream.

POSTERIOR PITUITARY HORMONE FUNCTIONS

Oxytocin affects pregnant and breast-feeding women.

★ It enhances the contraction of smooth muscle in the uterus during birth.

★ It stimulates milk ejection from the mammary glands (let-down reflex).

In nonpregnant women and men, oxytocin may enhance pair-bonding and care for young.

Antidiuretic hormone (**ADH**, also called vasopressin) acts on the kidney to enhance the reabsorption of water from the glomerular filtrate to the blood, thereby conserving water. In the absence of ADH (e.g., from pituitary damage), urine volume increases to as much as 20 liters (4.4 gallons) per day.

THYROID AND PARATHYROID GLANDS

The thyroid gland is a bilobed structure in the lower neck around the tracheal cartilages. A midline isthmus joins the two lobes.

Thyroid gland cells

Most of the thyroid gland consists of thyroid follicles. Follicular cells make up most of the wall of the follicle and are under control by TSH from the anterior pituitary. The center of each follicle contains a protein called **colloid** (thyroglobulin).

Parafollicular (or C) cells are located either between the follicles or embedded in the follicle wall. Parafollicular cells produce calcitonin in response to elevated blood calcium.

Calcitonin inhibits bone reabsorption by osteoclasts and accelerates the uptake of calcium and phosphate into bone.

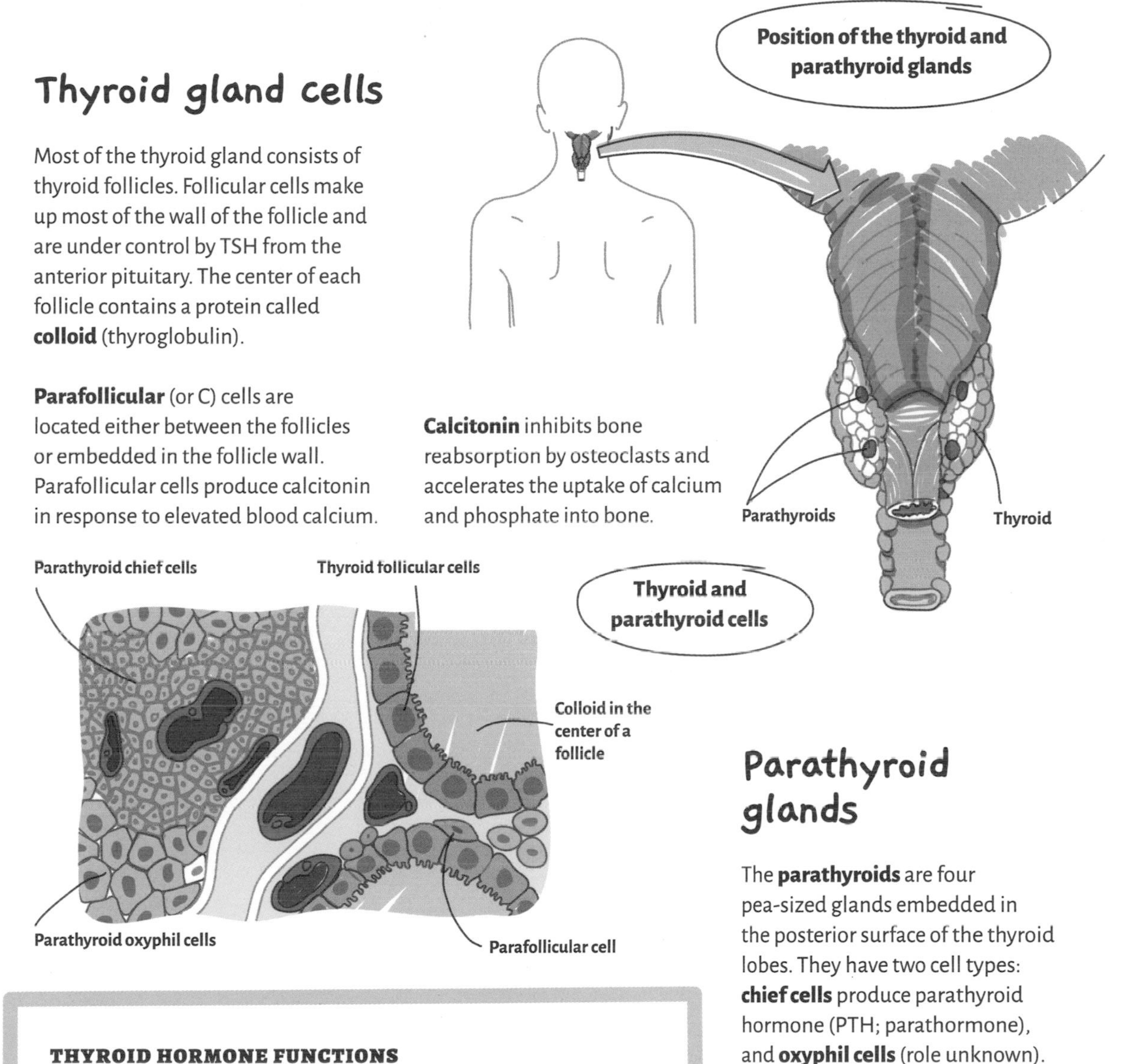

Position of the thyroid and parathyroid glands
Parathyroids
Thyroid
Thyroid and parathyroid cells
Parathyroid chief cells
Thyroid follicular cells
Colloid in the center of a follicle
Parathyroid oxyphil cells
Parafollicular cell

Parathyroid glands

The **parathyroids** are four pea-sized glands embedded in the posterior surface of the thyroid lobes. They have two cell types: **chief cells** produce parathyroid hormone (PTH; parathormone), and **oxyphil cells** (role unknown).

PTH increases blood calcium and magnesium and decreases blood phosphate. It increases reabsorption of bone by osteoclasts and reabsorption of calcium by the kidneys.

THYROID HORMONE FUNCTIONS

Follicular cells produce two types of iodinated thyroid hormone under the influence of TSH. **Thyroxine** (tetraiodothyronine) (T4) has four iodine atoms. **Triiodothyronine** (T3) has three iodine atoms. Both hormones regulate the following:

★Oxygen usage by cells and basal metabolic rate
★Cellular metabolism
★Growth and development

ENDOCRINE PANCREAS

The **endocrine pancreas** is composed of 1 to 2 million islets of Langerhans, which are globular clusters of cells embedded within the exocrine pancreas.

Structure of pancreatic islets

Pancreatic islets (islets of Langerhans) are spherical clusters of cells with vascular connections to the surrounding exocrine pancreas by the insuloacinar portal system. This connection allows islet hormones to control exocrine pancreatic function. Four types of hormone-secreting cells are known: alpha, beta, delta, and F cells.

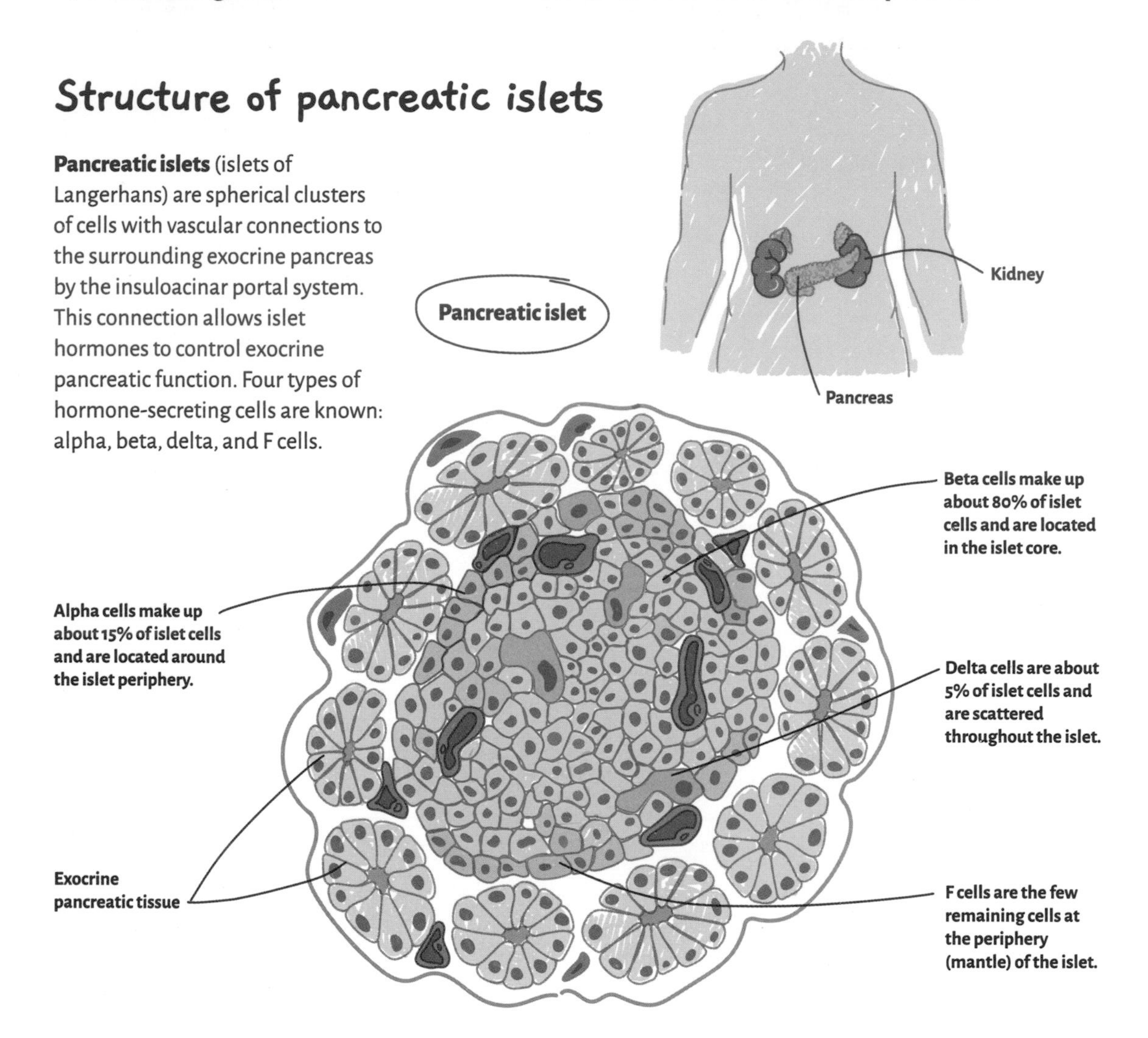

Pancreatic islet
Kidney
Pancreas
Beta cells make up about 80% of islet cells and are located in the islet core.
Alpha cells make up about 15% of islet cells and are located around the islet periphery.
Delta cells are about 5% of islet cells and are scattered throughout the islet.
Exocrine pancreatic tissue
F cells are the few remaining cells at the periphery (mantle) of the islet.

Alpha cells secrete glucagon in response to low blood glucose. Glucagon raises blood glucose by speeding up the breakdown of liver glycogen into glucose and by promoting liver production of glucose (gluconeogenesis).

Beta cells secrete insulin in response to elevated blood glucose. Insulin lowers blood glucose by quickening the uptake of glucose by cells, stimulating the conversion of glucose to glycogen in the liver and reducing glucose production in the liver.

Delta cells make gastrin and somatostatin. Gastrin stimulates stomach acid. Somatostatin inhibits secretion of both insulin and glucagon and slows the absorption of nutrients in the digestive tract.

F cells make pancreatic polypeptide. This hormone has a direct effect on the surrounding exocrine pancreas, inhibiting enzyme secretion. It also inhibits gallbladder contraction and the secretion of somatostatin by delta cells.

ADRENAL CORTEX AND MEDULLA

The adrenal (suprarenal) glands sit on the superior poles of each kidney against the posterior abdominal wall. Each gland has an outer cortex and inner medulla.

The adrenal cortex and medulla have different embryonic origins. The **adrenal cortex** is from the posterior abdominal wall, whereas the **adrenal medulla** is from the neural crest, a cell type that arises at the edge of the folding neural plate (see page 80).

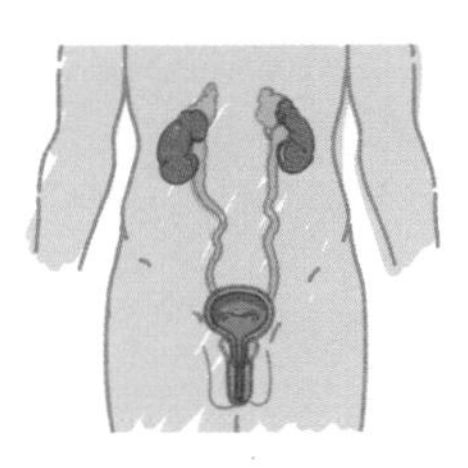

Adrenal cortex structure

The adrenal cortex consists of three layers.

Zona glomerulosa: Outermost layer of spherical and arched cell clusters; it is immediately deep to the gland capsule; 10% to 15% of the cortex.

Zona fasciculata: Middle zone composed of long straight cell cords, 75% of the cortex.

Zona reticularis: Inner zone of branching cords of cells; 5% to 10% of the cortex.

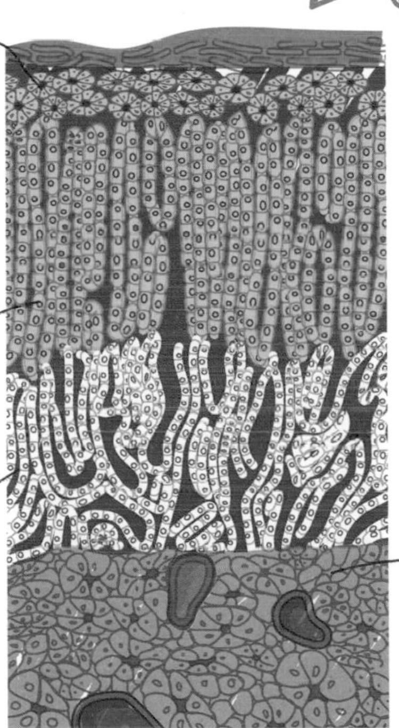
Chromaffin cells of the adrenal medulla

The hormones of the adrenal cortex include the following:

- ★**Mineralocorticoids** (e.g., aldosterone) made by zona glomerulosa. Aldosterone stimulates the retention of sodium by the kidney and renal secretion of potassium and hydrogen ions.
- ★**Glucocorticoids** (mainly cortisol) made by zona fasciculata under stimulation of ACTH from the anterior pituitary. Cortisol stimulates glucose production by the liver and has an anti-inflammatory effect, suppressing cellular and humoral immunity.
- ★**Androgens** (masculinizing hormones) made by zona reticularis. The main hormones are dehydroepiandrosterone (DHEA) and androstenedione. Both can be converted to testosterone or estrogen and stimulate the growth of pubic hair in girls during puberty.

ADRENAL MEDULLA HORMONES

The adrenal gland contains **chromaffin cells**, which are modified sympathetic neurons.

Chromaffin cells make the catecholamines epinephrine (also called adrenaline) and norepinephrine (also called noradrenaline) and some dopamine. Catecholamines are secreted into the bloodstream after sympathetic neural stimulus. They increase blood flow to skeletal muscle and increase heart rate and the force of contraction.

GONADS AND REPRODUCTIVE HORMONES

The gonads produce steroid hormones that not only regulate sexual function but also produce secondary sexual characteristics.

Growth in puberty

Humans are unusual among primates in having an adolescent growth spurt around the time of sexual maturation. This growth spurt rapidly increases the height of the individual by fostering growth of the long bones of the limbs. The growth spurt is controlled by growth, thyroid, and sex hormones. Other primates have a more gradual growth period. The peculiar human growth spurt may be associated with our upright walking and the need to lengthen limb bones rapidly.

Female reproductive cycle

During reproductive life, females undergo cyclical hormonal changes that produce changes in the ovary and uterus. Each cycle takes about 28 days and prepares the secondary oocyte for release and the uterine lining for possible implantation of an embryo.

	Ovarian cycle	Uterine cycle
Days 1 to 5	Follicular phase	Menstrual phase
Days 6 to 14	Follicular phase	Proliferative phase
Day 14	Ovulation	—
Days 15 to 28	Luteal phase	Secretory phase

The cycle is divided into three stages. The cycle stages have names for both the ovarian and uterine cycles (see above). (For stages of follicle development, see pages 171 and 172.)

The **menstrual phase** sees the shedding of the uterine epithelium and blood.

The **proliferative phase** involves the doubling of endometrial thickness, with the growth of endometrial glands and blood vessels.

The **secretory phase** sees the endometrial glands begin to secrete glycogen ready for implantation of the embryo.

Gonadotropin-releasing hormone (GnRH) from the hypothalamus controls the female reproductive cycle. GnRH stimulates the release of both FSH and LH from the anterior pituitary. FSH initiates follicle growth in the ovary and the secretion of estrogen by the growing follicles. LH stimulates further development of the follicles and estrogen secretion. LH also triggers ovulation at mid-cycle (day 14) and then promotes the formation of the corpus luteum, which in turn produces estrogen, progesterone, relaxin, and inhibin.

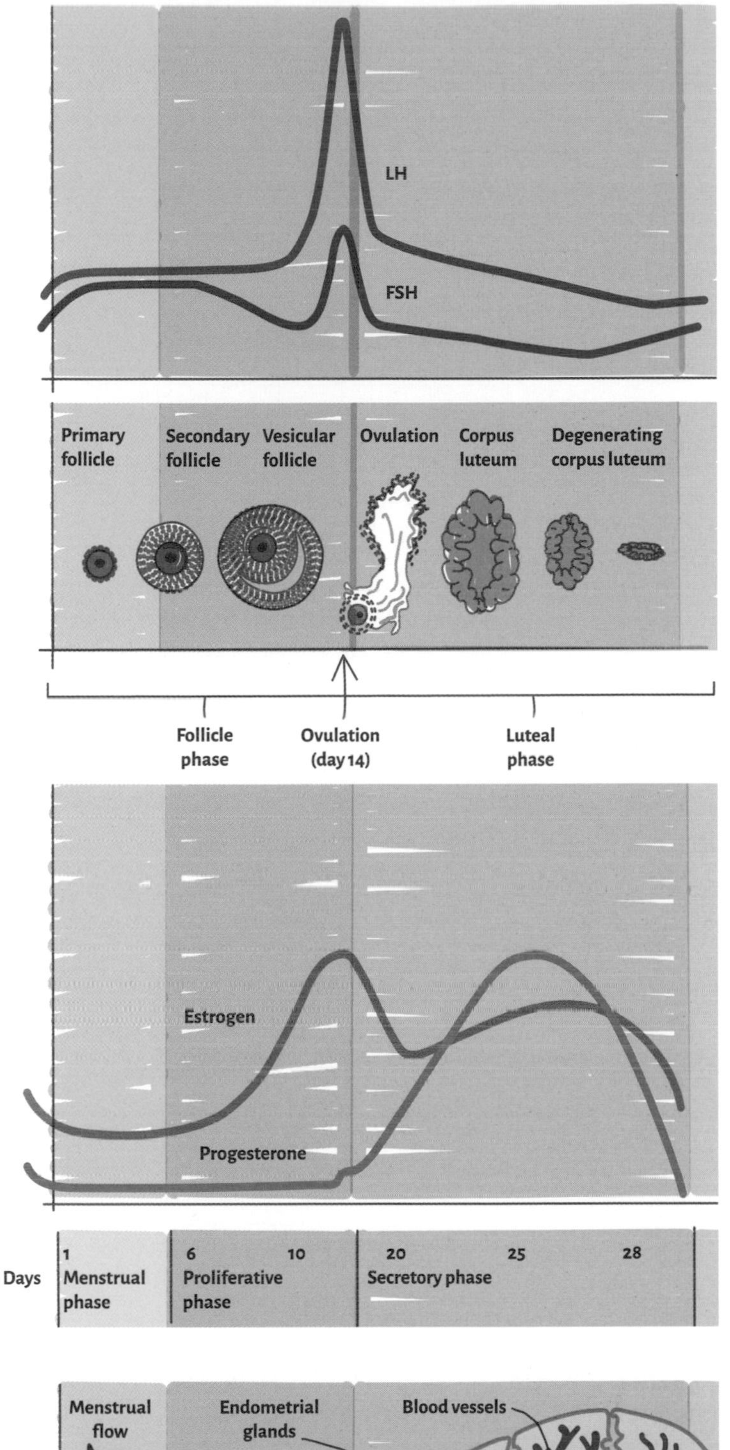
LH
FSH
Primary follicle
Secondary follicle
Vesicular follicle
Ovulation
Corpus luteum
Degenerating corpus luteum
Follicle phase
Ovulation (day 14)
Luteal phase
Estrogen
Progesterone
Days
1
Menstrual phase
6
10
Proliferative phase
20
25
28
Secretory phase

Menstrual cycle

The **menstrual cycle** involves rhythmic changes in pituitary hormones (FSH and LH), ovarian hormones (estrogen and progesterone), ovarian follicle development, and uterine endometrial structure and function.

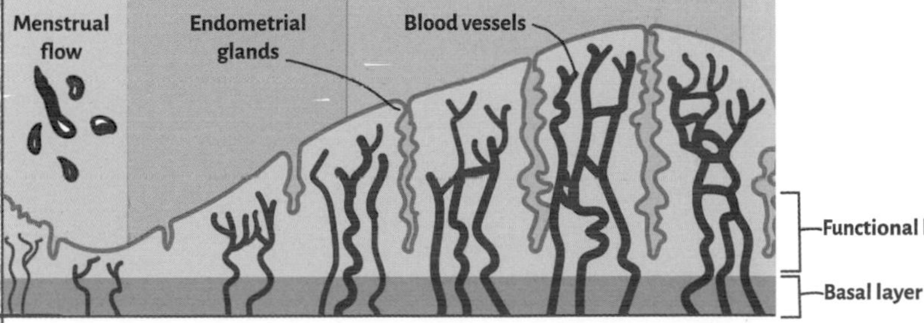
Menstrual flow
Endometrial glands
Blood vessels
Functional layer
Basal layer

Secondary sexual characteristics

Puberty is the period when secondary sexual characteristics begin to develop. Puberty is triggered by hypothalamic GnRH producing pulses of FSH and LH during sleep.

Secondary sexual characteristics are produced by androgens and estrogens and include for males the deepening of the voice from vocal ligament lengthening; increased muscle mass; facial, axillary, and pubic hair growth; elongation of the penis; and growth of the testes.

For females, it includes fat deposition in the breasts, thighs, and buttocks; the growth of pubic and axillary hair; and the onset of first menstruation (menarche).

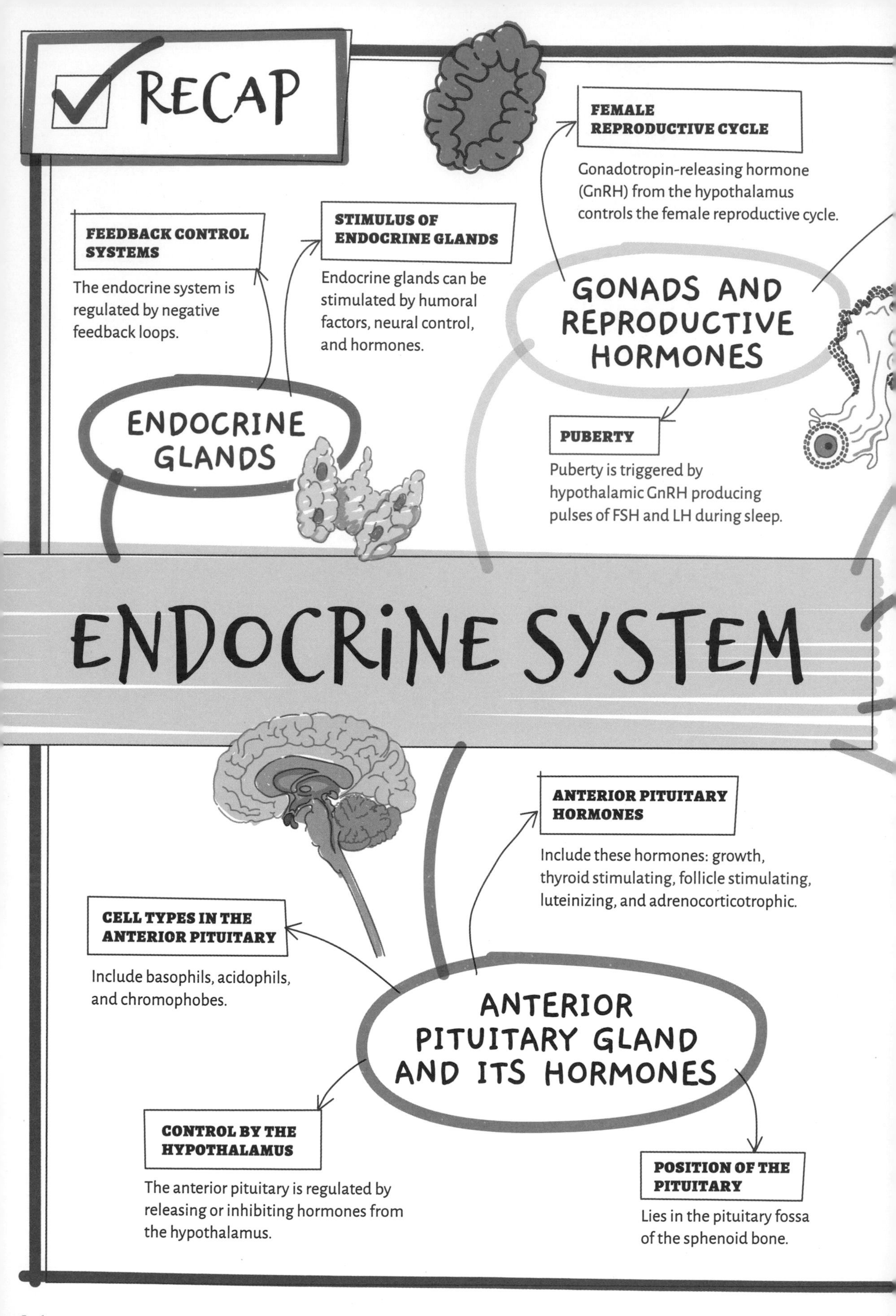

RECAP
FEEDBACK CONTROL SYSTEMS
The endocrine system is regulated by negative feedback loops.
STIMULUS OF ENDOCRINE GLANDS
Endocrine glands can be stimulated by humoral factors, neural control, and hormones.
ENDOCRINE GLANDS
FEMALE REPRODUCTIVE CYCLE
Gonadotropin-releasing hormone (GnRH) from the hypothalamus controls the female reproductive cycle.
GONADS AND REPRODUCTIVE HORMONES
PUBERTY
Puberty is triggered by hypothalamic GnRH producing pulses of FSH and LH during sleep.
ENDOCRINE SYSTEM
ANTERIOR PITUITARY HORMONES
Include these hormones: growth, thyroid stimulating, follicle stimulating, luteinizing, and adrenocorticotrophic.
CELL TYPES IN THE ANTERIOR PITUITARY
Include basophils, acidophils, and chromophobes.
ANTERIOR PITUITARY GLAND AND ITS HORMONES
CONTROL BY THE HYPOTHALAMUS
The anterior pituitary is regulated by releasing or inhibiting hormones from the hypothalamus.
POSITION OF THE PITUITARY
Lies in the pituitary fossa of the sphenoid bone.

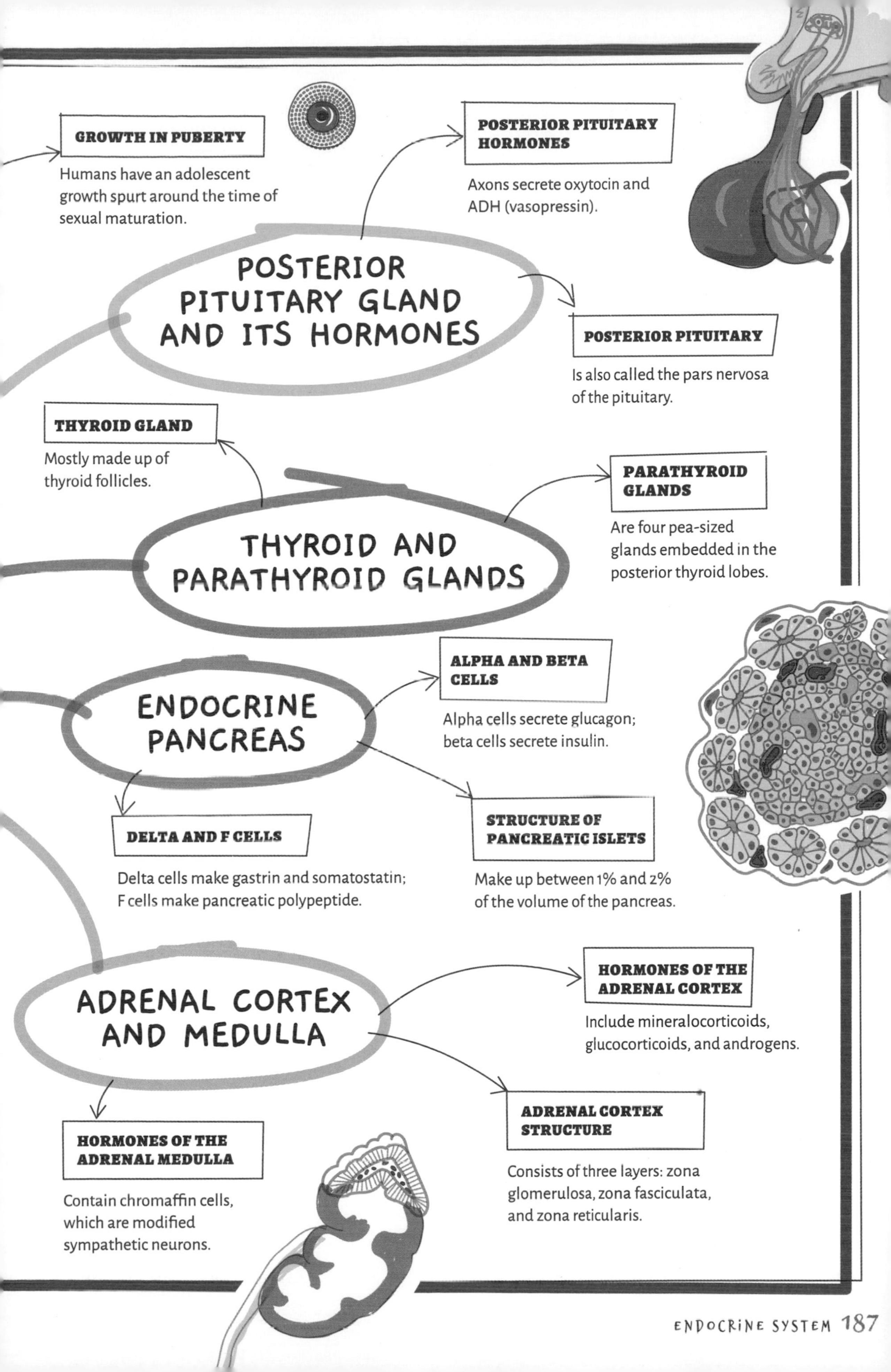

POSTERIOR PITUITARY GLAND AND ITS HORMONES

GROWTH IN PUBERTY

Humans have an adolescent growth spurt around the time of sexual maturation.

POSTERIOR PITUITARY HORMONES

Axons secrete oxytocin and ADH (vasopressin).

POSTERIOR PITUITARY

Is also called the pars nervosa of the pituitary.

THYROID AND PARATHYROID GLANDS

THYROID GLAND

Mostly made up of thyroid follicles.

PARATHYROID GLANDS

Are four pea-sized glands embedded in the posterior thyroid lobes.

ENDOCRINE PANCREAS

ALPHA AND BETA CELLS

Alpha cells secrete glucagon; beta cells secrete insulin.

DELTA AND F CELLS

Delta cells make gastrin and somatostatin; F cells make pancreatic polypeptide.

STRUCTURE OF PANCREATIC ISLETS

Make up between 1% and 2% of the volume of the pancreas.

ADRENAL CORTEX AND MEDULLA

HORMONES OF THE ADRENAL CORTEX

Include mineralocorticoids, glucocorticoids, and androgens.

HORMONES OF THE ADRENAL MEDULLA

Contain chromaffin cells, which are modified sympathetic neurons.

ADRENAL CORTEX STRUCTURE

Consists of three layers: zona glomerulosa, zona fasciculata, and zona reticularis.

INDEX

A

B

C

M

N

O

P

Acknowledgments

With thanks to Sarah Skeate for her informative illustrations,
Jane McKenna for the stylish page layout, and
Cynthia Pfirrmann for her expert advice.